Space–Time, Yang–Mills Gravity, and Dynamics of Cosmic Expansion

How Quantum Yang–Mills Gravity in the Super-Macroscopic Limit Leads to an Effective $G_{\mu\nu}(t)$ and New Perspectives on Hubble's Law, the Cosmic Redshift and Dark Energy

ADVANCED SERIES ON THEORETICAL PHYSICAL SCIENCE
A Collaboration between World Scientific and Institute of Theoretical Physics

ISSN: 1793-1495

Series Editors: Dai Yuan-Ben, Hao Bai-Lin, Su Zhao-Bin
(Institute of Theoretical Physics Academia Sinica)

Published

Advanced Series on Theoretical Physical Science **12**
Volume

Space–Time, Yang–Mills Gravity, and Dynamics of Cosmic Expansion

How Quantum Yang–Mills Gravity in the Super-Macroscopic Limit Leads to an Effective $G_{\mu\nu}(t)$ and New Perspectives on Hubble's Law, the Cosmic Redshift and Dark Energy

Jong-Ping Hsu
University of Massachusetts Dartmouth, USA

Leonardo Hsu
Santa Rosa Junior College, USA

World Scientific

NEW JERSEY · LONDON · SINGAPORE · BEIJING · SHANGHAI · HONG KONG · TAIPEI · CHENNAI · TOKYO

Published by

World Scientific Publishing Co. Pte. Ltd.

5 Toh Tuck Link, Singapore 596224

USA office: 27 Warren Street, Suite 401-402, Hackensack, NJ 07601

UK office: 57 Shelton Street, Covent Garden, London WC2H 9HE

British Library Cataloguing-in-Publication Data
A catalogue record for this book is available from the British Library.

Cover image credit: Tom Jarrett (IPAC/SSC)

Advanced Series on Theoretical Physical Science — Vol. 12
SPACE–TIME, YANG–MILLS GRAVITY, AND DYNAMICS OF COSMIC EXPANSION
How Quantum Yang–Mills Gravity in the Super-Macroscopic Limit Leads to an Effective
$G_{\mu\nu}(t)$ **and New Perspectives on Hubble's Law, the Cosmic Redshift and Dark Energy**

ISBN 978-981-120-043-4

For any available supplementary material, please visit
https://www.worldscientific.com/worldscibooks/10.1142/11283#t=suppl

Desk Editor: Ng Kah Fee

Typeset by Stallion Press
Email: enquiries@stallionpress.com

Printed in Singapore

Cosmic Relativity

Cosmic Relativity

Cosmic Relativity

Cosmic Relativity

To

The Physics Alumni, George Leung, Susumu Okubo, George Sudarshan and the University of Massachusetts Dartmouth. The research on space-time symmetry and quantum Yang–Mills gravity, as well as the dynamics of an expanding universe with cosmic time and cosmic relativity, was made possible by their support.

Cosmic Time

Cosmic Time

Cosmic Time

Cosmic Time

One new result of quantum Yang–Mills gravity is that the apparent curvature of space-time appears to be simply a manifestation of the flat space-time translational gauge symmetry for the motion of quantum particles in the classical limit.

Jong-Ping Hsu and Leonardo Hsu, "Overview," (2013).

The most glaring incompatibility of concepts in contemporary physics is that between Einstein's principle of general coordinate invariance and all the modern schemes for a quantum-mechanical description of nature.

Freeman J. Dyson, "Missed Opportunities," *Bull. Amer. Math. Soc.*, **78**, Sept. 1972.

Evidently, the usual statements about future positions of particles, as specified by their coordinates, are not meaningful statements in general relativity. This is a point which cannot be emphasized strongly enough.... Expressing our results in terms of the values of coordinates became a habit with us to such a degree that we adhere to this habit also in general relativity, where values of coordinates are not per se meaningful.

Eugene P. Wigner, *Symmetries and Reflections* (The MIT Press, 1967), pp. 52–53.

The important thing is to be able to make predictions about images on the astronomers' photographic plates,... and it simply doesn't matter whether we ascribe these predictions to the physical effect of gravitational fields on the motion of planets and photons or to a curvature of space and time. (The reader should be warned that these views are heterodox and would meet with objections from many general relativists.)

Steven Weinberg, *Gravitation and Cosmology* (John Wiley & Sons, 1972), p. 147.

Using Quantum Yang–Mills gravity with inertial frames in the super-macroscopic limit, one can accommodate the principle of relativity and the cosmological principle of homogeneity and isotropy. This feat is accomplished by introducing a new definition of time, which we call "cosmic time." We begin by choosing any inertial frame, called the F frame, in which to construct a system of synchronized clocks. The clocks are synchronized in such a way so that the speed of light is constant and isotropic in F. All observers then use these clocks to record time, regardless of in which inertial frame they are located. All observers then use, by definition, a cosmic time, which is an essential component of "cosmic relativity." Although the speed of light is anisotropic in all frames but F, the principle of relativity for physical laws still holds in cosmic relativity, enabling it to be consistent with experimental results....

Jong-Ping Hsu and Leonardo Hsu, Preface (this volume).

Preface

In this expanded edition, we expound on the intimate relationship between the space-time translation (T_4) gauge symmetry and the beautiful ideas of the Lie derivative and Pauli's variation. The Lie derivative (originally introduced by W. Ślebodziński in 1931) and Pauli's variation (1921) describe the coordinate invariant changes of a general tensor determined by an arbitrary vector function $\Lambda^\mu(x)$. In physics, Pauli's original idea of variation was motivated to obtain the invariant action and to derive field equations along with certain identities.

Within inertial and non-inertial frames of reference in flat space-time, Yang–Mills gravity reveals that the Pauli–Ślebodziński variation and the Lie derivative are identical to the gravitational T_4 gauge transformation. Furthermore, the vanishing of the Lie derivative of the gravitational action is precisely the T_4 gauge invariance of quantum Yang–Mills gravity. It appears that the most important application of the Lie derivative is to formulate a gravitational theory with external gauge group, which leads to Yang–Mills gravity. The Lie derivative and Cartan formula are the mathematical bases for treating the gravitational gauge transformation and symmetry with the external T_4 translation group. It seems to be non-trivial to treat the external space-time gauge group in the language of fiber bundles as it describes internal gauge group.

Furthermore, we present new views and implications of quantum Yang–Mills gravity and the cosmological principle in the super-macroscopic limit. In this regime, quantum Yang–Mills gravity provides a useful framework for modeling the motion of distant galaxies, even with a flat space-time. The combination of Yang–Mills gauge symmetry and the "effective" space-time metric tensors *a la* Einstein–Grossmann enables us to derive an equation, called Okubo equation, for the motion of distant galaxies and to model (and provide an upper limit for) Hubble recession velocities, as well as an exact cosmic redshift. Furthermore, the conserved baryon–lepton charges associated with a generalized Lee–Yang U_1 gauge symmetry lead to a new linear repulsive force between galaxies. Although this force is weaker than the gravitational force in the early universe, in the later epoch of evolution however, when the distances between typical galaxies are large enough, this repulsive force dominates over gravity. These properties are just right to understand the late-time accelerated cosmic expansion and could be considered a candidate for "dark energy."

Using quantum Yang–Mills gravity with inertial frames in the super-macroscopic limit, one can accommodate the principle of relativity and the cosmological principle of homogeneity and isotropy. This feat is accomplished by introducing a new definition of time, which we call cosmic time. We begin by choosing any inertial frame, called the F frame, in which to construct a system of synchronized clocks. The clocks are synchronized in such a way so that the speed of light is constant and isotropic in F. All observers then use these clocks to record time, regardless of in which inertial frame they are located. All observers then use, by definition, a cosmic time, which is an essential component of "cosmic relativity." Although the speed of light is anisotropic in all frames but F, the principle of relativity for physical laws still holds in cosmic relativity, enabling it to be consistent with experimental results such as the Michelson–Morley experiment, the relativistic Doppler shift and the decay-length (or life-time) dilatation of particle decaying in flight, etc. This distinguishes cosmic relativity from Galilean relativity. We stress that the speed of light is anisotropic only when measured in length/time. If space and time are measured using the same units, the speed of light remains isotropic in all frames.

New results for the dynamics of cosmic expansion are from recent publications and lecture notes of "particle cosmology" delivered in Fall 2018. Ideas exploring the birth of the universe based on a Big Jets model and the alternate dynamics of cosmic expansion based on Yang–Mills gravity and cosmic relativity originated from discussions among Leon Hsu, Daniel Katz and J.P. Hsu. Students' comments and questions in class were useful for determining how to present these ideas, and we would like to thank Joel Baer, Emma Klinkhamer, Hao Yun and Yi-yi Zhu. Some errors in the previous edition have been corrected. Finally, special acknowledgment is made to Bonnie C. Hsu for her resolute support of this project. J.P. would also like to thank Dana Fine for useful discussions.

Preface of "Space-Time Symmetry and Quantum Yang–Mills Gravity"

It began with the idea that the four-dimensional space-time for inertial frames could be formulated solely on the basis of a single postulate, i.e., the principle of relativity for physical laws. This idea, named "taiji relativity," was stimulated by a university term paper by one of the authors (LH) and discussed in the book *A Broader View of Relativity*. Although this foundational idea in and of itself was not new, it turned out that the framework of taiji relativity could also be generalized to derive explicit space-time transformations for non-inertial frames that simplified to the Lorentz transformations in the limit of zero acceleration, something never before accomplished. The taiji framework also supported the quantization of the physical fields of the strong and electroweak interactions, as well as all established conservation laws. Thus, with faith in the Yang–Mills idea and taiji relativity, it was not unreasonable to think that such a framework might also accommodate the gravitational interaction. This idea was additionally bolstered by the fact that the conservation of energy and momentum is a consequence of the translational symmetry of the action for a physical system, and that the energy–momentum tensor might be related to the source of the gravitational field.

The idea that the translational symmetry in flat space-time could be the gauge symmetry for gravitational potential fields has emerged slowly. At present, such an idea is unorthodox and is definitely a minority viewpoint, but it is an extremely intriguing one! Why should gauge symmetry in flat space-time be so successful for modeling all known interactions except gravity?

At first, deriving field equations and an equation of motion for classical objects that were logically and experimentally consistent seemed an insurmountable difficulty. Enlightenment came from the observation that, in the classical limit, translational gauge symmetry requires that all wave equations (with the exception of the gravitational wave equation) reduce to Hamilton–Jacobi type equations with an effective Riemannian metric tensor. In other words, even though the underlying physical space-time is flat, all objects behave as if they were in a curved space-time.

Many more years of search and research finally resulted in the discovery of a gauge invariant action within a generalized Yang–Mills framework that was consistent with all classical tests of gravity. Almost all the advantages of the Yang–Mills idea were then available to tackle such a quantum gravity, except that the gravitational coupling constant was not dimensionless.

Einstein's theory of gravity is formulated in the more complicated framework of curved space-time and as such appears to be too general to be compatible with the quantization of fields and the conservation of energy–momentum. Yang–Mills gravity represents an alternative approach that brings gravity back into the arena of gauge fields in flat space-time, in which the quantization of fields and the conservation of energy–momentum are already well-established.

The aim of this book is to provide a treatment of quantum Yang–Mills gravity with an emphasis on its ideas and the experimental evidence that might support it. The two big ideas of Yang–Mills gravity are as follows.

(1) There exists a fundamental space-time symmetry framework that can encompass all interactions in physics, including gravity, and is valid in both inertial and non-inertial frames of reference. This framework (the "taiji symmetry" framework) is based on a flat space-time with arbitrary coordinates for all reference frames and can accommodate a generalized Yang–Mills idea for external gauge groups and a totally unified model of all interactions, along the lines of the ideas of Glashow, Salam, Ward, and Weinberg.

(2) The gravitational field is the manifestation of the flat space-time translational gauge symmetry, which is the only exact gauge symmetry. All internal gauge symmetries are violated subtly by gravity, so that classical objects move in gravitational field as if they were in a curved space-time. Furthermore, the space-time translational gauge symmetry enables the unification of the Yang–Mills gravity with the electroweak theory and quantum chromodynamics.

This monograph summarizes our collaborative research and striving to bring forth new viewpoints and ideas over the past two decades since that university term paper. We hope the book will be useful for students and researchers interested in a unified picture of nature. The present work would not have been possible were it not for the publication of our earlier volumes, *A Broader View of Relativity:*

General Implications of Lorentz and Poincaré Invariance (World Scientific, 2006) and *100 Years of Gravity and Accelerated Frames: The Deepest Insight of Einstein and Yang–Mills* (with D. Fine, World Scientific, 2005).

We would like to thank Dana Fine, Howard Georgi and colleagues for their interest and discussions, and we are indebted to Kazuo Cottrell and Leslie Hsu for the reading and proofing of many chapters. One of us (JP) greatly benefitted from discussions with C. C. Chen, J. M. Wang, T. S. Cheng and W. T. Yang (Taida), T. Y. Wu (Tsing Hua and Academia Sinica), S. Okubo and R. E. Marshak (University of Rochester), H. W. Huang (Cornell), and E. C. G. Sudarshan, C. B. Chiu, D. A. Dicus and T. N. Sherry (Center for Particle Theory, Austin) in earlier years. Special acknowledgment is made to Bonnie Chiu Hsu for her tireless support throughout our lives.

Jong-Ping Hsu, *University of Massachusetts Dartmouth*

Leonardo Hsu, *University of Minnesota, Twin Cities*

Contents

Part II: Quantum Yang–Mills Gravity 91
Jong-Ping Hsu and Leonardo Hsu

Overview

0.1. Motivation: "The Most Glaring Incompatibility of Concepts in Contemporary Physics"

Despite the success of Einstein's theory of gravity,[a] there are difficult and long-standing problems with how general relativity fits into the big picture of physics. In the January 1972, Josiah Willard Gibbs Lecture, given under the auspices of the American Mathematical Society, Freeman Dyson pointed out one of the problems with the principle of general coordinate invariance,[2] saying "The most glaring incompatibility of concepts in contemporary physics is that between Einstein's principle of general coordinate invariance and all the modern schemes for a quantum-mechanical description of nature."

General coordinate invariance implies that space-time points can be given arbitrary labels.[b] Because the presence of gravity implies a change in the space-time metric, the physical separation of two bodies cannot be fixed in the sense that one can no longer talk about two particles having a fixed distance between each other.

---

[a] Hilbert elegantly embedded Einstein's idea of general coordinate invariance in an action to derive the gravitational field equation, independent of Einstein in 1915.[1]

[b] Wigner wrote "Evidently, the usual statements about future positions of particles, as specified by their coordinates, are not meaningful statements in general relativity. This is a point which cannot be emphasized strongly enough and is the basis of a much deeper dilemma than the more technical question of the Lorentz invariance of the quantum field equations. It pervades all the general theory, and to some degree we mislead both our students and ourselves when we calculate, for instance, the mercury [sic] perihelion motion without explaining how our coordinate system is fixed in space.... Expressing our results in terms of the values of coordinates became a habit with us to such a degree that we adhere to this habit also in general relativity, where values of coordinates are not per se meaningful."

Similarly, the relative rates of two clocks at two different locations are related in an arbitrary way and hence, time has no physical meaning. The metric tensor $g_{\mu\nu}$ is needed so that physical laws described by space-time can be independent of the coordinate systems.

In a lucid discussion of coordinates and momenta in general relativity, Wigner wrote "The basic premise of this theory [the general theory of relativity] is that coordinates are only auxiliary quantities which can be given arbitrary values for every event. Hence, the measurement of position, that is, of the space coordinates, is certainly not a significant measurement if the postulates of the general theory are adopted: the coordinates can be given any value one wants. The same holds for momenta. Most of us have struggled with the problem of how, under these premises, the general theory of relativity can make meaningful statements and predictions at all"[3] (See footnote b).

This property of coordinates is at odds with the physically meaningful and operationally defined space-time of quantum mechanics and gauge field theories with Poincaré invariance. Coordinate transformations in flat space-time form a much smaller group than that in curved space-time, and space and time coordinates in inertial frames can be operationally defined. Physics based on Poincaré invariance is endowed with great mathematical simplicity and elegance and supported by uncountable experiments. In particular, the concept of local commutativity, which is essential for quantum field theories, holds only if the space-time framework includes the Poincaré group.

A second problem with Einstein's principle of general coordinate invariance is that it does not include a conservation law for energy–momentum. From the group-theoretic viewpoint in mathematics, Einstein's gravity is based on the group formed by general coordinate transformations in curved space-time. This is a Lie group with a continuously infinite number of infinitesimal generators. In 1918, Noether's second theorem[4] suggested that energy could be conserved if and only if a physical theory is invariant under a finite or countably infinite number of infinitesimal generators, a result consistent with Hilbert's claim (1918) that the failure of the law of conservation for energy was characteristic of Einstein's theory of gravity.[4]

These serious problems have inspired physicists to search for a consistent and complete theory of classical and quantum gravity over the past century, a search that still goes on today.

0.2. The Taiji Symmetry Framework

The taiji symmetry framework, or "taiji space-time," is the maximally symmetric space-time with zero curvature.[c] It is capable of accommodating both inertial and non-inertial frames of reference and all known conservation laws, as well as all known interactions, including gravity, and the quantization of all fields.

Traditionally, discussions in particle physics, field theory and other branches of physics have all been based on inertial frames with the Lorentz and Poincaré transformations. However, to obtain a broader and more complete understanding of physics, it is desirable to generalize the physical framework to include non-inertial frames. Inertial frames represent only limiting and idealistic cases and moreover, there is now strong evidence that the observable universe is expanding with a non-zero acceleration.

In the first six chapters, we derive explicit coordinate transformations between inertial frames and non-inertial frames (including rotating frames and frames with a colinear acceleration and velocity where the acceleration is an arbitrary function of time) within the taiji symmetry framework. These transformations are derived on the basis of the "principle of limiting continuation," i.e., in the limit of zero acceleration, the transformations must simplify to the Lorentz transformations and the Poincaré metric tensor $P_{\mu\nu}$ of non-inertial frames must reduce to the Minkowski metric $\eta_{\mu\nu} = (1, -1, -1, -1)$. The existence of finite space-time transformations between inertial frames and such accelerated frames implies that the space-time associated with the non-inertial frames is flat,[5,6] i.e., the Riemann–Christoffel curvature tensor $R_{\lambda\mu\nu\alpha}$ is zero. The geometry of flat space-time with arbitrary coordinates is much simpler than the curved space-time of general relativity.

We discuss testable experimental predictions resulting from these coordinate transformations, some of which have already been performed, such as the lifetime dilation of unstable particles traveling in a circular storage ring, as well as new group properties of these transformations. Because inertial and non-inertial frames are not equivalent and no objects may travel faster than a light signal, transformations between inertial and non-inertial frames form only

---

[c]In ancient Chinese thought, the word "taiji" or "taichi" denotes "The Absolute," i.e., the ultimate principle or the condition that existed before the creation of the world.

pseudo-groups,[7] rather than the usual Lie groups, and map only a portion of the space-time in an accelerated frame to the entire space-time of inertial frames.

Finally, we note that, as stated by Noether's first theorem, the physical manifestations of symmetry are the conservation laws. Ideally, well-established laws such as the conservation of energy–momentum and angular momentum should be natural consequences of the symmetries of a physical framework. The taiji symmetry framework, which is based mathematically on a flat space-time with arbitrary coordinates and the Poincaré metric tensor, and accommodates both inertial and non-inertial frames, should be the most general framework for physical theories for which this is true. In contrast, a theory based on a curved space-time such as general relativity cannot automatically conserve the energy–momentum tensor, but only the "energy–momentum pseudo-tensor" $t^{\mu\nu}$ as expressed by $\partial_\mu t^{\mu\nu} = 0$. Physically, this is because in a curved space-time with arbitrary coordinates, one can remove the gravitational field from any given volume element so that unlike the energy associated with electromagnetic fields, it is meaningless to define an energy associated with a gravitational field at a given point, which is not completely satisfactory.

Within the taiji symmetry framework, Yang–Mills gravity reveals that the Lie derivative in coordinate expression in flat space-time is identical to the space-time translation (T_4) gauge transformation. The vanishing of the Lie derivative of the gravitational action is precisely the T_4 gauge invariance of quantum Yang–Mills gravity. It appears that the most important application of the Lie derivative is to formulate a gravitational theory with external gauge group, which leads to Yang–Mills gravity. The Lie derivative and Cartan formula are the mathematical bases for treating the gravitational gauge transformation and symmetry with the external space-time translation group.

0.3.   Quantum Yang–Mills Gravity

In the second part of this book (Chapters 7–13), we construct Yang–Mills gravity, which is a classical and quantum theory of gravity consistent with all known experimental results. It is based on a generalized Yang–Mills framework and translational gauge symmetry in the flat space-time of the taiji symmetry framework. As might be anticipated by the reader, perhaps the biggest challenge of the theory is to reproduce all the known experimental effects that are consistent with a curved space-time, within a framework based on a flat space-time.

Around 1920, Weyl realized that electromagnetic fields and the conservation of charge could be understood in terms of the symmetry of the U_1 gauge transformation.[8,d] Weyl's idea was subsequently generalized to the SU_2 gauge symmetry in 1954 by Yang and Mills, in order to understand isospin conservation in strong interactions.[9] This highly non-trivial generalization was Yang and Mills' deepest insight and was crucial to further progress in physics, especially in the unification of the electromagnetic and weak forces.

In 1955, Utiyama proposed a gauge-invariant interpretation of interactions and discussed a general gauge theory involving a Lie group of N generators, based on the work of Yang and Mills. Utiyama further formulated a gauge theory of gravity based on the Lorentz group and identified Einstein's field equations as the gauge field equations, which paved the way for future research on gauge theories of gravity.[10] Since then, quantum field theories with gauge symmetries of internal groups (which are independent of the external space-time symmetries) have been successfully applied to modeling all fundamental interactions except for the gravitational interaction. Decades later, gravity still stands alone as the last challenge to Yang, Mills, Utiyama, and Weyl's idea of gauge symmetry.

In tackling the challenge of creating a quantum theory of gravity, however, it appears that the original Yang–Mills framework for internal gauge groups is inadequate. In the approach we describe here, we generalize that framework to include external space-time gauge groups,[11] and we call this generalized Yang–Mills framework the Yang–Mills–Utiyama–Weyl (YMUW) framework.

The guiding principle in constructing Yang–Mills gravity is that there must be a gauge symmetry that is compatible with both the conservation law of energy–momentum and the quantization of gauge fields. One promising candidate in flat space-time is the global translational symmetry. Suppose a Lagrangian is invariant

---

[d]The physical meaning of the U_1 gauge invariance was analyzed by V. I. Oglevetski and I. V. Polubarinov. They first decomposed the field A_μ as

$$A_\mu = (A_\mu - \partial^{-2}\partial_\mu \partial^\nu A_\nu) + \partial^{-2}\partial_\mu \partial^\nu A_\nu \equiv A_\mu^1 + A_\mu^0,$$

showing that the physical spin 1 components, A_μ^1, of the massless vector field were unchanged under the U_1 gauge transformation, $A'_\mu = A_\mu + \partial_\mu \Lambda(x)$, and that only the unphysical spin 0 component, A_μ^0, was changed. The observable field strength $F_{\mu\nu}$ does not involve the unphysical spin 0 component. Thus, the physical predictions of the electromagnetic theory are gauge invariant.

under a global space-time translation,

$$x^\mu \to x'^\mu = x^\mu + \Lambda_o^\mu, \tag{0.1}$$

where the shifts of space-time coordinates Λ_o^μ are infinitesimal constants. This invariance implies the conservation of energy–momentum through Noether's theorem. The set of these transformations in flat four-dimensional space-time forms the T_4 group, which is an Abelian subgroup of the Poincaré group. However, the global transformations (0.1) are inadequate for the formulation of a gauge theory. We must enlarge the global group to include local translations in flat space-time,

$$x^\mu \to x'^\mu = x^\mu + \Lambda^\mu(x). \tag{0.2}$$

It suffices to consider the infinitesimal and arbitrary vector function $\Lambda^\mu(x)$. It is important to realize that the infinitesimal transformations (0.2) have two conceptual interpretations[11]: (i) They are local translations in flat space-time, and (ii) they are general coordinate transformations. This dual role of (0.2) implies that local translational gauge invariance is inseparable from general coordinate invariance in flat space-time.[e]

The invariance under the coordinate translations (0.2) is the mathematical basis of Yang–Mills gravity. In inertial frames, the fundamental space-time symmetry group is the Poincaré group, which contains two sub-groups: the global translational group T_4 and the Lorentz group (which contains rotations as a subgroup). The Abelian subgroup of space-time translations T_4 is our focus because it leads to the conservation of energy–momentum and it is the external T_4 gauge symmetry that enables us to construct a quantum theory of gravity that is consistent with experiments.

Although a full derivation and justification for Yang–Mills gravity must wait until later, we here give just enough detail to indicate how a theory of gravity based

---

[e]In this connection, it is interesting to note that Einstein followed the general coordinate invariance in *curved* space-time and hence, employed space-time curvature and a covariant derivative to formulate his theory. In contrast, Yang–Mills gravity follows the local translational invariance in flat space-time and hence, uses T_4 gauge curvature and a gauge covariant derivative to formulate a theory of the gravitational interaction. The difficulties of Einstein gravity discussed in Section 0.1 are thus avoided.

on a flat space-time can replicate the effects of a curved space-time. Consider a particle-physics approach to gravity based on the YMUW framework with external space-time translational gauge symmetry and for now, consider only inertial frames with the Minkowski metric tensor $\eta_{\mu\nu} = (1, -1, -1, -1)$. The formulations of electromagnetic and Yang–Mills-type gauge theories associated with internal gauge groups are all based on the replacement,

$$\partial_\mu \to \partial_\mu + ifB_\mu(x), \quad c = \hbar = 1, \tag{0.3}$$

in a Lagrangian. The field $B_\mu = B_\mu^a(x)t_a$ involves constant matrix representations of the generators t_a of internal gauge groups. However, in the YMUW framework, the generators of the external space-time translation group T_4 are the momentum operators $p_\mu = i\partial_\mu$ or the displacement operator $T_\mu = \partial_\mu$, where $p_\mu = iT_\mu$ ($c = \hbar = 1$). Hence, in Yang–Mills gravity, the replacement is assumed to take the form

$$\partial_\mu \to \partial_\mu - ig\phi_\mu^\nu p_\nu = \partial_\mu + g\phi_\mu^\nu \partial_\nu, \quad J_\mu^\nu = \delta_\mu^\nu + g\phi_\mu^\nu, \tag{0.4}$$

similar to (0.3) in non-Abelian gauge theories.

This replacement is the basic postulate for the local formulation of the Yang–Mills gravity. Since the generators of the T_4 group are $T_\mu = \partial_\mu$, we have a tensor gauge field $\phi^{\mu\nu} = \phi^{\nu\mu} = \eta^{\lambda\nu}\phi_\lambda^{\ \mu}$ (i.e., a spin-2 field) rather than a 4-vector field (i.e., a spin-1 field) in the T_4 gauge covariant derivative,

$$\Delta_\mu \equiv \partial_\mu + g\phi_\mu^\nu \partial_\nu \equiv J_\mu^\nu \partial_\nu. \tag{0.5}$$

Because the tensor field ϕ_μ^ν is associated with the most general flat space-time symmetry, as shown in the transformations (0.2), the gravitational tensor field $\phi_{\mu\nu}$ may be called the space-time translational gauge field or the space-time gauge field.

Two consequences of the replacements (0.4) and (0.5) can now be seen. First, we note that the replacement (0.4) in the Lagrangian differs from the U_1 replacement $\partial_\mu\psi \to (\partial_\mu + ieA_\mu(x))\psi$ in the free Lagrangian of the charged fermion field ψ. Thus, the covariant derivative of T_4 and U_1 gauge symmetries for a fermion (e.g., the electron with charge $e < 0$) is

$$\partial_\mu + ieA_\mu + g\phi_\mu^\nu \partial_\nu.$$

The anti-fermion field is associated with the complex conjugate, i.e.,

$$(\partial_\mu + ieA_\mu + g\phi_\mu^\nu \partial_\mu)^* = \partial_\mu - ieA_\mu + g\phi_\mu^\nu \partial_\nu.$$

The basic nature of the electric force between a charged fermion and an anti-fermion and that between a fermion and another fermion are respectively determined by

$$(ie)(-ie) = e^2 \quad (attractive) \quad and \quad (ie)(ie) = -e^2 \quad (repulsive).$$

In contrast, the basic nature of Yang–Mills gravitational force between a fermion and an anti-fermion and that between a fermion and another fermion are respectively determined by

$$(+g)(+g) = g^2 \quad (attractive) \quad and \quad (+g)(+g) = g^2 \quad (attractive).$$

Similarly, the gravitational force between an anti-fermion and another anti-fermion is also attractive. Therefore, we conclude that the gravitational force is always attractive because of the external space-time translation gauge symmetry in Yang–Mills gravity.

Second, the T_4 gauge covariant derivative Δ_μ in (0.5) has remarkable power in that it dictates that in flat space-time, all particle wave equations, including the electromagnetic wave equations, reduce to an equation of the Hamilton–Jacobi type with basically the same[f] "effective Riemannian metric tensor" $G_{\mu\nu}(x)$,

$$G^{\mu\nu}(x)(\partial_\mu S)(\partial_\nu S) - m^2 = 0, \quad m^2 \geq 0, \tag{0.6}$$

$$G^{\mu\nu}(x) = \eta_{\alpha\beta} J^{\alpha\mu}(x) J^{\beta\nu}(x), \quad J^{\mu\nu}(x) = \eta^{\mu\nu} + g\phi^{\mu\nu}(x), \tag{0.7}$$

in the classical (short-wavelength) limit.[g] Equation (0.6) shows the effects of a universal gravitational force acting on all classical objects. Because formally, it is the same as the corresponding equation in Einstein's gravity based on curved space-time, (0.6) implies that classical objects, such as planets and light rays, behave as if they were in a "curved space-time," even though the underlying physical space-time is actually flat.

Therefore, one new result of quantum Yang–Mills gravity is that the apparent curvature of space-time appears to be simply a manifestation of the flat

---

[f] There is a small difference between the effective Riemannian metric tensor of the fermions and that of the vector gauge bosons. See equations (8.36) and (8.40) in Chapter 8.

[g] For an early discussion of the geometric-optics limit of the Dirac equation with maximum four-dimensional gauge symmetry, see Ref. 12.

space-time translational gauge symmetry for the motion of quantum particles in the classical limit. In the first-order approximation, Yang–Mills gravity is consistent with the experimentally measured gravitational redshift. In the second-order approximation, it correctly predicts the perihelion shift of Mercury and the gravitational quadrupole radiation of binary pulsars.[13] Furthermore, Yang–Mills gravity allows a quantum theory of gravity to be formulated and Feynman–Dyson rules for the gravitational tensor fields and the associated "ghost particles"[13] to be derived, as we will show in Chapters 10 and 11. Thus, Yang–Mills gravity brings the gravitational interaction back into the experimentally established arena of gauge fields based on flat space-time.

In contrast, Einstein's approach involving the geometrization of physical fields in curved space-time appears to be successful only in the formulation of classical gravity, with the conservation of the energy–momentum pseudo-tensor, rather than the true energy–momentum tensor. If one were to adopt a similar approach to, say, the electromagnetic interaction, which is velocity-dependent, it would be natural to employ the Finsler geometry, where the fundamental metric tensors depend on both positions and velocities (i.e., the differentials of coordinates), rather than the Riemannian geometry.[14] However, all attempts thus far to geometrize classical electrodynamics have not even begun to approach the usefulness of quantum electrodynamics.

The classical theory of Einstein gravity is usually considered to be the most coherent mathematically and satisfying aesthetically.[2] However, there is little chance of making progress simply by going over its successes. Because of the success of Einstein gravity, many have made arguments that attempted to make it unique or necessary.[15,h] There is an old argument due to Feynman (and perhaps others) that the only consistent theory (with no more than two derivatives) of a spin-2 field coupled to itself is, in fact, general relativity. It is not necessary to analyze the detailed arguments one by one, all that is necessary is to present a counterexample.

---

[h] A century ago, Einstein observed "Concepts which have proved useful for ordering things easily assume so great an authority over us, that we forget their terrestrial origin and accept them as unalterable facts. They then become labeled as "conceptual necessities," "*a priori situations*," etc. The road to scientific progress is frequently blocked for long periods by such errors. It is therefore not just an idle game to exercise our ability to analyze familiar concepts, and to demonstrate the conditions on which their justification and usefulness depend, and the way in which these developed, little by little, from the data of experience. In this way they are deprived of their excessive authority."

Yang–Mills gravity, with its translational gauge symmetry in flat space-time and consistency with all known experiments, as well as other desirable features such as the preservation of the conservation law of energy–momentum, operational definitions of space-time coordinates, and quantizability of gravitational fields, is just such a counterexample from outside the curved space-time framework of general relativity.[i]

0.4.   Unification of All Interactions Based on the Yang–Mills–Utiyama–Weyl Framework

In Chapters 14–16, we consider a model for unifying all interactions in nature (including gravity) within the taiji symmetry and the generalized Yang–Mills frameworks, following the ideas of Glashow, Salam, Ward, and Weinberg for the unified electroweak interaction.[16] We here present a brief preview of the unification of the gravitational interaction with just the electroweak interaction.

The T_4 gauge covariant derivative (0.5) paves the way for a unification of the gravitational and electroweak ("gravity-electroweak") interactions based on $T_4 \times SU_2 \times U_1$. Consider a unified "gravity-electroweak" model based on a generalized covariant derivative d_μ,[17]

$$d_\mu \equiv J_\mu^v \partial_v - if W_\mu^a t^a - if' U_\mu t_o, \quad J_\mu^v = \delta_\mu^v + g\phi_\mu^v. \tag{0.8}$$

This is a linear combination of the usual gauge covariant derivative in the electroweak theory and the gravitational T_4 gauge covariant derivative (0.5). The gauge curvatures, $C_{\mu v\alpha}$, $W_{\mu v}^a$, and $U_{\mu v}$, associated with each group in the "gravity-electroweak" covariant derivative d_μ are given by

$$[d_\mu, d_v] = C_{\mu v\alpha} T^\alpha - if W_{\mu v}^a t^a - if' U_{\mu v} t_o, \quad [t^a, t^b] = i\epsilon^{abc} t^c, \tag{0.9}$$

and the $T(4)$ gauge curvature $C_{\mu v\alpha}$ in flat space-time is[17]

$$C_{\mu v\alpha} = J_{\mu\sigma} \partial^\sigma J_{v\alpha} - J_{v\sigma} \partial^\sigma J_{\mu\alpha} = \Delta_\mu J_{v\alpha} - \Delta_v J_{\mu\alpha} = -C_{v\mu\alpha}. \tag{0.10}$$

---

[i]For key features of Yang–Mills gravity and Einstein gravity, see a comparison at the end of this overview.

The new SU_2 and U_1 gauge curvatures in the presence of the gravitational gauge potential are then given by

$$W_{\mu\nu}^a = J_\mu^\sigma \partial_\sigma W_\nu^a - J_\nu^\sigma \partial_\sigma W_\mu^a + f \epsilon^{abc} W_\mu^b W_\nu^c \qquad (0.11)$$

and

$$U_{\mu\nu} = J_\mu^\sigma \partial_\sigma U_\nu - J_\nu^\sigma \partial_\sigma U_\mu = \Delta_\mu U_\nu - \Delta_\nu U_\mu, \qquad (0.12)$$

respectively. We observe that the T_4 gauge potential appears in SU_2 and U_1 gauge curvatures in (0.11) and (0.12), but the SU_2 and U_1 gauge potentials do not appear in the T_4 gauge curvature in (0.10). This reflects the universal nature of the gravitational interaction, in which all things participate. In the absence of gravity, i.e., $g = 0$, the gauge curvatures in (0.11) and (0.12) have the usual $SU_2 \times U_1$ gauge transformation properties in the Weinberg–Salam theory.[16]

One new implication of the preceding unification is that there is an extremely small violation of the local $SU_2 \times U_1$ gauge invariance due to gravity, i.e., the presence of J_μ^σ in the gauge curvatures (0.11) and (0.12). Such a violation of internal gauge symmetries can be traced to the presence of J_μ^ν in the unified covariant derivative d_μ in (0.8) and, in the case of the U_1 gauge symmetry, for example, appears to be just right for the emergence of the effective Riemannian metric tensors in the eikonal equations in the geometric-optics limit of the electromagnetic wave equations. Thus, in the gravity-electroweak model, the effective curved space-time for the motion of classical objects that enables its compatibility with classical tests of general relativity[17] and the violations of gauge symmetries of internal groups ($SU_2 \times U_1$) due to gravity are related.

It appears that all observed fundamental interactions in the physical world are manifestations of U_1, SU_2, SU_3, and T_4 gauge symmetries. The intimate relationship between interactions and gauge symmetries was first advocated by Utiyama, Yang, Mills and others.[18] The unification of all forces through our understanding of the profound and intertwined internal and external gauge symmetries based on the taiji symmetry framework may be the key to uncovering nature's deepest secrets.

---

[j] Utiyama (see Ref. 10) concluded that Einstein's equations are gauge field equations. But Yang believed that was an unnatural interpretation of gauge fields. He obtained gravitational equations, which are third-order differential equations for $g_{\mu\nu}$.

A Comparison of Key Features of Yang–Mills Gravity, Einstein Gravity and Electromagnetic Theory

Yang–Mills Gravity *(Flat space-time)*[$]

$\phi_{\mu\nu}$: *gauge potential fields,*

Physical frames of reference: defined,

Coordinates x^μ and momenta p_μ have operational meaning,

Invariance under local space-time translations in flat space-time:

$$x'^\mu = x^\mu + \Lambda^\mu(x), \quad \Lambda^\mu(x) = \textit{infinitesimal local translation,}$$

$$\phi'_{\mu\nu}(x) = \phi_{\mu\nu}(x) - \Lambda^\lambda(x)\partial_\lambda\phi_{\mu\nu}(x) - \phi_{\mu\alpha}(x)\partial_\nu\Lambda^\alpha(x) - \phi_{\alpha\nu}(x)\partial_\mu\Lambda^\alpha(x);$$

Space-time translation (T_4) group generators: iD_μ,[$$]

T_4 gauge covariant derivative:

$$(D_\mu + g\phi_{\mu\nu}D^\nu) = J_{\mu\alpha}D^\alpha, \qquad J_{\mu\alpha} = P_{\mu\alpha} + g\phi_{\mu\alpha};$$

Gauge field strength $(T_4$ gauge curvature$)$: $C_{\mu\nu\lambda}$,

$$[J_{\mu\alpha}D^\alpha, J_{\nu\beta}D^\beta] = C_{\mu\nu\lambda}D^\lambda, \quad C_{\mu\nu\lambda} = J_{\mu\alpha}D^\alpha J_{\nu\lambda} - J_{\nu\beta}D^\beta J_{\mu\lambda};$$

Lagrangian: $L = \left[\frac{1}{4g^2}(C_{\mu\nu\lambda}C^{\mu\nu\lambda} - 2C_{\mu\alpha}{}^\alpha C^{\mu\beta}{}_\beta)\right]\sqrt{-\det P_{\mu\nu}}$;

Maximum graviton self-coupling: 4-vertex.

[$]Inertial and non-inertial frames with the Poincaré metric tensors $P_{\mu\nu}$.

[$$]D_μ is the covariant derivative associated with the Poincaré metric tensors in a general frame of reference in flat space-time. It satisfies the relations $[D_\mu, D_\nu] = 0$.

Einstein Gravity (*Curved space-time*)

$\Gamma^\lambda_{\mu\nu}$: *gauge potential fields,*

Physical frames of reference: undefined,[$]

Coordinates x^μ and momenta p_μ have no operational meaning,[$]

Invariance under general coordinate transformations in curved space-time:

$x'^\mu = x^\mu + \xi^\mu(x), \quad \xi^\mu(x) = $ *arbitrary infinitesimal vector functions;*

$g'_{\mu\nu}(x) = g_{\mu\nu}(x) + g_{\alpha\beta}(x)[E_{\mu\nu}{}^{\alpha\beta}{}_\lambda]\xi^\lambda(x);$[$$]

"Infinite continuous group generators": $E_{\mu\nu}{}^{\alpha\beta}{}_\lambda = -(\delta^{\alpha\beta}_{\mu\nu}\overleftarrow{\partial}_\lambda + \delta^{\alpha\beta}_{\mu\lambda}\partial_\nu + \delta^{\alpha\beta}_{\lambda\mu}\partial_\mu),$

Covariant derivative: $\quad \Delta_\mu V_\nu = \partial_\mu V_\nu + \Gamma^\lambda_{\mu\nu} V_\lambda,$

Riemann–Christoffel (space-time) curvature: $R^\alpha_{\lambda\mu\nu},$

$[\Delta_\mu, \Delta_\nu]V_\lambda = R^\alpha_{\lambda\mu\nu} V_\alpha, \quad R^\alpha_{\lambda\mu\nu} = \partial_\nu\Gamma^\alpha_{\lambda\mu} - \partial_\mu\Gamma^\alpha_{\lambda\nu} - \Gamma^\alpha_{\mu\beta}\Gamma^\beta_{\nu\lambda} + \Gamma^\alpha_{\nu\beta}\Gamma^\beta_{\mu\lambda};$

Lagrangian: $\quad L = \left\{\frac{R}{16\pi G}\right\}\sqrt{-\det g_{\mu\nu}}, \quad R = g^{\mu\nu}R^\beta_{\mu\beta\nu};$

Maximum graviton self-coupling: ∞-*vertex.*

[$]See Wigner's comments in Ref. 3.

[$$] The expression for $E_{\mu\nu}{}^{\alpha\beta}{}_\lambda$ may be used with some caution. It is merely to show the difficulty of interpreting Einstein's gravitational fields as gauge fields in the Yang–Mills approach.[19,j] For rigorous discussions of the infinite continuous group and its generators, see Noether's paper in Ref. 4. The transformation $\delta g_{\mu\nu} = g'_{\mu\nu} - g_{\mu\nu}$ can also be expressed as follows:

$$\delta g_{\mu\nu} = E_{\mu\nu}{}^{\alpha'\beta'}{}_{\lambda''}\, g_{\alpha'\beta'}\xi^{\lambda''}$$

$$\equiv -\int [(\delta^{\alpha\beta}_{\mu\nu}\partial_\lambda\delta(x-x')\delta(x-x'')$$

$$+ \delta(x-x')\{\delta^{\alpha\beta}_{\mu\lambda}\partial_\nu\delta(x-x'') + \delta^{\alpha\beta}_{\lambda\nu}\partial_\mu\delta(x-x'')\}]g_{\alpha\beta}(x')\xi^\lambda(x'')dx'\,dx'',$$

$$\delta(x-x') \equiv \delta^4(x-x'), \quad dx' \equiv d^4x', \text{ etc.}$$

<u>Electromagnetic Theory</u> (*Minkowski space-time*) [$]

A^μ: *gauge potential fields,*

Physical frames of reference: defined,

Coordinates x^μ and momenta p_μ have operational meaning,

Invariance under U_1 gauge transformations,

$A'_\mu = A_\mu + \partial_\mu \Lambda(x),$ $\Lambda(x) = $ *arbitrary infinitesimal function,*

U_1 group generator: 1,

U_1 gauge covariant derivative: $\Delta_\mu = \partial_\mu - ieA_\mu,$

Gauge field strength (U_1 gauge curvature): $F_{\mu\nu},$

$[\Delta_\mu, \Delta_\nu] = ieF_{\mu\nu},$ $F_{\mu\nu} = \partial_\mu A_\nu - \partial_\nu A_\mu;$

Lagrangian: $L = \left[-\tfrac{1}{4}F_{\mu\nu}F^{\mu\nu}\right];$

Photon self-coupling: none.

[$]*Inertial frames.*

0.5. The Big Picture of the Universe Revealed by Quantum Yang–Mills Gravity

Since only long-range forces play an important role at all length scales, let us use gravity to illustrate the big picture of nature. It is interesting that quantum Yang–Mills gravity reveals qualitatively different laws of motion at different length scales. Such a physical picture emerges naturally through the combination of Yang–Mills gauge symmetry in flat space-time and the effective metric tensor a la Einstein–Grossmann.

(I) **Microscopic World:** In quantum gravity, the key role is played by the space-time translational T_4 gauge group with the generators $p_\mu = i\partial_\mu$ so that we have the T_4 gauge covariant derivative Δ_μ, as shown in (0.5)

$$\Delta^\mu \equiv \partial^\mu - ig\phi^{\mu\nu}p_\nu \equiv J^{\mu\nu}\partial_\nu, \tag{0.13}$$

where $J^{\mu\nu}(x) = \eta^{\mu\nu} + g\phi^{\mu\nu}(x)$, $\eta^{\mu\nu} = (1, -1, -1, -1)$. In contrast to all other gauge theories with internal gauge groups, this external space-time T_4 gauge covariant derivative Δ^μ possess three distinct properties: (i) The coupling constant g has the dimension of length rather than being dimensionless, so that it can be related to the Newtonian gravitational constant G, i.e., $g^2 = 8\pi G$. (ii) The coupling term involving g in the gauge covariant derivative Δ_μ does not contain the imaginary number "i" because the T_4 group generators take the form $p_\mu = i\partial_\mu$. Its complex conjugate does not change sign, so that there is only one kind of gravitation force between matter (it is always attractive). In other gauge theories such as quantum electrodynamics (QED) and quantum chromodynamics (QCD), the coupling terms contain "i" and hence, admit both attractive and repulsive forces between, say, electric charges. (iii) The T_4 gauge field $\phi_{\mu\nu}(x)$ is a tensor field rather than a vector field, in contrast to QED and QCD. This tensor field $\phi_{\mu\nu}$ is crucial for Yang–Mills gravity to be connected to an effective metric tensor a la Einstein–Grossmann. This connection can be demonstrated below by the Dirac wave equation of quantum particles in the presence of gravity, i.e.,

$$i\gamma_\mu\Delta^\mu\psi - m\psi + \frac{i}{2}[\partial_\nu(J^{\mu\nu}\gamma_\mu)]\psi = 0. \qquad (0.14)$$

(II) **Macroscopic World:** The equation describing the propagation of a light ray can be derived from Maxwell wave equations by taking the geometric-optics limit (i.e., letting wavelength approach zero). In analogy, when we take the geometric-optics limit of the Dirac wave equation (0.14), we obtain a classical Hamilton–Jacobi type equation (0.6),

$$G^{\mu\nu}(x)(\partial_\mu S)(\partial_\nu S) - m^2 = 0, \qquad (0.15)$$

$$G^{\mu\nu}(x) = \eta_{\alpha\beta}J^{\alpha\mu}(x)J^{\beta\nu}(x), \qquad (0.16)$$

where $G^{\mu\nu}(x)$ appears in (0.15) as if it were the metric tensor of space-time. However, it is an effective metric tensor because it is really just a function of the T_4 gauge field $\phi_{\mu\nu}$ in flat space-time, as shown in (0.16), which is determined by the Yang–Mills gravity with T_4 gauge symmetry.[k]

(III) **Super-Macroscopic World with a Standard Cosmic Time** t_c: All previous equations in the macroscopic and microscopic worlds do not help us to understand the Hubble recession velocity and the cosmic redshift. Fortunately, the combination

---

[k]The equation of motion (0.15) is formally the same as that in general relativity. We called it the Einstein–Grossmann equation in memory of their collaborations.

of Yang–Mills gravity and the cosmological principle of homogeneity and isotropy enable us to tackle these problems. However, homogeneity implies that $G^{\mu\nu}$ in (0.15) and (0.16) can only depend on time, and not on space, i.e., $G^{\mu\nu}(t)$. So space and time are no longer on an equal footing, contrary to the requirement of the four-dimensional symmetry of space-time and special relativity. Thus, how to formulate a relativistically invariant cosmology model becomes a non-trivial problem within the framework of Lorentz invariant Yang–Mills gravity.

It turns out that we can resolve this difficulty by introducing a scalar cosmic time t_c for all observers. We can choose any one inertial frame, called the F frame, to set up a grid of synchronized clocks.[20] The clocks are synchronized in such a way so that the speed of light is constant and isotropic in F. All observers in all inertial frames use the nearest F clock to record time, so that this represents a cosmic time system. Although all other inertial frames now have a non-isotropic speed of light, measured in units of length per time,[1] such a theory of cosmic relativity can be formulated on the basis of the principle of relativity and as a result, the theory turns out to be consistent with all known experiments.[21]

Such an operationally defined cosmic time t_c within the four-dimensional symmetry framework is particularly useful for the super-macroscopic world. The reason is that all measurements by observers in a great many inertial frames in the physical world then have a simple standardized time to measure recession velocities, expansion rates, distances, the ages of star clusters, and the age of the universe.[22]

Quantum Yang–Mills gravity and the cosmological principle in the super-macroscopic world lead to a new effective metric tensor,

$$G^{\mu\nu}(x) \rightarrow G^{\mu\nu}(t_c) = \eta^{\mu\nu} U^{-2}(t_c), \quad U(t_c) \propto t_c^{1/2}, \qquad (0.17)$$

for a matter-dominated cosmos. Thus, equation (0.15) for macroscopic objects takes a new simple form

$$\eta^{\mu\nu} U^{-2}(t_c)(\partial_\mu S)(\partial_\nu S) - m^2 = 0. \qquad (0.18)$$

This is called the Okubo equation, which is Lorentz invariant and governs the motion of distant galaxies in the super-macroscopic world. Naturally, the massive Okubo equation can be derived from a Lorentz invariant cosmic action involving

---

[1]When one expresses both space and time using the same unit (i.e., measuring time in meters or distances in seconds), the speed of light is again isotropic in all inertial frames.

the super-macroscopic effective metric tensor (0.17)

$$S_{\cos} = \int (-mds), \quad ds^2 = G_{\mu\nu}(t_c)dx^\mu dx^\nu, \qquad (0.19)$$

where $G^{\mu\nu}(t_c)G_{\nu\lambda}(t_c) = \delta^\mu_\lambda$. Thus, we have a relativistic-cosmology model based on the Okubo equation (0.18) or, equivalently, the Lorentz invariant cosmic action (0.19). As usual, the equation describing the propagation of a light ray and the cosmic redshifts are assumed to be described by the massless Okubo equation (0.18) with $m = 0$.

0.6.   Baryon–lepton Charges, Late-Time Accelerated Cosmic Expansion and Dark Energy

Let us consider only conserved baryon charges for simplicity since the same discussions and equations can be applied to conserved lepton charges. The original idea for understanding the conservation of baryon number (or baryon charges) was to associate it with the usual U_1 gauge symmetry, as discussed by Lee and Yang in 1955.[23,m] Their formulation of dynamics for baryon charges is identical to that of QED. Based on Eötvös' experiments, they estimated that the coupling strength of such an inverse-square baryon force is about 10^{-5} smaller than that of the gravitational force. As a result, there have been no observable physical effects of Lee–Yang's U_1 fields (or gauge bosons). Some physicists did not consider such a U_1 symmetry corresponding to conserved baryon charges to be local gauge symmetry.

However, in view of the success of the unified electroweak theory and QCD, it is natural to assume that gauge symmetry is indeed a universal symmetry principle.[24,n] Thus, baryon and lepton charges must correspond to local U_1 gauge symmetries. The reason that these massless gauge fields are not observed in the laboratory could be due to their extremely small coupling strength. Suppose Lee–Yang's usual U_1 symmetry (associated with a scalar gauge function and phase function) is generalized to a general U_1 symmetry (associated with a vector gauge function and Hamilton's characteristic phase function). Then, baryon–lepton charges can produce a linear repulsive force between galaxies. Modeling a supernova as a baryonic sphere of radius R_s located within the universe, modeled

---

[m]The baryon conservation law was first postulated by Stueckelberg.
[n]The fundamental interactions of particles are determined by gauge invariance. This idea was stressed by Utiyama, Yang and others.

as a baryonic sphere with radius R_o, this linear repulsive force is roughly,

$$F \propto \left[\frac{r}{R_o} - \frac{rR_s^2}{5R_o^3} \right]. \tag{0.20}$$

The properties of this force turn out to be just right for baryon–lepton charges to play an important and dominant role in the late-time cosmic acceleration,[25] even though they cannot be detected in high energy laboratories and in our Milky Way galaxy.

In the dynamics of cosmic expansion with general U_1 symmetry,[26] we shall often use "phase symmetry" instead of the usual term "gauge symmetry" in order to emphasize that phase symmetries involve Lorentz vector gauge functions $\Lambda_\mu(x)$, rather than the usual scalar gauge function $\Lambda(x)$). Furthermore, phase fields satisfy fourth-order field equations rather than the usual second-order gauge field equations.

To explore the late-time accelerated cosmic expansion, we suggest that the universal space-time scale factor $U(t)$ in the relativistic-cosmology model could be further modified by taking into account all the long-range cosmic forces acting on galaxies, including gravitational and baryon-lepton forces. The novel repulsive linear forces between distant galaxies produced by baryon and lepton charges could provide a new quantitative understanding of the mysterious "dark energy" related to the late-time accelerated cosmic expansion. This enables us to estimate an effective vacuum energy density as

$$\rho^b \approx (10^{-66}\text{--}10^{-48})\,\text{GeV}^4, \quad c = \hbar = 1, \tag{0.21}$$

where the experimental value is $\approx 10^{-47}\,\text{GeV}^4$.[o]

In summary, Yang–Mills gravity, the cosmological principle and the cosmic Okubo equation suggest new views and understandings of our universe:

(i) The linear Hubble law is the low-velocity approximate solution to the Okubo equation for distant galaxies.

(ii) The Okubo equation leads to a generalized Hubble law for all recession velocities, which has an upper limit, i.e., the speed of light in vacuum.

(iii) The universe began with the maximum recession velocity $\dot{r} = c$, $r = 2p^2c^3/(m^2\beta^2) \equiv r_o$, and cosmic redshift $z = \infty$, at time $t = 0$ for a matter-dominated universe; and the universe will end with $\dot{r} \to 0$, $r \to \infty$, $z \to 0$,

---

[o]For cosmological parameters in Planck 2018 results, see Ref. 27.

as time $t \to \infty$. A more realistic model could be constructed by removing the assumption that the universe is matter-dominated.

(iv) The Okubo equations with $m \geq 0$ are the basic equations of motion for all distant galaxies and for the redshift in the super-macroscopic limit.

(v) The Okubo equation with $m > 0$ suggests that the universe underwent an initial expansion, which resembled some sort of "detonation" at time $t = 0$ with the speed c.

References

1. A. Einstein and M. Grossman, in *100 Years of Gravity and Accelerated Frames: The Deepest Insights of Einstein and Yang–Mills* eds. J. P. Hsu and D. Fine (World Scientific, 2005), p. 48; A. Einstein, *ibid.*, p. 65. D. Hilbert, *ibid.*, p. 120.

2. F. J. Dyson, *Bull. Amer. Math. Soc.* **78** (1972).

3. E. P. Wigner, in *Symmetries and Reflections, Scientific Essays* (The MIT Press, 1967), pp. 52–53.

4. E. Noether, *Goett. Nachr.* **235** (1918). English translation of Noether's paper by M.A. Tavel is online. Google search: M.A. Tavel, Noether's paper.

5. T.-Y. Wu and Y. C. Lee, *Int. J. Theor. Phys.* **5**, 307 (1972); T.-Y. Wu, in *Theoretical Physics, Vol. 4, Theory of Relativity* (Lian Jing Publishing Co., Taipei, 1978), pp. 172–175.

6. J. P. Hsu and L. Hsu, in *A Broader View of Relativity, General Implications of Lorentz and Poincaré Invariance* (World Scientific, 2006), Chapters 18, 19, 24, 25 and 26. Online: "google books, a broader view of relativity, Hsu."

7. O. Veblen and J. H. C. Whitehead, in *The Foundations of Differential Geometry* (Cambridge University Press, 1954), pp. 37–38.

8. H. Weyl, in *Space-Time-Matter*, 4th ed. (Dutton, 1922), Trans. H. L. Brose (Dover, 1930); F. London, *Z. Phys.* **37**, 375 (1927); V. Fock, *Z. Phys.* **39**, 226 (1927); V. I. Oglevetski and I. V. Polubarinov, *Nuovo Cimento* **23**, 173 (1962).

9. C N. Yang and R. L. Mills, *Phys. Rev.* **96**, 191 (1954); T. D. Lee and C. N. Yang, *Phys. Rev.* **98**, 155 (1955); R. Shaw, PhD thesis (1955).

10. R. Utiyama, *Phys. Rev.* **101**, 1597 (1956); T. W. B. Kibble, *J. Math. Phys.* **2**, 212 (1961); *100 Years of Gravity and Accelerated Frames: The Deepest Insights of Einstein and Yang–Mills*, eds. J. P. Hsu and D. Fine (World Scientific, 2005), Chapters 7 and 8 (available online: "google books, 100 years of gravity, J. P. Hsu."); T. W. Hehl, P. von der Heyde, G. D. Kerlick and J. M. Nester, *Rev. Mod. Phys.* **48**, 393 (1976); G. Grignani and G. Nardelli, *Phys. Rev. D* **45**, 2719 (1992); D. Cangemi and R. Jackiw, *Phys. Rev. Lett.* **69**, 233 (1992); T. Strobl, *Phys. Rev. D* **48**, 5029 (1993); G. Grignani and G. Nardelli, *Nucl.*

Phys. B **412**, 320 (1994); N. A. Batakis, *Phys. Lett. B* **391**, 59 (1997); Y. Ne'eman, *Acta Physica Polonica, B* **29**, 827 (1998); V. Aldaya, J. L. Jaramillo and J. Guerrero, *J. Math. Phys.* **44**, 5166 (2003); T. Kawai, *Phys. Rev. D* **62**, 104014/1 (2000).

11. J. P. Hsu, *Int. J. Mod. Phys. A* **21**, 5119 (2006).

12. L. Landau and E. Lifshitz, *The Classical Theory of Fields*, Trans. by M. Hamermesh (Addison-Wesley, Cambridge, MA, 1951), pp. 312–323, pp. 136–137 and p. 319. J. P. Hsu, *Phys. Rev. Lett.* **42**, 934 (1979).

13. J. P. Hsu, *Int. J. Mod. Phys. A* **24**, 5217 (2009); *Eur. Phys. J. Plus* **126**, 24 (2011).

14. J. P. Hsu, *Nuovo Cimento B* **108**, 183 (1993); **108**, 949 (1993); **109**, 645 (1994).

15. A. Einstein, *Phys. Zeitschr.* **17**, 101 (1916).

16. K. Huang, in *Quarks, Leptons and Gauge Field* (World Scientific, 1982), pp. 108–117 and pp. 152–156.

17. J. P. Hsu, *Mod. Phys. Lett. A* **26**, 1707 (2011); *Chin. Phys. C* **36**, 403 (2012).

18. J. P. Hsu and D. Fine (eds.), in *100 Years of Gravity and Accelerated Frames: The Deepest Insights of Einstein and Yang–Mills* (World Scientific, 2005), Chapters 4(C) and 10(C).

19. C. N. Yang, in *100 Years of Gravity and Accelerated Frames: The Deepest Insights of Einstein and Yang–Mills* (World Scientific, 2005), pp. 387–389.

20. J. P. Hsu Editorial, *Nature* **303**, 129 (1983); J. P. Hsu, *Found. Phys.* **8**, 371 (1978); **6**, 317 (1976); J. P. Hsu and T. N. Sherry, *Found. Phys.* **10**, 57 (1980); J. P. Hsu, *Il Nuovo Cimento* **74B**, 67 (1983).

21. J. P. Hsu and L. Hsu, in *A Broader View of Relativity* (World Scientific, 2006), Chapters 7, 9, 10.

22. J. P. Hsu, L. Hsu and D. Katz, *Mod. Phys. Lett. A* **36**, 1850116 (2018).

23. T. D. Lee and C. N. Yang, in *100 Years of Gravity and Accelerated Frames: The Deepest Insights of Einstein and Yang–Mills* eds. J. P. Hsu and D. Fine (World Scientific, 2005), p. 155; E. C. G. Stueckelberg, *Helv. Phys. Acta* **11**, 299 (1938).

24. Y. Utiyama, *Phys. Rev.* **101**, 1597 (1956); C. N. Yang, *Phys. Today* **33**, 42 (1980).

25. S. Perlmutter *et al.*, *Astrophys. J.* **517**, 565 (1999); A. G. Riess *et al.*, *Astron. J.* **116**, 1009 (1998).

26. J. P. Hsu, *Chin. Phys. C.* **41**, 015101 (2017); arXiv: 1609.00227 [physics.gen-ph]; J. P. Hsu, *Eur. Phys. J. Plus* **129**, 108 (2014).

27. S. Weinberg, in *Cosmology* (Oxford University Press, 2008), p. 39, p. 57, p. 510; N. Aghanim *et al.* (Planck Collaboration), arXiv:1807.06209.

About the Authors

Jong-Ping Hsu received his B.S. from the National Taiwan University and M.S. from the National Tsing Hua University. He earned his Ph.D. degree in 1969 studying particle physics at the University of Rochester (New York) with Professor S. Okubo. He has done research at McGill University, Rutgers, the University of Texas at Austin, the Marshall Space Flight Center, NASA, and the University of Massachusetts Dartmouth. He has been a visiting scientist at Brown, MIT, Taiwan University, Beijing Normal University, and the Academy of Science, China. His research is concentrated in the areas of gauge field theories, Yang–Mills gravity, total-unified model of interactions and broad views of four-dimensional symmetry. He has published more than 175 papers and articles, and two books, *A Broader View of Relativity: General Implications of Lorentz and Poincaré Invariance* (with L. Hsu, World Scientific, 2006) and *Lorentz and Poincaré Invariance* (with Y. Z. Zhang, World Scientific, 2001). He has also co-edited four conference proceedings and the book *100 Years of Gravity and Accelerated Frames: The Deepest Insights of Einstein and Yang–Mills* (with D. Fine, World Scientific, 2005). He is currently a chancellor professor and the Director of the Jing Shin Research Fund at the University of Massachusetts Dartmouth.

Leonardo Hsu earned his Ph.D. doing both experiment and theoretical work on the physics of semiconductors at the University of California, Berkeley. After doing a postdoc in physics education research at the Center for Innovation in Learning at Carnegie Mellon University, he joined the faculty of the University of Minnesota, Twin Cities from 2000–2017. His primary research interest is in studying how students learn to solve problems in introductory physics courses. He also studies transport processes in semiconductors and their alloys, as well as the implications of the principle of relativity for different relativity theories. He was a co-editor of the *JingShin Theoretical Physics Symposium in Honor of*

Professor Ta-You Wu (World Scientific, 1998) and the monograph *A Broader View of Relativity: General Implications of Lorentz and Poincaré Invariance* (World Scientific, 2006). His work has been supported by the Potz Science Fund and by the National Science Foundation. He now teaches physics at Santa Rosa Junior College, applying knowledge from his own work and the work of others in physics education research.

Part I

A More General Space-Time Symmetry Framework

Leonardo Hsu and Jong-Ping Hsu

Chapter 1

Space-Time Symmetry, Natural Units and
Fundamental Constants

1.1. Underpinnings

Before discussing the development of the taiji relativity framework, it is worth
taking some time to examine some of the supporting structures of the theory
that are often taken for granted, namely the units used for specifying physical
quantities and the fundamental constants that do or do not appear in the theory. In
developing taiji relativity, we hope to create the simplest space-time framework that
can accommodate all the laws of physics in a unified manner, where by "simplest,"
we mean that it is based on the smallest number of postulates. Correspondingly,
in our discussions, we choose to use a system of units that is also, in a way, the
"simplest," in the sense that it has only one unit. This is the system of "natural
units," in which all physical quantities are specified in terms of a power of the
same single unit.[1-3,a] One desirable feature of natural units is that the equations
expressing physical laws are simpler to write down and have fewer constants that
can obscure the essential physics they embody. For example, both the speed of
light c and Planck's constant $\hbar$ have unit magnitude with no dimensions and thus
do not appear in any of the equations. The system of natural units is widely used
by particle physicists for just this reason, although it is far from being the most
convenient or useful choice for many everyday applications.

---

[a] Pauli began his paper on "The Connection Between Spin and Statistics" with a discussion of 'Units and
Notations': 'Since the requirements of the relativity theory and the quantum theory are fundamental
for every theory, it is natural to use as units the vacuum velocity of light c, and Planck's constant divided
by 2π which we shall simply denote by $\hbar$. This convention means that all quantities are brought to the
dimension of a power of a length by multiplication with the power of $\hbar$ and c.'

3

One might ask however, if the natural unit system is simply a calculational convenience or whether there is a physical justification for, in effect, setting c and $\hbar$ equal to one. If there is to be a physical justification, one must be able to operationally define each of the base units in a well-established unit system, say, the SI system of units, in terms of one single unit in a physically meaningful way. For example, one could arbitrarily decide that 1 meter is equivalent to 5 kelvin, but such an equivalence would not be physically meaningful. On the other hand, saying that 1 second is equivalent to 299 792 458 meter is physically meaningful, as we shall see.

Below, we argue that there is indeed a physical basis for the natural unit system and thus its use in constructing a framework for physical laws, by presenting definitions for the most relevant base SI units in terms of a single unit. We also discuss the implications of these definitions for the nature of physical constants such as the speed of light c, Planck's constant $\hbar$ and Boltzmann's constant k_B.

1.2.   Physical Basis for the System of Natural Units

1.2.1.   *Equivalence of length and time*

Prior to the twentieth century, space and time seemed to be completely different entities and thus two different units, the meter and second, were invented to quantify them. From a modern viewpoint however, the space-time symmetry of special relativity implies that space and time are not independent and separate, but parts of a four-dimensional space-time. This view is reflected in the fact that in the SI system, the unit of length meter, is defined in terms of the unit of time, second.[3]

Such a definition is possible because in our physical laws, four coordinates, three spatial and one temporal, are necessary to specify the space-time location of an event. Although there is nothing wrong logically with using different units to express these four quantities, it is mathematically simpler to use the same single unit, either meter or second. As an analogy, there is nothing wrong with expressing north–south distances in meters and east–west distances in miles. Doing so, however, introduces an extra conversion constant into physical equations that is clearly artificial. The mathematical form of physical laws is much simpler when distances in both directions are expressed using the same unit.[4] The same is true for intervals of both space and time.

In order to establish a physically meaningful equivalence between the meter and the second, we must choose a physical phenomenon that involves the dimensions of both length and time. The constant speed c of a particle with a zero rest mass provides a convenient conversion factor between the two units. For practical reasons of precision and reproducibility, the unit of time second is chosen to be the base unit in the SI system. The modern (1983) definition of the meter is then "The meter is the length of the path travelled by light in vacuum during a time interval of 1/299 792 458 of a second."[3]

One could imagine that had the four-dimensional symmetry of our physical universe been understood from the very beginning, we might now only talk about the single dimension of length (or time), instead of the two dimensions of length and time, and only a single unit for both spatial and temporal intervals might have arisen, rather than the two separate units of second and meter. In fact, some textbooks on special relativity[4,5] purposely write the Lorentz transformation equations with both spatial and temporal coordinates expressed using the same unit to emphasize this point. Thus, we see that the constant $c = 299\ 792\ 458$ m/s originates from an incomplete understanding of physics and is merely a conversion factor between the human-defined units of meter and second.[6] (See also Ref. 4, p. 5)

1.2.2. *Equivalence of mass and length*

Similarly, before the development of quantum mechanics and Einstein's understanding of the equivalence of mass and energy, mass seemed to be a completely separate quantity from time and space. As such, the present definition of the kilogram, the SI unit of mass, is based on a physical artifact, the international prototype of the kilogram. However, just as the four-dimensional symmetry of our universe indicates that a single unit is sufficient to express both spatial and time intervals, the dual wave-particle nature of physical entities provides a physical basis for establishing a physical equivalence between the unit of mass and time. Using the relationships between energy and frequency $E = \hbar\omega$ (one could also use de Broglie relation $p = \hbar k$) and Einstein's relationship $E = mc^2$ (which becomes $E = m$ when expressing both length and time using the same units), we can then write $E = m = \hbar\omega$ to establish a relationship between a particle's mass and the angular frequency of its corresponding wave, where Planck's constant $\hbar$ acts as a the conversion factor between mass and time. Using the fact that 1 meter is equivalent to 1/299 792 458 of a second, we have

$$\hbar = 1.054 \times 10^{-34} \mathrm{J} \cdot \mathrm{s} = 1.054 \times 10^{-34} \left(\frac{1}{299\ 792\ 458} \right)^2 \mathrm{kg} \cdot \mathrm{s} = 1, \quad (1.1)$$

so that

$$1\,\text{kg} = 8.511\ldots \times 10^{50}\text{s}^{-1} = 2.839\ldots \times 10^{42}\text{m}^{-1}. \tag{1.2}$$

Following this line of thought, one serious proposal for a future definition of the kilogram that is being considered by the General Conference on Weights and Measures (CGPM) is "The kilogram is the mass of a body whose equivalent energy is equal to that of a number of photons whose frequencies sum to exactly $[(299\,792\,458)^2/66\,260\,693] \times 10^{41}$ hertz."[7]

In the system of natural units used by particle physicists, it is more usual to set $\hbar = h/2\pi = 1$ rather than $h = 1$, so that $1\,\text{kg}$ is equivalent to $[2\pi(299\,792\,458)^2/66\,260\,693] \times 10^{41}\,\text{s}^{-1}$ since $E = \hbar\omega$. In this case, one could define the kilogram as "The kilogram is the mass of a body whose equivalent energy is equal to that of a number of photons whose **angular** frequencies sum to exactly $[2\pi(299\,792\,458)^2/66\,260\,693] \times 10^{41}\,\text{s}^{-1}$." We will use this alternative definition. It is important to note that although the numerical value of the equivalence between the kilogram and inverse second is a matter of choice and not unique, the main result, that there is a physical basis for expressing masses and time intervals using the same unit (albeit different powers of that single unit), still stands.

At present, the proposed definition given above for the SI unit kilogram has not yet been adopted because the stability and reproducibility of the mass of the international prototype of the kilogram is still better than that of any atomic standard that we can yet achieve.[8,9] However, if the kilogram were to be redefined in terms of the frequency of a collection of photons, one consequence would be the fixing of Planck's constant $\hbar$ to an exact value, analogous to the case where the redefinition of the meter in 1983 had the effect of fixing the speed of light c to an exact value, with no uncertainty. Planck's constant would then become analogous to the speed of light, a conversion factor between the human defined units of kilogram and second.

Although the ability to express lengths, masses, and time intervals using the same single unit implies that all physical quantities can be expressed in terms of powers of that unit,[b] one can also develop corresponding operational equivalences for the other base SI units. Let us consider current ampere and temperature as examples.

---

[b]The reason is that all units in SI can be expressed in terms of the units meter, kilogram and second. For example, the unit Newton for force can be expressed as $\text{kg} \cdot \text{m} \cdot \text{s}^{-2}$.

1.2.3.   *Equivalence of current and length*

The present definition of the SI unit of current ampere is based on the force between two parallel current carrying wires

$$\frac{F}{l} = \frac{2I_1 I_2}{c^2 d} \tag{1.3}$$

written in Gaussian units where F/l is the force per unit length between two infinite current-carrying wires, I_1 and I_2 are the currents in the two wires, c is the speed of light, and d is the perpendicular distance between the wires. Using the equivalences previously developed between the units of second, meter, and kilogram, we can insert 2×10^{-7} newton per meter for F/l, $c = 1$, 1 ampere for both I_1 and I_2, and 1 meter for d, to obtain the conversion 1 ampere $= 5.331 \ldots \times 10^{17}$ second$^{-1} = 1.778 \ldots \times 10^9$ meter^{-1}. Thus the definition of the ampere can be rewritten in terms of the unit meter as follows: "The ampere is that constant current which, if maintained in two straight parallel conductors of infinite length, of negligible circular cross-section, and placed 1 meter apart in vacuum, would produce between these conductors a force equal to $2.109 \ldots \times 10^{10}$ meter^{-2} per meter of length (or $5.683 \ldots \times 10^{35}$ second^{-2} per second of length)."

1.2.4.   *Equivalence of temperature and length*

The present definition of the SI unit of temperature is "The kelvin, unit of thermodynamic temperature, is the fraction 1/273.16 of the thermodynamic temperature of the triple point of water."[3] However, just as special relativity shows that space and time are both parts of a four-dimensional space-time and quantum mechanics shows that mass and frequency (or wavelength) are simply different aspects of the same quantity, the kinetic theory of gases implies that temperature is not an independent characteristic of a system, but simply one type of energy scale. For example, the temperature of a monatomic ideal gas is related to the average kinetic energy of its individual gas molecules through $E = (3/2)k_B T$. In natural units, both energy and temperature are dimensionally equivalent to the kilogram, the inverse meter, or the inverse second.

To establish a physically meaningful conversion factor between the kelvin and these other units, we note that the Boltzmann constant k_B already provides a conversion between the energy and the temperature of a system. In analogy with using c and $\hbar$ as conversion factors between the meter, second, and kilogram, a conversion which fixes the value of the Boltzmann constant k_B to unity ($k_B = 1$) is convenient and results in the equivalence 1 kelvin = $1.309\ldots \times 10^{11}$ second^{-1} = $436.7\ldots$ meter^{-1}. A new definition of the unit kelvin might then be "The kelvin, unit of thermodynamic temperature, is the thermodynamic temperature of a system whose energy is equal to the energy of a collection of photons whose angular frequencies sum to $1.309\ldots \times 10^{11}$ second^{-1} (or, equivalently, $4.367\ldots \times 10^{2}$ meter^{-1})."

Were this definition to be adopted, then, as was the case for the speed of light c and as may become the case for Planck's constant $\hbar$, the Boltzmann constant k_B would take on the status of a conversion constant between kelvin and second, a historical artifact resulting from an incomplete understanding of the atomic basis of temperature and energy. In this sense, the Boltzmann constant would become similar to Joule's constant, which was used to relate mechanical energy (in joules) and heat (in calories). When it was realized that both were the same type of quantity, the unit joule was used to express both and Joule's constant of 4.184 J/cal became merely another conversion constant with an equivalent value of one.

Similarly, other SI base units such as mole and luminous intensity can also be written in terms of the unit second.[2]

1.3.   Nature of the Fundamental Constants

In the equations of physics embodying our theories that allow us to predict and explain the diverse phenomena we observe in the universe, certain quantities appear repeatedly. These quantities, conventionally called fundamental constants or fundamental physical constants, typically either have their origins in the mathematical formulation of physical theories (such as $\hbar$ and Newton's gravitational constant G) or are properties of particles that are currently viewed as the basic building blocks of matter (such as the charge and mass of the electron). Levy-Leblond[10] grouped the physical constants into two categories, (1) constants characterizing whole classes of physical phenomena (such as the electric charge e and the universal gravitational constant G), and (2) universal constants (such as c and h) which act as concept or theory synthesizers (for example, Planck's constant h synthesizes the concepts of momentum and wavelength through the relation $p = h/\lambda$).

In this section, we use the natural system of units to divide the fundamental constants into two categories of our own, (A) those constants that are inherent descriptors of our universe, and (B) those constants that are historical products of an incomplete understanding of our universe and thus have values that are determined by human convention (or more precisely, by the initial definitions of the diverse units that were developed to describe our world, such as meter, second, and kilogram). Our reason for doing so is that in our formulation of what we hope to be the simplest space-time framework and theory that can accommodate all known interactions, the only constants we would like to appear are those from category (A), whose values are inherent properties of the universe and cannot be explained by any known theory. This is not to say that constants from category (B) are not useful. Indeed, the values of those constants are important for interpreting the results of experiments that can check the consistency and explanatory power of our theories and for making connections between physics and other disciplines. However, it should not be necessary for them to appear in a fundamental theory of physics.[11,c] Thus, papers in metrology that discuss the measurement of various physical constants[12,13] actually describe two conceptually different types of experiments. One is an experiment which seeks to determine more precisely one of the fundamental numbers that characterizes our universe. The second is an experiment that seeks to determine more precisely the conversion factor between two historically independent definitions of two units.

One criterion that can be used to determine the category to which a particular constant belongs is whether or not the value of that constant can be made exact or unity by a suitable physically grounded (re-)definition of units. For example, quantities such as c, $\hbar$, and k_B are category (B) constants because they can be made to have the value unity by redefining the second, meter, kilogram, and kelvin, as previously discussed.[14,d] Likewise, the vacuum permittivity ϵ_0 and permeability μ_0 are category (B) constants. Although they are not equal to unity under present definitions of the units, their values are exact, and they could be made unity by a suitable redefinition of the ampere. As a final example, the Stefan–Boltzmann constant σ is related to other constants by $\sigma = 2\pi^5 k_B^4/(15h^3 c^2)$ and

---

[c]Wilczek states that parameters such as e, $\hbar$, and the mass of the electron m_e can be eliminated by a suitable choice of a system of units for length $(\hbar^2/m_e e^2)$, time $(\hbar^3/m_e e^4)$, and mass (m_e). This is much the same as choosing a system of natural units, in which parameters such as c, $\hbar$ and k_B are eliminated. As Wilczek mentions later in the article, for a more complete theory, additional parameters such as the fine structure constant α inevitably come in.

[d]It is interesting to compare this result with Dirac's outlook on the physics of the future: "The physics of the future, of course, cannot have the three quantities $\hbar$, e and c all as fundamental quantities. Only two of them can be fundamental, and the third must be derived from those two. It is almost certain that c will be one of the two fundamental ones. The velocity of light, c, is so important in the four-dimensional picture, and it plays such a fundamental role in the special theory of relativity, correlating our units of space and time, that it has to be fundamental... I think one is on safe ground if one makes the guess that in the physical picture we shall have at some future stage e and c will be fundamental quantities and $\hbar$ will be derived."

in natural units (where $c = \hbar = k_B = 1$), has the exact value $\pi^2/60$. By a suitable redefinition of the kelvin (or meter, second, or kilogram), its value can be made unity.

Constants whose values cannot be made equal to unity by any physically grounded choice of units, and are therefore category (A) constants, inherent characteristics of our universe, include the electron charge e, the weak mixing (or the Weinberg) angle θ_W (found in the unified electroweak theory), and the strong coupling constant α_s (in quantum chromodynamics). Constants such as the Josephson constant $K_J = 2e/h$ and the von Klitzing constant $R_K = h/e^2$, by virtue of their relationship to category (A) fundamental constants, are themselves also fundamental constants. However, e, α_e, J_K, and R_K are not four independent fundamental constants, since they are all powers or multiples of each other. At present, the only quantities that qualify as category (A) fundamental constants under this criterion are the coupling constants for each of the four fundamental forces G_N (gravitational), $\alpha_e{}^e$ and θ_W (electroweak), and α_s (strong).[15,f]

In summary, natural units are not merely a calculational convenience, but have a conceptual basis rooted in the nature of our physical universe. Furthermore, the system of natural units provides a means of separating out the constants that should or should not appear in a fundamental physical theory. We shall use natural units throughout this book.

References

1. W. Pauli, *Phys. Rev.* **58**, 716 (1940).
2. L. Hsu and J. P. Hsu, *Eur. Phys. J. Plus* **127**, 11 (2012). DOI 10.1140/ epjp/i2012-12011-5.
3. B. N. Taylor (ed.), *NIST Special Publication 330: The International System of Units (SI)* (National Institute of Standards and Technology, Gaithersburg, MD, 2001), pp. 5–9; J. D. Jackson, *Classical Electrodynamics*, 3rd ed. (John Wiley & Sons, New York, 1999), pp. 775–776.
4. E. F. Taylor and J. A. Wheeler, *Spacetime Physics*, 2nd. ed. (W. H. Freeman, New York, 1992), pp. 1–4.

---

[e]We shall use Heavyside–Lotentz rationalized units with $\hbar = c = 1$, $\alpha_e = e^2/(4\pi) = 1/137.036$, where $-e = -0.3028$ is the charge of the electron in natural units. ($\alpha_e = e^2/(\hbar c)$ in cgs [Gaussian] units and $\alpha_e = e^2/(4\pi\hbar c\epsilon_0)$ in mks [SI] units.)

[f] That the gravitational coupling constant G is not dimensionless is intimately related to the fact that the gravitational interaction is related to the external space-time translational gauge symmetry. See Ref. 6. In the conventional theory of gravity, it may be attributed to the property that energy–momentum tensor is the source of the gravitational field.

5. T. A. Moore, *A Traveler's Guide to Spacetime: An Introduction to the Special Theory of Relativity* (McGraw-Hill, New York, 1995).

6. J. P. Hsu and L. Hsu, *A Broader View of Relativity, General Implications of Lorentz and Poincaré Invariance*, 2nd ed. (World Scientific, 2006), pp. 434–436. Available online: "google books, a broader view of relativity, Hsu."

7. I. M. Mills, P. J. Mohr, T. J. Quinn, B. N. Taylor and E. R. Williams, *Metrologia* **43**, 227 (2006).

8. B. N. Taylor and P. J. Mohr, *Metrologia* **36**, 63 (1999).

9. E. R. Williams, R. L. Steiner, D. B. Newell and P. T. Olson, *Phys. Rev. Lett.* **81**, 2404 (1998).

10. J.-M. Levy-Leblond, *Riv. Nuovo Cimento* **7**, 1897 (1977).

11. F. Wilczek, *Phys. Today* **56**, 10 (2003).

12. P. J. Mohr and B. N. Taylor, *Rev. Mod. Phys.* **72**, 351 (2000).

13. V. Kose and W. Woeger, *Metrologia* **22**, 177 (1986).

14. P. A. M. Dirac, *Sci. Am.* **208**, 48 (1963).

15. J. P. Hsu, *Eur. Phys. J. Plus* **126**, 24 (2011), [arXiv: 1102.2253].

Chapter 2

The Taiji Relativity Framework

2.1. A New Space-Time Framework

The primary goal of physics is to devise the most logically simple and parsimonious theories that can predict and explain the wide range of observed phenomena in the universe. A fundamental physical theory should be based on the smallest and simplest set of postulates. Einstein formulated special relativity on the basis of two postulates:

(I) the principle of relativity, which states that the laws of physics have the same form in every inertial frame, and
(II) the universality of the speed of light, which states that the speed of light is independent of the motion of the source and of the observer.

In this and the next few chapters, we shall show that from a modern viewpoint, one can formulate a relativity theory that is consistent with the results of all known experiments, solely on the basis of the principle of relativity (postulate I). The idea of formulating a relativity theory using only the principle of relativity is not new. In the literature, one can find statements to the effect that the universality of the speed of light is a direct consequence of the principle of relativity and that the second postulate is not needed.[1,2] However, such discussions typically end there, without further elaboration. In our case, we shall show that a relativity theory based solely on the first postulate, while calculationally equivalent to special relativity, has a very different conceptual foundation, particularly when viewed in light of the physical basis of natural units.

In a previous book, we showed how this different conceptual foundation led to a fresh view of old physics and new results, including making more tractable the solution of the relativistic many-body problem and helping to develop more convenient definitions for thermodynamics quantities in a relativistic framework.[3] In this volume, we demonstrate how this conceptual difference allows us to construct a flat four-dimensional space-time with arbitrary coordinates that can serve as a framework for both a classical and quantum theory of gravity. Unlike the arbitrary coordinates of the curved space-time in general relativity, those in this flat space-time retain their physical significance and can be operationally defined. Furthermore, because the resulting theory of gravity is a gauge theory in a flat space-time, it can be unified with the other known interactions in a way not allowed by the curved space-time of Einstein gravity.

We call this flat four-dimensional space-time framework "taiji relativity" because it is based on what we believe to be the smallest and simplest set of postulates and system of units (natural units) and "taiji," in ancient Chinese thought, symbolizes "The Absolute," or the ultimate principle or condition that existed before the creation of the world.[4]

2.2.   Taiji Relativity

We begin by deriving the taiji coordinate transformations between two inertial frames using natural units. Although these equations will look identical to those in special relativity (the Lorentz transformations), there is an important conceptual difference between them that we will discuss at the end of this chapter. Moreover, in Chapters 3–6 we will show how these transformations can be generalized to develop explicit and physically meaningful coordinate transformations between non-inertial frames with arbitrary linear accelerations or rotational motion, a feat that general relativity cannot accomplish, despite its ability to predict physical phenomena such as the perihelion shift of planets and the change in the rate of ticking of clocks in different gravitational fields.

As usual, we describe an event using the four coordinates

$$(w, x, y, z) = x^{\mu} \quad \text{and} \quad (w', x', y', z') = x'^{\mu}, \quad \mu = 0, 1, 2, 3, \qquad (2.1)$$

in inertial frames $F(w, x, y, z)$ and $F'(w', x', y', z')$, respectively. The quantities w and w' are the evolution variables and have the dimension of length. We use the letter w to distinguish it from t, which we will take to be the evolution variable expressed using the conventional unit second and also refer to w and w' as the taiji time. In special relativity, the quantities w and t are related by $w = (299\ 792\ 458\ \text{m/s})t$ in every inertial frame. Such is not necessarily the case in taiji relativity.

Using the usual simplifying conventions that the relative motion between F and F' frames is along the parallel x and x' axes, that the origins of F and F' coincide at $w = w' = 0$, and that the transformation equation between the two inertial frames is linear,[a] the transformations between the primed and unprimed coordinates can be written as

$$w' = a_1 w + a_2 x, \quad x' = b_1 w + b_2 x, \quad y' = y, \quad z' = z, \tag{2.2}$$

where a_1, a_2, b_1, and b_2 are constants to be determined. The principle of relativity implies that physical laws must have the same form in any inertial frame. This statement is equivalent to demanding that the four-dimensional interval, $s^2 = \eta_{\mu\nu} x^\mu x^\nu = x_\mu x^\mu = w^2 - r^2, \eta_{\mu\nu} = (1, -1, -1, -1)$, be invariant so that[b]

$$w'^2 - r'^2 = w^2 - r^2. \tag{2.3}$$

Substituting the expressions for w and x from (2.2) into (2.3) and equating the coefficients of w^2, wx, and x^2 in addition to using the condition that when $dx'/dw' = 0$, $dx/dw = \beta$, we can then determine all four unknown constants in terms of β, where β is a dimensionless constant that characterizes the magnitude of the relative motion between F and F'. The four-dimensional coordinate transformations can then be written as

$$w' = \gamma(w - \beta x), \quad x' = \gamma(x - \beta w), \quad y' = y, \quad z' = z; \tag{2.4}$$

$$\gamma = \frac{1}{\sqrt{1 - \beta^2}}, \tag{2.5}$$

with a corresponding inverse transformation, derived from (2.4) and (2.5)

$$w = \gamma(w' + \beta x'), \quad x = \gamma(x' + \beta w'), \quad y = y', \quad z = z'. \tag{2.6}$$

---

[a]We can demonstrate that the principle of relativity demands that the transformation be linear in the following way: Let us consider the relations between (w, x) and (w', x'). Suppose they are related by nonlinear relations: $w' = aw + bx + cwx + ew + fx + \cdots$, and $x' = Ax + Bw + Cwx + \cdots$, where $a, b, c, \ldots, A, B, C, \ldots$ are constant. After taking differentiations, one has $dw' = a^* dw + b^* dx$ and $dx' = A^* dw + B^* dx$, where a^*, b^*, A^* and B^* are, in general, functions of w and x. For an object at rest in the inertial frame $F'(w', x')$, i.e., $dx' = 0$, its velocity as measured in the inertial frame $F(w, x)$ is $dx/dw = \beta = $ constant. This relation, together with the infinitesimal four-dimensional interval, leads to the result $a^* = A^* = \gamma$ and $b^* = B^* = -\gamma\beta$, where γ is given by (2.5). This implies that a^*, b^*, A^* and B^* are constants. It follows that the relations between (w', x') and (w, x) are linear.

[b]For a free particle, $r^2/w^2 = \beta^2 = $ constant. Equation (2.3) can be written as $m^2 w^2/s^2 - m^2 r^2/s^2 = m^2$, which is exactly the same as the "energy–momentum" relation $(p_0)^2 - p^2 = m^2$ because $p_0 = m/\sqrt{1 - \beta^2}$ and $\mathbf{p} = m\boldsymbol{\beta}/\sqrt{1 - \beta^2}$.

From the solution of the Maxwell equations for a plane wave, it follows that the propagation of a light signal emitted from any source is described by $s^2 = 0$ or equivalently $ds^2 = dw^2 - dr^2 = 0$.[c] In that case, the speed of light β_L implied by $ds = 0$ is thus

$$\beta_L = |dr/dw| = 1. \tag{2.7}$$

Since ds^2 is invariant, this result, which is a consequence of the principle of relativity, holds in all inertial frames: the speed of a light signal is isotropic and independent of the motion of the source and of the observer.

From equations (2.4)–(2.5), we can also derive the velocity transformation,

$$u'_x = \frac{u_x - \beta}{1 - u_x\beta}, \quad u'_y = \frac{u_y}{\gamma(1 - u_x\beta)}, \quad u'_z = \frac{u_z}{\gamma(1 - u_x\beta)};$$

$$\mathbf{u}' = \left(\frac{dx'}{dw'}, \frac{dy'}{dw'}, \frac{dz'}{dw'}\right), \quad \mathbf{u} = \left(\frac{dx}{dw}, \frac{dy}{dw}, \frac{dz}{dw}\right), \tag{2.8}$$

where $\mathbf{u}$ is the velocity of a particle as measured by an F frame observer and $\mathbf{u}'$ is the velocity of the same particle as measured by an F' observer. Since we use natural units, these velocities are dimensionless quantities, the components of which range in value from -1 to 1.

2.3.   Operationalization of Taiji Time

If taiji relativity is to be a physical theory, then the taiji time w must have an operational definition. As we are accustomed to the usual kinds of clocks that measure time in seconds, one might ask how the measurement of time as expressed by w can be physically realized. The answer is straightforward. Because taiji relativity and special relativity are mathematically identical if one sets $c = 1$ in special relativity, one can use the usual procedure for synchronizing clocks in special relativity to set up a clock system in taiji relativity. In order to obtain taiji time measurements in units of meters, we merely re-label the clocks from seconds to meters after synchronization and scale the readings so that a time interval Δt of 1 second corresponds to a time interval Δw of 299 792 458 m.

---

[c]The speed of a plane wave in vacuum can be determined by the motion of a particular constant value of its phase, $k_0 w - \vec{k} \cdot \vec{r} = constant$. Since the wave 4-vector of light satisfies $(k_0)^2 - (\vec{k})^2 = 0$, the speed is $dr/dw = \pm k_0/|\vec{k}| = \pm 1$, which implies $dw^2 - dr^2 = 0$ or $ds^2 = 0$. The finite transformation (2.4) is equivalent to the transformation of the differentials, $dw' = \gamma(dw - \beta dx)$, $dx' = \gamma(dx - \beta dw)$, $dy' = dy$, $dz' = dz$. In the same sense, the finite four-dimensional interval (2.3) is equivalent to the differential interval $ds^2 = dw'^2 - dr'^2 = dw^2 - dr^2$.

A taiji-time Δw can also be understood in terms of an optical path length. For a finite interval, the propagation of a light signal starting from the origin $\vec{r} = 0$ at $w = 0$ is described by

$$w^2 - r^2 = 0, \tag{2.9}$$

where r is the distance traveled by the light signal during the taiji-time interval $\Delta w = w$. Since equation (2.9) implies that $r = w$ and the speed of light $\beta_L \equiv r/w = 1$ is invariant, w is also the distance traveled by the light signal. Thus, the time interval Δw between two events can be interpreted as the distance traveled by a light signal between the occurrence of the two events.

This interpretation of w suggests another way of operationalizing the taiji time, which is to use clocks whose rates of ticking are based on the motion of a light signal. Such clocks, which have been discussed in many books on special relativity, consist of a light bulb and light detector mounted next to each other facing in the same direction, and a mirror placed half a meter away facing the two devices.[5] The light bulb flashes, sending out a light signal that reflects off the mirror and returns to the detector. When the detector receives the signal, it triggers the light bulb to flash again. Because a light signal travels a total distance of 1 meter between flashes, the time interval Δw marked by the clock between flashes is $\Delta w = 1$ meter.

2.4. Conceptual Difference Between Taiji Relativity and Special Relativity

As noted previously, the taiji coordinate transformation equations (2.4) and (2.5) are identical to the conventional Lorentz transformation equations if one sets $c = 1$ so that $t = w$. Therefore, taiji relativity shares all of the mathematical and operational characteristics of special relativity. For example, clocks for measuring the taiji-time w can be synchronized in the same way as clock systems for relativistic time, the taiji coordinate transformations share all of the Lorentz and Poincaré group properties of special relativity, and all of the experimental predictions of special relativity are duplicated by taiji relativity. However, there is an important conceptual difference between the two theories.

In special relativity (and as defined by the 1983 meeting of the General Conference on Weights and Measures), the two quantities, which are directly proportional and related by $w = (299\ 792\ 458\ m/s)t$, are merely two different ways of expressing the same idea, namely a quantification of the evolution of a system. As Taylor and Wheeler pointed out,[6] the number 299 792 458 is simply a conversion factor whose value is the consequence of the choices made by humans in defining the units of meter and second.

  However, in taiji relativity, we propose a new way of thinking about the relationship between w and t. If the current relationship between them is merely a product of the human perceptual system and the choices made in selecting standards for the dimensions of time and length, why not explore the implications of a different choice in the relationship between w and t? Just because $w = ct$ leads to a relativity theory that is consistent with all experiments does not imply that it is the only relationship that is consistent with all experiments. Because changing the relationship between w and t is equivalent to changing the definition of the meter in terms of the second, a human convention, none of the experimental predictions of the theory verified over the last century would change. Some types of new relationships, such as setting $w = (9.83571 \times 10^8 \text{ ft/s})t$ are trivial. However, what if we were to adopt a relationship that is more radically different, such as

$$w = bt, \tag{2.10}$$

where b is not a constant but a function itself of space and time?[d,e]

  At first glance, such a relationship may seem impractical. Not only would the speed of light measured in units of meter per second no longer be a universal constant (although the dimensionless speed as measured in natural units would still be a universal constant as demonstrated in (2.7)), clocks set up to display the time t in seconds would appear to run in some crazy non-uniform way.

  However, as we have shown in a previous book,[8] such a relationship can, in fact, result in a relativity theory that

(a) keeps all inertial frames equivalent,
(b) still correctly predicts experimental results,
(c) still has all of the Lorentz and Poincaré group properties of special relativity, and most importantly
(d) can be more convenient for certain types of calculations than special relativity (though consequently, it is less convenient than special relativity for other types of calculations).

---

[d]These papers discuss a four-dimensional symmetry framework based on the usual first postulate and a different second postulate, namely, a common time $t' = t$ (or $w = bt$ and $w' = b't$) for all observers in different inertial frames.[7]

[e]Such a radically different relationship, $w = bt$, also implies a change in the concept of simultaneity in terms of the time t within the four-dimensional symmetry framework. For a comprehensive discussion of the arbitrariness of the concept of simultaneity (see Ref. 2).

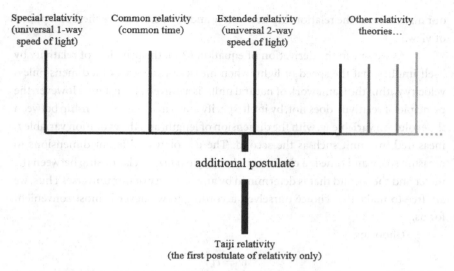

Fig. 2.1.　"Tree" of relativity theories showing the logical connection between taiji relativity, based solely on the principle of relativity, and other relativity theories for which an additional postulate has been made.

A useful analogy for thinking about such alternative relationships between w and t is the different choices one can make for coordinate systems: Cartesian, spherical, cylindrical, hyperbolic, etc. One can think of special relativity, with its particularly simple relationship between w and t, as analogous to the Cartesian coordinate system. Changing the scaling factor between w and t while keeping it constant is analogous to adopting a different Cartesian coordinate system with a different grid spacing. Moving to a relationship $w = bt$ where b is a function of space and time is analogous to changing to a coordinate system such as spherical coordinates, with its complex relationships between (x, y, z) and (r, θ, ϕ). Neither system is more correct than any other and all predict the same physics. However, depending on the problem to be analyzed, certain coordinate systems are more convenient and are better suited to revealing symmetries and invariants of the physical systems under consideration (see Fig. 2.1).

2.5.　A Short Digression: The Role of a Second Postulate

Although some authors claimed that the universality of the speed of light is a direct consequence of the principle of relativity and thus that a separate second postulate is unnecessary, Einstein evidently thought that a second postulate WAS necessary to establish the universality of the speed of light.[1,2] In light of equation (2.7) and our discussion of the relationship between w and t, we can resolve these two points of view.

As we saw in the derivation of equation (2.7), the principle of relativity by itself implies that the speed of light, when measured in terms of a dimensionless velocity within the framework of natural units, is a universal constant. However, the principle of relativity does not, by itself, specify any particular relationship between the evolution variable w, with the dimension of length, and the evolution variable t, measured in a unit such as the second. The use of two different dimensions to measure space and time is a choice and there is no *a priori* relationship between the meter and the second that is determined by any property of our universe. Thus, we are free to make that choice ourselves, according to whatever is most convenient for us.

Choosing

$$w = ct$$

in every inertial frame, as Einstein did, corresponds to making a second postulate that "The speed of light, measured in units of meter per second, is independent of the motion of the source and of the observer" and is equivalent to defining the unit second as "The interval of time that passes when a light signal travels a distance 299 792 458 m in a vacuum." On the other hand, choosing

$$w = bt,$$

where b is a function, results in a different second postulate and a different definition of the second. Although the speed of light measured in units of meter per second is no longer a universal constant under this choice, this is purely because we have chosen a different definition of the "second." *The dimensionless speed of light expressed in natural units is still a universal constant.*

Thus, whether or not a second postulate is necessary for the formulation of special relativity can be seen as a difference in one's viewpoint of space and time. In Einstein's era, space and time were thought to be two very different entities as evidenced by the use of two different dimensions and two independently defined units to quantify them. Einstein (and to a lesser extent, Poincaré) was able to free himself sufficiently from that mode of thought to develop special relativity, in which space and time are unified in a four-dimensional space-time. However, Einstein also did not have the decades of experience working with special relativity as a fundamental part of physics that modern physicists have, and thus was not able to see all of the implications of his theory, such as that spatial and temporal quantities could then be expressed using the same units. For him, it was necessary to introduce a second postulate in order to obtain the universality of the speed of light in the system of units in use at the time. From our modern point of view in which both spatial and temporal intervals can be expressed using the same units, it is revealed that the universality of the speed of light (when expressed as a dimensionless velocity in natural units) is indeed a direct consequence of the principle of relativity.

References

1. L. Landau and E. Lifshitz, *The Classical Theory of Fields*, Trans. M. Hamermesh (Addison-Wesley, 1957), p. 2; E. F. Taylor and J. A. Wheeler, *Spacetime Physics*, 2nd ed. (W. H. Freeman and Company, 1992), p. 60.
2. A. A. Tyapkin, *Sov. Phys. Usp.* **15**, 205 (1972).
3. J. P. Hsu and L. Hsu, *A Broader View of Relativity, General Implications of Lorentz and Poincaré Invariance*, 2nd ed. (World Scientific, 2006), Chapters 13–15. Available online: google books, a broader view of relativity, Hsu; J.-P. Hsu and L. Hsu, *Phys. Lett. A* **196**, 1 (1994); L. Hsu and J.-P. Hsu, *Nuovo Cimento B* **111**, 1283 (1996).
4. M. Lai and T.-Y. Lin (eds), *The New Lin Yutang Chinese–English Dictionary* (Panorama Press, Hong Kong, 1987), p. 90.
5. J. P. Hsu and Y. Z. Zhang, *Lorentz and Poincaré Invariance, 100 Years of Relativity* (World Scientific, New Jersey, 2001), p. xxvi.
6. E. F. Taylor and J. A. Wheeler, *Spacetime Physics*, 2nd ed. (W. H. Freeman and Company, 1992); T. Moore, *Six Ideas that Shaped Physics* (McGraw- Hill Science, 2002); H. Poincaré, *C. R. Acad. Sci. Paris* **140**, 1504 (1905); *Rend. Circ. Mat. Palermo* **21**, 129 (1906); H. M. Schwartz, *Am. J. Phys.* **39**, 1287 (1971); **40**, 862 (1972); **40**, 1282 (1972).
7. J. P. Hsu, *Nuovo Cimento B* **74**, 67 (1983); *Phys. Lett. A* **97**, 137 (1983); Editorial, *Nature* **303**, 129 (1983); J. P. Hsu, *Found. Phys.* **8**, 371 (1978); **6**, 317 (1976); J. P. Hsu and T. N. Sherry, *ibid* **10**, 57 (1980); J. P. Hsu and C. Whan, *Phys. Rev. A* **38**, 2248 (1988), Appendix.
8. J. P. Hsu and L. Hsu, *A Broader View of Relativity, General Implications of Lorentz and Poincaré Invariance*, 2nd ed. (World Scientific, 2006), Chapters 7–16. Available online: google books, a broader view of relativity; Editorial, *Nature* **303**, 129 (1983).

Chapter 3

The Principle of Limiting Continuation of Physical Laws and Coordinate Transformations for Frames with Constant Accelerations

3.1. The Principle of Limiting Continuation

One of the most powerful methodologies in physics for understanding a complex system or situation is that of stripping out its complications and reducing it to its bare essentials in order to create a tractable problem that can be modeled using known principles. As the system becomes better understood, the various complications can be restored, perhaps using perturbation theory or the like, to create progressively more sophisticated representations until the predictive and explanatory power of the model satisfies some criteria of precision or adequacy relative to the real-world situation. In this tradition, physical theories have all been formulated for the case of inertial frames even though, strictly speaking, all physically realizable frames are non-inertial due to the gravitational interaction and the accelerated expansion of the universe. Although the inertial frame approximation is adequate for a great number of situations, ideally, our understanding of physical principles should not be restricted to inertial frames.

Thus far, general relativity represents our best theory for understanding physics in non-inertial frames. Nevertheless, it suffers from a number of drawbacks that limit its predictive and explanatory power. One of the most prominent of these drawbacks is that the mathematical framework of general relativity appears to be too general to develop an explicit metric tensor for accelerated frames.[a] Thus, in order to write down an explicit coordinate transformation between a non-inertial and an inertial frame, or between two non-inertial frames, one must make

---

[a]Space-time coordinates in general relativity have no operational or physical meaning, E. P. Wigner, *Symmetries and Reflections, Scientific Essays* (The MIT Press, 1967), pp. 52–53; see also Section 0.1 in this volume.

additional assumptions or impose additional conditions that are not part of the framework of general relativity.

Even then, one must be careful to impose the "right" additional conditions (in effect, to choose a "correct" coordinate system) or else non-physical terms may appear in the calculations.[1] This is fundamentally different from the case in classical physics, where an injudicious choice of coordinates can lead to a mathematically intractable, but still physically correct, set of equations. This demand for the selection of a "correct" coordinate system appears to be inconsistent with the principle of general coordinate invariance and, although this problem has been recognized for decades, no completely satisfactory resolution has yet been found.

In the meantime, there have been several attempts to develop explicit coordinate transformations involving non-inertial frames,[2] none of which has been completely satisfactory. For example, Møller's transformation, sometimes referred to as Rindler coordinates, reduces to the identity transformation in the limit of zero acceleration, rather than the Lorentz transformation.

In this chapter and in Chapters 4 and 5, we describe our approach to generalizing the Lorentz transformation to both inertial and non-inertial frames. Unlike many of the previous attempts, which begin with general relativity and choose a particular gauge (such as the harmonic gauge), we base our derivation on what we call the principle of limiting continuation of physical laws. Simply put:

> The laws of physics in a reference frame F_1 with an acceleration a_1 must reduce to those in a reference frame F_2 with an acceleration a_2 in the limit where a_1 approaches a_2.[b]

As a practical matter, the space of possibilities of explicit metric tensors for frames with arbitrary accelerations is too large for the principle of limiting continuation to be of much use. However, we can accomplish this task by making the generalization of the metric tensors from inertial frame to accelerated frames in a series of small steps. In this chapter, we first make the minimum generalization by using the principle of limiting continuation to develop an explicit coordinate transformation between an inertial frame and another frame with a constant

---

[b]Since the accelerations a_1 and a_2 are not specified, in the special case that $a_1 \to a_2 = 0$, this principle of limiting continuation of physical laws reduces to the principle of relativity in the zero acceleration limit, provided the limit exists. In this sense, the principle of limiting continuation of physical laws includes the "limiting Poincaré and Lorentz invariance" as a special case, and appears to be a simple generalization of the principle of relativity from inertial frames to non-inertial frames.

acceleration point along the same line as its relative velocity. We will call this motion constant linear acceleration (CLA) and will discuss later in this chapter an operational definition for "constant acceleration." We then derive the metric tensor for CLA frames, explore the group properties of the coordinate transformation and use it to make a number of specific experimental predictions. In Chapter 4, we use our results for CLA frames to develop an explicit coordinate transformation between an inertial frame and another frame with an arbitrary linear acceleration parallel pointing along the same line as its relative velocity (ALA). Finally, in Chapter 5, we use our results to develop an explicit coordinate transformation between an inertial frame and another frame that rotates about a fixed point. In all cases, the space-time coordinates retain their traditional operational definitions, unlike the arbitrary coordinates in general relativity.

All inertial and non-inertial frames will be treated within the taiji symmetry framework, i.e., using a flat space-time with arbitrary coordinates and the Poincaré metric tensor, which is the Minkowski metric tensor generalized from inertial to non-inertial frames. As mentioned previously, the taiji symmetry framework has the advantages that it has the maximum space-time symmetry, guaranteeing the law of conservation of the energy–momentum tensor and enabling the quantization of all physical fields (including gravity) when formulated as a (generalized) Yang–Mills gauge field. These subjects will be discussed later in this book.

3.2. Constant Linear Acceleration: The Wu Transformations

Our first step in constructing a coordinate transformation for accelerated frames is to make a minimal generalization from the Poincaré transformations[c,3] to a set of coordinate transformations between an inertial frame and a frame with a constant linear acceleration (CLA), where both the acceleration and velocity of the non-inertial frame point along the same line (they may be either parallel or antiparallel). Postulating a minimum departure[d,4] of the CLA transformations from the Poincaré transformations, we write the relationship between the infinitesimal space-time intervals of an inertial frames $F_I(x_I) = F_I(w_I, x_I, y_I, z_I)$ and a CLA

---

[c]The name "Poincaré group" was probably first used in a paper by E. Wigner and in a set of lecture notes by A. S. Wightman (informed by R. F. Streater). Wigner said that the geometrical principles of invariance "were recognized by Poincaré first, and I like to call the group formed by these invariables the Poincaré group."

[d]This is analogous to the case in which gauge symmetry does not uniquely determine the electromagnetic action and one must also postulate a minimal electromagnetic coupling.

frame $F(x) = F(w, x, y, z)$ as

$$ds^2 = W^2 dw^2 - dx^2 - dy^2 - dz^2 = dw_I^2 - dx_I^2 - dy_I^2 - dz_I^2. \tag{3.1}$$

In general, W is a function of w and x as long as the motion of the CLA frame $F(x)$ is along the parallel x and x_I axes. We note that if one were to postulate the form $ds^2 = W^2 dw^2 - X^2 dx^2 - Y^2 dy^2 - Z^2 dz^2$ or $ds^2 = W^2 dw^2 - R^2(dx^2 + dy^2 + dz^2)$, then there are too many unknown functions to be determined by the principle of limiting continuation. In addition, if one were to postulate $ds^2 = dw^2 - R^2(dx^2 + dy^2 + dz^2)$, then there is no physical solution for the accelerated transformation. Equation (3.1) leads to the following transformations for the coordinate differentials:

$$dw_I = \gamma(Wdw + \beta dx), \quad dx_I = \gamma(dx + \beta Wdw),$$
$$dy_I = dy, \quad dz_I = dz; \quad \gamma = \sqrt{1 - \beta^2}, \tag{3.2}$$

where $W = W(x, w)$ and $\beta = \beta(w)$ are unknown functions to be determined. In order for a finite space-time transformation between F_I and F to exist, these unknown functions must satisfy the integrability conditions[e,5]

$$\frac{\partial(\gamma W)}{\partial x} = \frac{\partial(\gamma \beta)}{\partial w}, \quad \frac{\partial \gamma}{\partial w} = \frac{\partial(\gamma \beta W)}{\partial x}. \tag{3.3}$$

Assuming the usual linear relationship between the velocity β and the evolution variable w, i.e., $\beta = \alpha_o w + \beta_o$, leads to the finite transformation[f]

$$w_I = \gamma\beta\left(x + \frac{1}{\alpha_o \gamma_o^2}\right) - \frac{\beta_o}{\alpha_o \gamma_o} + w_o,$$

$$x_I = \gamma\left(x + \frac{1}{\alpha_o \gamma_o^2}\right) - \frac{1}{\alpha_o \gamma_o} + x_o, \quad y_I = y + y_o, \quad z_I = z + z_o; \tag{3.4}$$

$$\gamma = \frac{1}{\sqrt{1 - \beta^2}}, \quad \gamma_o = \frac{1}{\sqrt{1 - \beta_o^2}}, \quad \beta = \alpha_o w + \beta_o,$$

---

[e]The integrability conditions lead to a general solution for $W(w, x)$, i.e., $W(w, x) = \gamma^2 \alpha_o x + f(w)$, where $f(w)$ is an arbitrary function of w and must approach 1 in the limit of zero acceleration. For example, if one chooses $f(w) = 1$, one has the following transformation

$$w_I = \gamma\beta x + \frac{1}{\alpha_o}(\sin^{-1}\beta - \sin^{-1}\beta_o), \quad x_I = \gamma x - \frac{1}{\alpha_o}\left(\frac{1}{\gamma} - \frac{1}{\gamma_o}\right), \quad y_I = y, \ldots$$

where $\beta = \alpha_o w + \beta_o$. Thus far, it appears that no physical principle can determine $f(w)$ or the Wu factor $W(w, x)$ and $\beta(w)$ in (3.3) uniquely.

[f]It is possible to make a different assumption for the function $\beta(w)$. In this case, one has a different time in CLA frames. See Ref. 5, Section 20a in Chapter 20.

which satisfies the initial conditions $x_I = x_o$ and $w_I = w_o$ when $w = x = 0$. We call (3.4) the Wu transformations in honor of T. Y. Wu's idea of a kinematical approach to deriving transformations between accelerated frames in a space-time with a vanishing Riemann–Christoffel curvature.[5] As expected, in the special case $\beta_o = 0$ and $w_o = x_o = y_o = z_o = 0$, the Wu transformation (3.4) reduces to Møller's accelerated transformation,[g,2,6,7]

$$w_I = \left(x + \frac{1}{\alpha_o}\right) \sinh(\alpha_o w^*), \quad x_I = \left(x + \frac{1}{\alpha_o}\right) \cosh(\alpha_o w^*) - \frac{1}{\alpha_o},$$

$$y_I = y, \quad z_I = z, \tag{3.5}$$

with a change of the time variable, $w = (1/\alpha_o) \tanh(\alpha_o w^*)$. It furthermore reduces smoothly to the Poincaré transformation in the limit of zero acceleration, $\alpha_o \to 0$,

$$w_I = \gamma_o(w + \beta_o x) + w_o, \quad x_I = \gamma_o(x + \beta_o w) + x_o,$$

$$y_I = y + y_o, \quad z_I = z + z_o, \tag{3.6}$$

in accordance with the principle of limiting continuation.[h]

This smooth connection with the Lorentz transformation in the limit of zero acceleration when $w_o = x_o = y_o = z_o = 0$ in (3.6) is a crucial property of the accelerated transformations not found in other proposed transformations[2] between non-inertial frames. The inverse Wu transformation derived from (3.4) is

$$w = \frac{w_I - w_o + \beta_o/(\alpha_o \gamma_o)}{\alpha_o(x_I - x_o) + 1/\gamma_o} - \frac{\beta_o}{\alpha_o},$$

$$x = \sqrt{(x_I - x_o + 1/\gamma_o \alpha_o)^2 - (w_I - w_o + \beta_o/\alpha_o \gamma_o)^2} - \frac{1}{\alpha_o \gamma_o^2}, \tag{3.7}$$

$$y = y_I + y_o, \quad z = z_I + z_o,$$

and differentiation of (3.4) leads to (3.2) with

$$W = \gamma^2(\gamma_o^{-2} + \alpha_o x), \quad \gamma = \frac{1}{\sqrt{1 - \beta^2}}, \quad \beta = \alpha_o w + \beta_o. \tag{3.8}$$

---

[g]The authors also made an exact calculation of the clock paradox problem, including the effects of linear accelerations and decelerations.

[h]In (3.4), we use the approximation $\gamma \approx \gamma_o(1 + \beta_o \alpha_o w \gamma_o^2)$, which is obtained by expanding γ to the first order in α_o (which is assumed to be small), while keeping $\beta_o < 1$ to all orders.

We can now write the invariant infinitesimal interval (3.1) in terms of the Minkowski metric tensor $\eta_{\mu\nu}$ for F_I and $P_{\mu\nu}$ for CLA frame F,

$$ds^2 = \eta_{\mu\nu}dx_I^\mu \, dx_I^\nu = P_{\mu\nu}dx^\mu \, dx^\nu,$$

$$\eta_{\mu\nu} = (1, -1, -1, -1), \quad P_{\mu\nu} = (W^2, -1, -1, -1), \quad W = \gamma^2(\gamma_o^{-2} + \alpha_o x).$$

$$(3.9)$$

Thus, the physical space-time of a CLA frame is characterized by $P_{\mu\nu}$, which we call the Poincaré metric tensor, which reduces to the Minkowski metric $\eta_{\mu\nu}$ in the limit of zero acceleration. We will refer to W in the metric tensors $P_{\mu\nu}$ and the differential form of the Wu transformation (3.2) as the Wu factor, whose physical implications will be discussed later.

One can also generalize the Wu transformation to the case where the time-dependent velocity $\boldsymbol{\beta}$ and the CLA are in the same arbitrary direction $\boldsymbol{\beta}/\beta$ not necessarily along the x/x_I-axis. In this case, the finite Wu transformation (3.4) becomes

$$w_I = \gamma\beta\left[\frac{\boldsymbol{\beta} \cdot \mathbf{r}}{\beta} + \frac{1}{\alpha_o\gamma_o^2}\right] - \frac{\beta_o}{\alpha_o\gamma_o} + w_o,$$

$$\mathbf{r}_I = \mathbf{r} + \left[(\gamma - 1)\frac{\boldsymbol{\beta} \cdot \mathbf{r}}{\beta} + \left(\frac{\gamma}{\gamma_o} - 1\right)\frac{1}{\alpha_o\gamma_o}\right]\frac{\boldsymbol{\beta}}{\beta} + \mathbf{r}_o, \qquad (3.10)$$

$$\gamma = \frac{1}{\sqrt{1 - \beta^2}}, \quad \gamma_o = \frac{1}{\sqrt{1 - \beta_o^2}}, \quad \boldsymbol{\beta} = \alpha_o w + \boldsymbol{\beta}_o, \quad \boldsymbol{\alpha}_o \| \boldsymbol{\beta}_o.$$

3.3. Operational Meaning of the Space-Time Coordinates and "CLA"

It is important to define what "physical space-time coordinates" are in accelerated frames and how one can realize them operationally. In inertial frames, the space-time coordinates (w_I, x_I, y_I, z_I) can be pictured as a rigid and uniform grid of meter sticks and identical clocks that have been synchronized using isotropic light signals.[i] Let us use the term "space-time clock" to denote a clock (on the grid) that shows both the time w_I and the position r_I of the clock, i.e., (w_I, x_I, y_I, z_I). One can always use the space-time clock closest to an event to record the space-time coordinates of that event.

---

[i]Cf. Ref. 5, Sections 7d and 18c.

In CLA frames, since the Wu transformation between an inertial and a CLA frame is nonlinear, one can no longer picture the space-time coordinates of a CLA frame as a uniform grid of space-time clocks. However, consider the inverse Wu transformation (3.7). The space-time clocks in the CLA frame can be synchronized using only this transformation. A "clock" is simply a device that shows a time and position and, if based on, say, a computer chip, it can run at an arbitrary rate and show an arbitrary time and position. In order to synchronize clocks in the CLA frame, we can imagine that the computer clocks associated with that frame have the ability to measure their position r_I relative to the F_I frame, to obtain w_I from the nearest F_I clock, and then to compute and display w and $r = (x, y, z)$ using (3.7) (with known velocity β_o and acceleration α_o) on a readout. If these clocks in the CLA frame are uniformly spaced on a regular grid, the differences in their time readings will not be uniform. However, this grid of computer clocks will automatically become the more familiar Einstein clocks of inertial frames as the acceleration α_o approaches zero, provided that $w = ct$ and $w' = ct'$. This indicates that the coordinates (w, x, y, z) in (3.7) or (3.4) for a CLA frame play the same role and have an analogous physical meaning to the space-time coordinates in the transformations for inertial frames.

We now examine the operational definition of "constant linear acceleration." Consider the invariant action S_f for a "free particle" in a CLA frame, which can be associated with the Wu transformations with time variable w

$$S_f = -\int_a^b mds = \int_{w_a}^{w_b} L_w dw,$$

$$ds^2 = W^2 dw^2 - dx^2 - dy^2 - dz^2, \quad P_{\mu\nu} = (W^2, -1, -1, -1), \qquad (3.11)$$

$$L_w = -m\sqrt{P_{\mu\nu} u^\mu u^\nu} = -m\sqrt{W^2 - (\beta^i)^2}, \quad u^\mu = \frac{dx^\mu}{dw} = (1, \beta^i),$$

where $W(w, x) = \gamma^2(g_o^{-2} + \alpha_o x)$.

As usual, the covariant momentum p_i and the corresponding energy p_0 (or the Hamiltonian H) with dimensions of mass are given by

$$p_i = -\frac{\partial L_w}{\partial \beta^i} = \left(\frac{-m\Gamma\beta_x}{W}, \frac{-m\Gamma\beta_y}{W}, \frac{-m\Gamma\beta_z}{W} \right), \quad p_i = P_{ik}p^k = -p^i, \quad (3.12)$$

$$p_0 = \left(\frac{\partial L_w}{\partial \beta^i}\beta^i - L_w\right) = m\Gamma W = P_{00}p^0, \quad p^0 = m\frac{dx^0}{ds} = m\frac{dw}{ds};$$

$$\tag{3.13}$$

$$\Gamma = \frac{1}{\sqrt{1 - \beta^2/W^2}}, \quad \beta^2 = \beta_x^2 + \beta_y^2 + \beta_z^2 = -\beta_i\beta^i, \quad \beta^i = \frac{dx^i}{dw} = u^i,$$

where $i = 1, 2, 3$. Thus, the covariant momentum $p_\mu = (p_0, p_1, p_2, p_3) = (p_0, -\mathbf{p})$ transforms like the covariant coordinate $dx_\mu = P_{\mu\nu}dx^\nu$. We have

$$p_{I0} = \gamma\left(\frac{p_0}{W} - \beta p_1\right), \quad p_{I1} = \gamma\left(p_1 - \beta\frac{p_0}{W}\right), \quad p_{I2} = p_2, \quad p_{I3} = p_3;$$

$$\tag{3.14}$$

$$p_{I0}^2 - \mathbf{p}_I^2 = (p_0/W)^2 - \mathbf{p}^2 = m^2. \tag{3.15}$$

Now consider a particle at rest in the CLA frame ($\mathbf{p} = 0$). From (3.14) and (3.15), we have $p_0 = mW$ and $p_{I0} = m\gamma$. Thus, when $\mathbf{r}$ is constant,

$$\left(\frac{dp_{I0}}{dx_I}\right)_x = m\gamma^3\frac{d\beta}{dx_I} = \frac{m\alpha_0}{(\gamma_o^{-2} + \alpha_o x)}; \quad \beta = \alpha_o w + \beta_o. \tag{3.16}$$

Thus, we see that an object moving with a CLA undergoes a uniform change per unit length of its energy p_{I0}, as measured in an inertial frame F_I. Similarly, a CLA frame is a reference frame that is comoving with such an object. An object at rest in F and located at, say, $\mathbf{r} = (x, y, z)$ will have a constant change of energy p_{I0} per unit distance traveled as measured from an inertial frame F_I. The magnitude of this constant change depends on the initial velocity β_o of the CLA frame and the location of the object in a highly non-trivial manner, as shown in (3.16).

The principle of limiting continuation of physical laws (where we treat space-time coordinate transformations as physical laws) is critical in obtaining the result (3.16). Serendipitously, what we mean by CLA is precisely the type of acceleration that is realized in high energy particle accelerators, facilitating comparison between the theory and experiment, as will be discussed later in this chapter.

3.4.   Singular Walls and Horizons in Accelerated Frames

Let us now examine the mapping of physical coordinates between an inertial frame and a CLA frame. First, we consider the inverse Wu transformations (3.7) with

$x_o^\mu = 0$, the case in which the origins of the inertial and CLA frame coincide at $w = w_I = 0$. From the transformation equation for w, we see that lines of constant w correspond to straight lines in F_I, and that all such lines pass through the point $(w_I, x_I) = (-\beta_0/(\alpha_0\gamma_0), -1/(\alpha_0\gamma_0))$. In contrast, from the transformation equation for x, lines of constant x correspond to hyperbolic lines that satisfy the equation

$$\left(x + \frac{1}{\alpha_0\gamma_0^2}\right)^2 = \left(x_I + \frac{1}{\gamma_0\alpha_0}\right)^2 - \left(w_I + \frac{\beta_0}{\alpha_0\gamma_0}\right)^2. \tag{3.17}$$

The graphs in Fig. 3.1 show lines of constant x (hyperbolic lines) and of constant w (straight lines) in the (w_I, x_I) plane for the Wu transformations with non-zero inertial velocity.

Second, from an examination of the Wu factor W in (3.9), one can see that P_{00} vanishes when $x = -1/(\alpha_0\gamma_0^2)$ (with arbitrary y and z coordinates). This corresponds to a wall singularity in the CLA frame. All points in the inertial frame F_I are mapped into only a portion of the CLA frame F. The location of the wall depends on the magnitude of the acceleration and as would be expected, as the acceleration α_0 approaches zero, the distance between the wall and the origin of the CLA frame approaches infinity.

In the portion of space beyond the wall in the CLA frame ($x < -1/(\alpha_0\gamma_0^2)$), the differentials dw_I and Wdw have different signs, which is unphysical. This situation resembles the negative energy solutions for a classical particle in some ways. This portion of space is inaccessible to observers in inertial frames. The inverse Wu transformation (3.7) implies that as the coordinate x approaches $-1/(\alpha_0\gamma_0^2)$:

(a) clocks slow down and stop at the singular wall

$$\left(\frac{dw}{dw_I}\right)_{dx=0} = \frac{1}{\gamma W} = \frac{(x_I + 1/\gamma_0\alpha_0)^2 - (w_I + \beta_0/\alpha_0\gamma_0)^2}{\alpha_0(x_I + 1/\alpha_0\gamma_0)^3} \rightarrow 0; \tag{3.18}$$

(b) the speed of light as measured by observers in the CLA frame F increases without bound

$$\left(\frac{dx}{dw}\right)_{ds=0} = \pm|W| = \pm\frac{\alpha_0(x_I + 1/\gamma_0\alpha_0)^2}{\sqrt{(x_I + 1/\gamma_0\alpha_0)^2 - (w_I + \beta_0/\alpha_0\gamma_0)^2}} \rightarrow \pm\infty.$$

$$\tag{3.19}$$

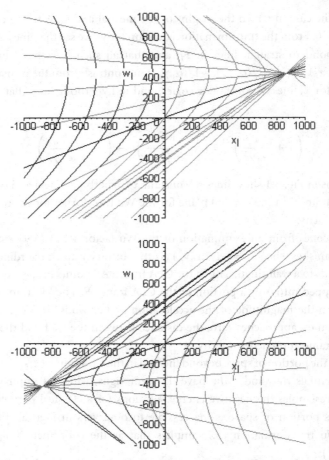

Fig. 3.1.   Lines of constant w and x in the F_I (inertial) frame. The CLA frame has velocity $\beta_o = 0.5$ and accelerations $\alpha_o = -0.001$ (top graph) and $\alpha_o = +0.001$ (bottom graph).

Result (a) shows that the rate of ticking of a clock at rest relative to F at $(x, 0, 0)$ is slower than that of a clock at rest relative to F_I at $(x_I, 0, 0)$. Result (b) is due to the fact that the law for the propagation of light is given by $ds^2 = 0$, and is related to rate of clock ticking (a).

3.5.   The Wu Pseudo-Group

It is well established that the set of Poincaré transformations constitutes a unique one-to-one mapping between the whole of space-time in two inertial frames. The set of Poincaré transformations satisfies all group properties and forms the Poincaré group. However, when the Poincaré transformations are generalized to the Wu transformations that map coordinates between inertial and non-inertial frames, the whole of space-time in the inertial frame is mapped to only a portion of the space-time in the accelerated frame, as shown in Section 3.4, and the usual group properties are not completely satisfied. Nevertheless, as we will show, the Wu transformations satisfy a weaker set of criteria and form a "pseudo-group," as defined by Veblen and Whitehead.[8]

We first demonstrate explicitly that the subset of Wu transformations consisting of the transformations for all CLA frames with relative velocities along a particular direction (such as parallel x/x'-axes) satisfies the properties of a normal group. These transformations can be defined as nonlinear transformations that preserve the quadratic form,

$$ds^2 = P_{\mu\nu}dx^\mu dx^\nu = P'_{\mu\nu}dx'^\mu dx'^\nu = P''_{\mu\nu}dx''^\mu dx''^\nu,$$

$$P_{\mu\nu} = (W^2, -1, -1, -1), \quad W = \gamma^2(\gamma_o^{-2} + \alpha_o x), \tag{3.20}$$

$$P'_{\mu\nu} = (W'^2, -1, -1, -1), \quad W' = \gamma'^2(\gamma'^{-2}_o + \alpha'_o x'), \quad \text{etc.}$$

where $P_{\mu\nu}$, $P'_{\mu\nu}$ and $P''_{\mu\nu}$ are Poincaré metric tensors in the CLA frames $F(x)$, $F'(x')$ and $F''(x'')$, respectively. Each frame has its own constant acceleration, α_o, α'_o, and α''_o with velocities $\beta(w) = \alpha_o w + \beta_o$, $\beta'(w') = \alpha'_o w' + \beta'_o$, and $\beta''(w'') = \alpha''_o w'' + \beta''_o$.[j] With the help of (3.4), the transformations between an inertial frame F_I and the CLA frames $F(x)$, $F'(x')$ and $F''(x'')$ can be written as

$$w_I = \gamma\beta\left(x + \frac{1}{\alpha_o\gamma_o^2}\right) - \frac{\beta_o}{\alpha_o\gamma_o} = \gamma'\beta'\left(x' + \frac{1}{\alpha'_o\gamma'^2_o}\right) - \frac{\beta'_o}{\alpha'_o\gamma'_o}$$

$$= \gamma''\beta''\left(x'' + \frac{1}{\alpha''_o\gamma''^2_o}\right) - \frac{\beta''_o}{\alpha''_o\gamma''_o},$$

$$x_I = \gamma\left(x + \frac{1}{\alpha_o\gamma_o^2}\right) - \frac{1}{\alpha_o\gamma_o} = \gamma'\left(x' + \frac{1}{\alpha'_o\gamma'^2_o}\right) - \frac{1}{\alpha'_o\gamma'_o} \tag{3.21}$$

$$= \gamma''\left(x'' + \frac{1}{\alpha''_o\gamma''^2_o}\right) - \frac{1}{\alpha''_o\gamma''_o},$$

---

[j]The transformations for the differentials in accelerated frames F and F' take the form $\gamma(Wdw + \beta dx) = \gamma'(W'dw' + \beta'dx')$, $\gamma(dx + \beta Wdw) = \gamma'(dx' + \beta'W'dw')$, $dy = dy'$, and $dz = dz'$. (cf. equation (3.2)).

where $\gamma = 1/\sqrt{1-\beta^2}$, $\gamma' = 1/\sqrt{1-\beta'^2}$, and $\gamma'' = 1/\sqrt{1-\beta''^2}$ and we have ignored the trivial transformations of y and z for simplicity. From (3.21), the coordinate transformations between the CLA frames F and F' are then

$$w = \frac{1}{\alpha_o}\left[\frac{\gamma'\beta'Q' - \beta_o'/(\alpha_o'\gamma_o') + \beta_o/(\alpha_o\gamma_o)}{\gamma'Q' - 1/(\alpha_o'\gamma_o') + 1/(\alpha_o\gamma_o)} - \beta_o\right],$$

$$x = \sqrt{\left(\gamma'Q' - \frac{1}{\alpha_o'\gamma_o'} + \frac{1}{\alpha_o\gamma_o}\right)^2 - \left(\gamma'\beta'Q' - \frac{\beta_o'}{\alpha_o'\gamma_o'} + \frac{\beta_o}{\alpha_o\gamma_o}\right)^2 - \frac{1}{\alpha_o\gamma_o^2}},$$

$$Q' \equiv \left(x' + \frac{1}{\alpha_o'\gamma_o'^2}\right), \quad \beta' = \alpha_o' + \beta_o', \tag{3.22}$$

with the inverse transformations

$$w' = \frac{1}{\alpha_o'}\left[\frac{\gamma\beta Q - \beta_o/(\alpha_o\gamma_o) + \beta_o'/(\alpha_o'\gamma_o')}{\gamma Q - 1/(\alpha_o\gamma_o) + 1/(\alpha_o'\gamma_o')} - \beta_o'\right],$$

$$x' = \sqrt{\left(\gamma Q - \frac{1}{\alpha_o\gamma_o} + \frac{1}{\alpha_o'\gamma_o'}\right)^2 - \left(\gamma\beta Q - \frac{\beta_o}{\alpha_o\gamma_o} + \frac{\beta_o'}{\alpha_o'\gamma_o'}\right)^2 - \frac{1}{\alpha_o'\gamma_o^2}},$$

$$Q \equiv \left(x + \frac{1}{\alpha_o\gamma_o^2}\right), \quad \beta = \alpha_o w + \beta_o. \tag{3.23}$$

Similarly from (3.21), the coordinate transformations between F and F'' are

$$w = \frac{1}{\alpha_o}\left[\frac{\gamma''\beta''Q'' - \beta_o''/(\alpha_o''\gamma_o'') + \beta_o/(\alpha_o\gamma_o)}{\gamma''Q'' - 1/(\alpha_o''\gamma_o'') + 1/(\alpha_o\gamma_o)} - \beta_o\right],$$

$$x = \sqrt{\left(\gamma''Q'' - \frac{1}{\alpha_o''\gamma_o''} + \frac{1}{\alpha_o\gamma_o}\right)^2 - \left(\gamma''\beta''Q'' - \frac{\beta_o''}{\alpha_o''\gamma_o''} + \frac{\beta_o}{\alpha_o\gamma_o}\right)^2 - \frac{1}{\alpha_o\gamma_o^2}},$$

$$Q'' \equiv \left(x'' + \frac{1}{\alpha_o''\gamma_o''^2}\right), \quad \beta'' = \alpha_o''w'' + \beta_o''. \tag{3.24}$$

From (3.23) and (3.24), the transformations between $F'(x')$ and $F''(x'')$ are

$$w' = \frac{1}{\alpha_o'}\left[\frac{\gamma''\beta''Q'' - \beta_o''/(\alpha_o''\gamma_o'') + \beta_o'/(\alpha_o'\gamma_o')}{\gamma''Q'' - 1/(\alpha_o''\gamma_o'') + 1/(\alpha_o'\gamma_o')} - \beta_o'\right],$$

$$\tag{3.25}$$

$$x' = \sqrt{\left(\gamma''Q'' - \frac{1}{\alpha_o''\gamma_o''} + \frac{1}{\alpha_o'\gamma_o'}\right)^2 - \left(\gamma''\beta''Q'' - \frac{\beta_o''}{\alpha_o''\gamma_o''} + \frac{\beta_o'}{\alpha_o'\gamma_o'}\right)^2 - \frac{1}{\alpha_o'\gamma_o'^2}}.$$

We see explicitly from (3.23), (3.24), and (3.25) that the transformations between any two CLA frames with relative motion along the $x/x'/x''$-axes have the same form, satisfying the group property of closure. Equation (3.23) demonstrates the existence of an inverse transformation that has the same form as the transformations themselves. Substituting (3.23) into (3.22) results in the identity transformation, $w = w$ and $x = x$. Finally, one can also show that this subset of Wu transformations satisfies the associative rule. Thus, this subset of Wu transformations satisfies all the properties of the usual continuous group.

However, the situation is different when transformations for non-parallel directions are considered. In the case of the Poincaré transformations, the combination of two Lorentz boosts in different directions is equivalent to a suitable combination of a boost and a rotation, and thus this compound boost still exists within the (usual) finite-dimensional set of Poincaré transformations. In the case of the Wu transformations however, the combination of two transformations for CLA frames with accelerations in different directions cannot be expressed as a combination of a single such transformation with a rotation. Thus, it does not belong to the usual finite-dimensional set of Wu transformations. In fact, the set obtained from generators of the Wu transformations using the Lie bracket, and the brackets of brackets, and so on, never closes. That is, the Wu transformations define a group with infinitely many generators.[9] This difference is likely related to the distortion of space-time coordinates in accelerated frames relative to an inertial frame, as shown by the metric tensors $P_{\mu\nu}$ in (3.9). Also, the Wu transformations map only a portion of the space in a CLA frame F to a portion of the space in another CLA frame F'.

To deal with these types of transformations, mathematicians O. Veblen and J. H. C. Whitehead developed the concept of a pseudo-group.[8] A set of transformations forms a pseudo-group if

(i) the resultant of two transformations in the set is also in the set,
(ii) the set contains the inverse of every transformation in the set,

where "set" is defined as an infinite-dimensional set, rather than the usual Lie set. The complete set of Wu transformations for CLA (in all directions) forms a pseudo-group, which we will call the Wu pseudo-group.[k,l] These pseudo-group

---

[k]The Wu transformations appear to be also the same as the pseudo-group discussed by Cartan, who required the identity transformation to exist in the pseudo-group he defined. (Cf. online: 'wiki, pseudogroup.')

[l]In fact, the properties of the Wu pseudo-group are richer than those of traditional pseudo-groups because the complete set of Wu transformations also satisfies the associative property of normal groups.

properties are determined by the physical requirements of non-inertial frames and their accelerations, including the limiting continuation of physical laws and space-time transformations, within the taiji symmetry framework.

3.6.   Relationship Between the Wu and Møller Transformations

The solutions of the integrability conditions (3.3) are not unique. Instead of assuming (or defining) the velocity β to be a linear function of time w (i.e., $\beta = \alpha_o w + \beta_o$) to derive (3.4), suppose one assumes that $W(w, x)$ is independent of time, i.e., $W = W_x$ and that the velocity β is a function of time $\beta = \beta(w)$ only. We obtain a different solution,

$$W_x = k_1 x + k_2, \quad \beta = \tanh(k_1 w + k_3), \tag{3.26}$$

in which the velocity β is a complicated hyperbolic tangent function of time. In this case, the time is, in general, different from that in the Wu transformations (3.4), so we will use w^* to denote time in the new solution (3.26). If the initial velocity is β_o, $\beta(0) = \beta_o$, then we have $k_3 = \tanh^{-1} \beta_o$.

Suppose we follow the approach of Wu and Lee[6] to impose a boundary condition that the velocity β is related to a constant acceleration α^* by the relation $\beta = \alpha^* w_I + \beta_o$ at $x_I = 0$. This condition enables us to determine $k_1 = \gamma_o \alpha^*$. In the limit of zero acceleration, where α^* or k_1 vanishes, W_x must be 1 for inertial frames. Thus, we have $k_2 = 1$ in (3.26). With the help of the principle of limiting continuation, we obtain

$$w_I = \left(x + \frac{1}{\gamma_o \alpha^*} \right) \sinh(\gamma_o \alpha^* w^* + \tanh^{-1} \beta_o) - \frac{\beta_o}{\alpha^*}$$

$$x_I = \left(x + \frac{1}{\gamma_o \alpha^*} \right) \cosh(\gamma_o \alpha^* w^* + \tanh^{-1} \beta_o) - \frac{1}{\alpha^*}, \tag{3.27}$$

$$y_I = y, \quad z_I = z,$$

which we call the generalized Møller–Wu–Lee (MWL) transformation for an inertial frame F_I and a CLA frame F with non-zero initial velocity. The Møller time w^* in the transformation (3.27) can take all values from $-\infty$ to $+\infty$. From the preceding discussions, we have $k_1 = \gamma_o \alpha^*$, $k_2 = 1$, $k_3 = \tanh^{-1} \beta_o$ and (3.27), so W_x and the velocity β in (3.26) (with w replaced by w^*) take the form

$$W_x = \gamma_o \alpha^* x + 1,$$

$$\beta = \tanh(\gamma_o \alpha^* w^* + \tanh^{-1} \beta_o) = \frac{w_I + \beta_o/\alpha^*}{x_I + 1/\alpha^*} \to \alpha^* w_I + \beta_o \quad \text{for } x_I \to 0. \tag{3.28}$$

In the special case of zero initial velocity, $\beta_o = 0$, the transformations (3.27) reduce to Møller's accelerated transformation (3.5) with $\alpha^* = \alpha_o$. This transformation was also derived by Wu and Lee using a purely kinematical approach in flat space-time. One can verify that in the limit of zero acceleration $\alpha^* \to 0$, (3.27) reduces to the Lorentz transformation

$$w_I = \gamma_o(w^* + \beta_o x), \quad x_I = \gamma_o(x + \beta_o w^*), \quad y_I = y, \quad z_I = z, \qquad (3.29)$$

where

$$\sinh(\tanh^{-1}\beta_o) = \gamma_o\beta_o, \quad \cosh(\tanh^{-1}\beta_o) = \gamma_o.$$

Although the generalized MWL transformations (3.27) appear to be quite different from the Wu transformations (3.4), they are in fact related by a simple change of time variables. From (3.4), we have

$$\alpha_o w + \beta_o = \frac{w_I + \beta_o/(\gamma_o\alpha_o)}{x_I + 1/(\gamma_o\alpha_o)}. \qquad (3.30)$$

Comparing (3.30) and (3.28) with the identification $\alpha^* = \alpha_o\gamma_o$, we see the following unique relation between the times w and w^*:

$$w = \frac{1}{\alpha_o}\left[\tanh(\gamma_o\alpha^* w^* + \tanh^{-1}\beta_o) - \beta_o\right], \quad \alpha^* = \alpha_o\gamma_o. \qquad (3.31)$$

The generalized MWL transformation (3.27) can be obtained by substituting (3.31) into the Wu transformations (3.4). Both (3.27) and (3.4) reduce to the Lorentz transformation in the limit of zero acceleration and both have the same group properties. Although it may seem that only one of the two transformations can be correct because the readings on the clocks in the CLA frame will differ depending on whether one chooses to use (3.27) or (3.4), in actuality, both transformations are correct, in the sense of predicting the same physical outcomes. For example, let us rederive the result for the change in energy per unit length of a particle undergoing CLA by following the steps from (3.11) to (3.16), but this time using the generalized MWL transformations (3.27). The metric tensor of the CLA frame is still $P_{\mu\nu} = (W^2, -1, -1, -1)$, but this time, the Wu factor $W = W_x = (1+\alpha^*\gamma_o x)$ is different. If a particle is at rest in the CLA frame ($\mathbf{p} = 0$), located at $\mathbf{r} = (x, y, z) = $ constant, then its change in energy per unit length, as measured in an inertial frame, would be given by

$$\left(\frac{dp_{I0}}{dx_I}\right)_x = m\gamma^3 \frac{d\beta}{dx_I} = \frac{m\gamma_o\alpha^*}{(1+\alpha^*\gamma_o x)} = \frac{m\alpha_0}{(\gamma_o^{-2}+\alpha_o x)}; \tag{3.32}$$

$$\beta = \tanh(\gamma_o\alpha^* w^* + \tanh^{-1}\beta_o), \quad \alpha^* = \alpha_o\gamma_o, \tag{3.33}$$

which is the same as the result (3.16) obtained using the Wu transformations (3.4) because we have the relation $\alpha^* = \alpha_o\gamma_o$. Thus, the Wu and the generalized MWL transformations have the same physical implications. The only quantities that would be different between the two are those that depend directly on clock readings, such as velocities and time intervals. However, physical predictions, such as scattering cross-sections, energies, and even the lifetime dilations of accelerated particles, remain the same. This is a reflection of the fact that the choice of clock system for measuring physical quantities is purely a human convention and has no bearing on the physics itself. As long as the theory retains the essential limiting Lorentz and Poincaré invariance, its predictions will be consistent with experimental results. Within an experimentally consistent theoretical framework, a choice of clock synchronizations is analogous to a choice of coordinate systems. Different choices are convenient for different situations, but all lead to the same physics and physical predictions.

If one uses the generalized MWL transformations instead of the Wu transformations, one can setup a corresponding computer clock system in a CLA frame to show (w^*, x, y, z). Is this time w^* physical time for CLA frames?

It is reasonable to define "physical time" for any clock system in CLA frames which is consistent with the accelerated lifetime dilation or decay-length dilation of an accelerated unstable particle in flight. In this sense, the "times" or "evolution variables" in both inertial and non-inertial frames have equal physical significance, as long as they can describe the physical phenomena of lifetime dilations of particles in flight. As we shall demonstrate in Section 3.7, the Wu transformation and the generalized MWL transformation predict the same lifetime dilation, as measured in an inertial frame. Therefore, both times w and w^* in the Wu and the generalized MWL transformations, respectively, are physical time for CLA frames. The situation is similar to relativistic and common times in relativity theory.[m] Our results suggest that any space-time transformations that preserve the fundamental metric tensor

---

[m] See Ref. 5, Chapter 8.

of the form $(W^2, -1, -1, -1)$ are physical space and time coordinates for CLA frames.

3.7.   Experimental Tests of the Wu Transformations

We now discuss two possible experimental tests of the Wu transformations. The first is the lifetime dilation of accelerated particles. The second is the Doppler effect on radiation from an accelerated source.

Suppose a particle is at rest in an inertial frame F_I' and has a lifetime $\tau_I'(rest) = w_{I2}' - w_{I1}'$, as measured by an observer in F_I'. Its lifetime is $\tau_I = w_{I2} - w_{I1}$, as measured by observers in F_I. The Lorentz transformation (3.6) with $x^\mu = 0$ gives the usual lifetime dilation

$$\tau_I = \gamma_o \tau_I' \ (rest). \tag{3.34}$$

Experimentally, it is difficult to measure $\tau_I'(rest)$ because it is almost impossible to arrange for an observer to be in a frame that is comoving with unstable particles in high-energy laboratory. Fortunately, according to the principle of relativity, all inertial frames are equivalent. Thus, the lifetime $\tau_I(rest)$ of a particle at rest in F_I and measured by observers in F_I is the same as the lifetime $\tau_I'(rest)$ of the same kind of particle at rest in F_I' and measured by observers in F_I',

$$\tau_I(rest) = \tau_I'(rest). \tag{3.35}$$

Thus, the lifetime dilation (3.34) in standard special relativity can be expressed in terms of observable quantities in an inertial frame F_I,

$$\tau_I = \gamma_o \tau_I(rest),$$

which has been confirmed by high-energy experiments in the laboratory frame F_I.

An inertial frame F_I and a CLA frame F are not equivalent. However, in order to calculate the lifetime of an accelerated particle to compare it with experiment, one needs to postulate a relation similar to (3.35). Let us consider observable vectors such as space-time differential dx^μ, the wave vector k_μ, or a momentum vector. All of these physically observable vector quantities (e.g., energy–momentum of a particle, wave vector of the radiation emitted from a source, etc.) can be denoted by contravariant or covariant vectors, $V^\mu(rest)$ and $V_\mu(rest)$, if the particle or the source of light waves is "at rest" in a CLA frame F and these quantities are measured by observers in the same CLA frame F. Following the principle of limiting continuation for physical laws, it is natural to postulate the following relation for two different CLA frames, F and F',

$$V^0(\text{rest})\sqrt{P_{00}} = V'^0(\text{rest})\sqrt{P'_{00}},$$

$$\text{or} \quad V^0(\text{rest})W = V'^0(\text{rest})W', \tag{3.36}$$

which can also be expressed in terms of a covariant component

$$V_0(\text{rest})\sqrt{P^{00}} = V'_0(\text{rest})\sqrt{P'^{00}},$$

$$\text{or} \quad \frac{V_0(\text{rest})}{W} = \frac{V'_0(\text{rest})}{W'}, \tag{3.37}$$

where we have used $V_\mu = P_{\mu\nu}V^\nu$ or $V_0 = W^2 V^0$. This postulate (3.36) or (3.37) may be termed the "weak equivalence of non-inertial frames." Relationships (3.36) and (3.37) cannot be derived from the Wu transformations (3.21). However, they are consistent with the principle of limiting continuation of physical laws.

Now, let us consider a specific example of the lifetime of a particle at rest in a CLA frame $\tau(\text{rest})$. Equation (3.36) implies that this lifetime is related to the lifetime of the same kind of particles at rest in inertial frame F_I (for which the Wu factor W is 1) by

$$\tau(\text{rest})W = \tau_I(\text{rest}), \quad W = \gamma^2(\gamma_o^{-2} + \alpha_o x), \tag{3.38}$$

where the lifetime of a particle is usually sufficient small and can be considered as the zeroth component of a contravariant differential vector dx^μ. This relationship is consistent with the principle of limiting continuation and can be tested indirectly by experiments of measuring the lifetime dilation of accelerated particles.

Since the lifetime of unstable particles is usually very short, those lifetimes τ_I and τ, as measured by observed in F_I and F, can be identified with the infinitesimal time intervals dw_I and dw in (3.2), respectively. Suppose a particle is at rest with $\mathbf{r} = 0$ and $d\mathbf{r} = 0$ in a CLA frame, and has a position $x_I = L$ as measured in F_I. From (3.2) with $d\mathbf{r} = 0$ and (3.38), we have the lifetime τ_I decaying in flight with constant acceleration α_o,

$$\tau_I = \frac{1}{\sqrt{1-\beta^2}}\tau(\text{rest})W = \gamma_o(1 + \alpha_o\gamma_o L)\tau_I(\text{rest}), \tag{3.39}$$

where all quantities are measured in the inertial laboratory frame F_I.

According to the Wu transformations, a CLA α_o corresponds physically to a uniform increase in the energy of the particle per unit length traveled. This prediction (3.39) for the dilation of the lifetime of an unstable particle undergoing such a CLA can be tested experimentally by measurements of the decay-length (or equivalently, the "lifetime") of an unstable particle in a linear accelerator.[10]

Given the discussion regarding the relationship between the Wu transformations and the generalized MWL transformations in Section 3.6, one might wonder whether the generalized MWL transformations predict the same result (3.39). The answer is yes. Differentiating the generalized MWL transformations (3.27), one obtains (3.2) with $W = W_x = 1 + \alpha^* \gamma_o x$ and β given by (3.28). The Wu factor W in (3.38) would be replaced by W_x. Substituting in these results and simplifying, one again obtains the same result (3.39), supporting the equivalence of the Wu and the generalized MWL transformations.

The Wu transformations (3.4) can also be tested experimentally by measuring the shift of the wavelength of light emitted from a source undergoing a CLA. From equation (3.14), one can obtain the Wu transformations of the covariant wave 4-vector $k_\mu = p_\mu$ between an inertial frame F_I and a CLA frame F:

$$k_{I0} = \gamma \left(\frac{k_0}{W} - \beta k_1 \right), \quad k_{I1} = \gamma \left(k_1 - \frac{\beta k_0}{W} \right), \tag{3.40}$$

and

$$k_{I0}^2 - k_{I1}^2 = \left(\frac{k_0}{W} \right)^2 - (k_1)^2 = 0, \tag{3.41}$$

where $k_{I2} = k_2 = 0$, $k_{I3} = k_3 = 0$. The quantities k_{I0} and k_0 are the frequencies (or equivalently, energies) of the same wave measured by observers in F_I and F respectively.

According to the weak equivalence of non-inertial frames (3.37), $k_0(\text{rest})$ and $k_{I0}(\text{rest})$ satisfy

$$\frac{k_0(\text{rest})}{W} = k_{I0}(\text{rest}), \quad W = \gamma^2 (\gamma_o^{-2} + \alpha_o x), \tag{3.42}$$

where k_0 is the zeroth component of the covariant wave vector. From (3.40), (3.42) and $\beta = \alpha_o w + \beta_o$, we obtain the shift in frequency k_{I0} and wavelength λ_I

$$k_{I0} = k_{I0}(\text{rest})\gamma(1 - \beta) \approx k_{I0}(\text{rest})\gamma_o(1 - \beta_o)(1 - \alpha_o w \gamma_o^2), \tag{3.43}$$

$$\frac{1}{\lambda_I} = \frac{1}{\lambda_I(\text{rest})}\gamma(1 - \beta) \approx \frac{1}{\lambda_I(\text{rest})}\gamma_o(1 - \beta_o)(1 - \alpha_o w \gamma_o^2), \tag{3.44}$$

for radiation emitted from a source undergoing CLA. In (3.43) and (3.44), we have again used the approximation $\gamma \approx \gamma_o(1 + \beta_o \alpha_o w \gamma_o^2)$, obtained by expanding γ to the first order in α_o (which is assumed to be small), while keeping $\beta_o < 1$ to all orders. We call the results (3.43) and (3.44) the accelerated Wu–Doppler shift or the accelerated Wu–Doppler effect.[11]

For comparison with experiments, it is more convenient to express the quantities in (3.43) and (3.44) in terms of distances, rather than the time w of the CLA frame. For this purpose, the following relation may be used to express $\alpha_o w$ in (3.43) and (3.44) in terms of distances,

$$\alpha_o w + \beta_o = \frac{w_I + \beta_o/(\alpha_o\gamma_o)}{x_I + 1/(\alpha_o\gamma_o)} = \sqrt{1 - \left(\frac{\gamma_o^{-1} + \alpha_o\gamma_o x}{1 + \alpha_o\gamma_o x_I}\right)^2}. \tag{3.45}$$

Suppose that the radiation source is located at $\mathbf{r} = 0$ in the CLA frame, and enters an accelerated potential at $x_I = 0$ in the laboratory frame F_I with an initial velocity β_o (measured from F_I). The radiation is then measured when the source has been accelerated to the point $x_I = L$. In this case, equation (3.45) with small α_o can be approximated as

$$\alpha_o w \approx \sqrt{2\alpha_o L + \beta_o^2} - \beta_o. \tag{3.46}$$

From equations (3.44) and (3.46), we obtain

$$\delta\lambda_I = \lambda_I - \lambda_I(\text{rest}) \approx \lambda_I(\text{rest})\left[-\frac{3}{2}\beta_o^2 - (1 - 2\beta_o)\sqrt{2\alpha_o L + \beta_o^2}\right], \tag{3.47}$$

to first order in α_o and to second order in β_o. As expected, in the limit of zero acceleration, $\alpha_o \to 0$, the Wu–Doppler shift (3.47) reduces to the usual relativistic formula (to order β_o^2) for the Doppler shift.

As with our result for the lifetime dilation of an accelerated particle, we can also check our result (3.47) using the generalized MWL transformations rather than the Wu transformations. Following the derivation from (3.40) to (3.44) using the generalized MWL transformations (3.27), one again has the same equations for (3.40) through (3.42), provided the Wu factor W is replaced by W_x and the velocity β is given by (3.28), where $\alpha^* = \alpha_o\gamma_o$. Using (3.28), we also obtain (3.43) and (3.44), but with $\alpha_o w\gamma_o$ replaced by $\alpha^*\gamma_o w^*$. Following steps (3.45) through (3.47), one can express the time w^* in the expression $\alpha^*\gamma_o w^*$ in terms of the distance L, and again arrive at the same expression (3.47) for the Wu–Doppler effect. Thus, the generalized MWL transformations are physically equivalent to the Wu transformations.

The predictions (3.44) and (3.47) can be tested in the laboratory frame F_I using a method similar to that in the Ives–Stilwell experiment.[12]

References

1. V. A. Brumberg and S. M. Kopeikin, in *Reference Frames*, Astrophysics and Space Science Library, Vol. 154, eds. J. Kovalevsky, I. I. Mueller and B. Kolaczek (Springer, 1989), pp. 115–141.

2. C. Møller, *Danske Vid. Sel. Mat.-Fys.* **20**, (1943); *The Theory of Relativity* (Oxford University Press, London, 1952), pp. 253–258; H. Lass, *Am. J. Phys.* **31**, 274 (1963); W. Rindler, *Am. J. Phys.* **34**, 1174 (1966); R. A. Nelson, *J. Math. Phys.* **28**, 2379 (1987); S. G. Turyshev, O. L. Minazzoli and V. T. Toth, *J. Math. Phys.* **53**, 032501 (2012).

3. E. P. Wigner, *Proc. Natl. Acad. Sci.* **51**, (1964); *Symmetries and Reflections, Scientific Essays* (MIT Press, 1967), pp. 18–19; H. Poincaré, *C. R. Acad. Sci. Paris* **140**, 1504 (1905); *Rend. Circ. Mat. Palermo* **21**, 129 (1906).

4. J. J. Sakurai, *Invariance Principles and Elementary Particles* (Princeton University Press, 1964), p. v, pp. 3–5, and p. 10.

5. J.-P. Hsu and L. Hsu, *Nuovo Cimento* **112**, 575 (1997) and *Chin. J. Phys.* **35**, 407 (1997); in *A Broad View of Relativity: General Implications of Lorentz and Poincaré Invariance* (World Scientific, Singapore, 2006), Chapters 18–19. (Available online at "google books, Leonardo Hsu.")

6. T.-Y. Wu and Y. C. Lee, *Int. J. Theor. Phys.* **5**, 307 (1972); T.-Y. Wu, *Theoretical Physics, Vol. 4, Theory of Relativity* (Lian Jing Publishing Co., Taipei, 1978), pp. 172–175; J.-P. Hsu and L. Hsu, in *JingShin Theoretical Physics Symposium in Honor of Professor Ta-You Wu*, eds. J. P. Hsu and L. Hsu (World Scientific, Singapore, 1998), pp. 393–412.

7. J. P. Hsu and S. M. Kleff, *Chin. J. Phys.* **36**, 768 (1998); S. Kleff and J. P. Hsu, in *JingShin Theoretical Physics Symposium in Honor of Professor Ta-You Wu*, eds. J. P. Hsu and L. Hsu (World Scientific, Singapore, 1998), pp. 348–352. (Available online at "google books, Leonardo Hsu.")

8. O. Veblen and J. H. C. Whitehead, in *The Foundations of Differential Geometry* (Cambridge University Press, 1953), pp. 37–38.

9. D. Fine, Private correspondence, (2012).

10. J. P. Hsu and L. Hsu, in *A Broad View of Relativity: General Implications of Lorentz and Poincaré Invariance* (World Scientific, Singapore, 2006), pp. 310–317.

11. L. Hsu and J. P. Hsu, *Nuovo Cimento B* **112**, 1147 (1997). Appendix.

12. H. E. Ives and G. R. Stilwell, *J. Opt. Soc. Am.* **28**, 215 (1938).

Chapter 4

Coordinate Transformations for Frames with Arbitrary Linear Accelerations and the Taiji Pseudo-Group

4.1. Arbitrary Linear Accelerations: The Taiji Transformations

In Chapter 3, we used the principle of limiting continuation of physical laws to develop the Wu transformations, a minimal generalization of the Lorentz transformations that give the coordinate transformations between an inertial frame and a non-inertial frame with a constant acceleration, where the acceleration and the velocity are both along the same line (they may point in the same or opposite directions).[a,b] In this chapter, we make a further generalization, again using the principle of limiting continuation, to develop a set of coordinate transformations between an inertial frame and a non-inertial frame with an "arbitrary linear acceleration (ALA)," i.e., where the acceleration and velocity are both along the same line, but the acceleration is an arbitrary function of time. We will call these non-inertial frames "arbitrary linear acceleration frames" or ALA frames, and the transformations the "taiji transformations." As we shall see, the set of taiji transformations, like the Wu transformations, also forms a pseudo-group[3] and reduces to the Wu transformations in the limit of constant acceleration. For the space-time transformations between inertial and accelerated frames from least to most general, see Figure 4.1.

Consider a non-inertial frame $F(w, x, y, z)$ that has an acceleration $\alpha(w)$ along the x-axis that is an arbitrary function of time, rather than a constant.

---

[a] For early discussions of space-time transformations for accelerated frames, see Ref. 1.

[b] The Wu transformations also lead to space-time transformations between any two frames with a constant linear accelerations, where the limit of zero acceleration exists. For more recent discussions based on limiting Lorentz and Poincaré invariance (which is a special case of the principle of limiting continuation of physical laws), see Ref. 2.

Space-time transformation between two inertial frames
$$(\beta = \beta_0 = \text{constant}; \quad \gamma_0 = (1 - \beta_0^2)^{-1/2})$$

$$w_I = \gamma_0(w - \beta_0 x), \quad x_I = \gamma_0(x - \beta_0 w), \quad y' = y, \quad z' = z;$$

$$\Big\downarrow \quad \beta = \beta_0 \rightarrow \beta = \beta_0 + \alpha_0 w$$

Space-time transformation between an inertial frame and a frame with a
Constant-Linear-Acceleration
$$[\beta = \beta_0 + \alpha_0 w, \quad \gamma = (1 - \beta^2)^{-1/2}]$$

$$w_I = \gamma\beta\left(x + \frac{1}{\alpha_0\gamma_0^2}\right) - \frac{\beta_0}{\alpha_0\gamma_0}, \quad x_I = \gamma\left(x + \frac{1}{\alpha_0\gamma_0^2}\right) - \frac{1}{\alpha_0\gamma_0}, \quad y_I = y, \quad z_I = z;$$

$$\Big\downarrow \quad \beta = \beta_0 + \alpha_0 w \rightarrow \beta = \beta(w) = \text{arbitrary}$$

Space-time transformation between an inertial frame and a frame with an
Arbitrary-Linear-Acceleration
$$[\beta = \beta(w) = \text{arbitrary}, \quad \gamma = (1 - \beta^2)^{-1/2}]$$

$$w_I = \gamma\beta\left(x + \frac{1}{\alpha(w)\gamma_0^2}\right) - \frac{\beta_0}{\alpha_0\gamma_0}, \quad x_I = \gamma\left(x + \frac{1}{\alpha(w)\gamma_0^2}\right) - \frac{1}{\alpha_0\gamma_0}, \quad y_I = y, \quad z_I = z$$

Fig. 4.1.   The hierarchy of space-time transformations between inertial and accelerated frames from least to most general.

We consider the case where the velocity $\beta(w)$ is also taken to be along the x-axis,

$$\beta(w) = \beta_1(w) + \beta_0, \quad \alpha(w) = \frac{d\beta(w)}{dw} = \frac{d\beta_1}{dw}, \quad \beta(0) = \beta_0, \quad \alpha(0) = \alpha_0.$$
$$(4.1)$$

The last two conditions in (4.1) imply that when $w = x = 0$, w_I and x_I need not also be zero, in general.

One of the simplest generalizations of the Wu transformations for the constant-acceleration case is to write the infinitesimal coordinate transformations between $F(x)$ and $F_I(x_I)$ in the following form:

$$dw_I = \gamma(W_a dw + \beta dx), \quad dx_I = \gamma(dx + \beta W_b dw), \quad dy_I = dy, \quad dz_I = dz; \tag{4.2}$$

$$\gamma = \sqrt{1 - \beta^2}, \quad \beta = \beta_1(w) + \beta_o, \quad \beta^2 < 1, \tag{4.3}$$

where $W_a = W_a(w, x)$ and $W_b = W_b(w, x)$ are two different unknown functions, in contrast to (3.2). For the same reasons as in the development of (3.2), the terms involving dx in (4.2) do not gain an extra coefficient.

The principle of limiting continuation requires that the two unknown functions W_a and W_b in (4.2), (a) reduce to the same function W in (3.8) in the limit of constant acceleration, and (b) satisfy the following two integrability conditions for the differential equations in (4.2):

$$\frac{\partial(\gamma W_a)}{\partial x} = \frac{\partial(\gamma\beta)}{\partial w}, \quad \frac{\partial(\gamma\beta W_b)}{\partial x} = \frac{\partial\gamma}{\partial w}. \tag{4.4}$$

Analogous to the derivation of the Wu transformations, γ and β are assumed to be functions of w only, leading to the results

$$W_a = \gamma^2\alpha(w)x + A(w), \quad W_b = \gamma^2\alpha(w)x + B(w). \tag{4.5}$$

Substituting (4.5) into (4.2), we obtain

$$w_I = \gamma\beta x + \int \gamma A(w)dw, \quad x_I = \gamma x + \int \gamma\beta B(w)dw, \tag{4.6}$$

while the transformations of the y and z coordinates remain

$$y_I = y + y_o, \quad z_I = z + z_o. \tag{4.7}$$

The principle of limiting continuation of physical laws now dictates that (4.6) must reduces to (3.4) in the limit,

$$\alpha(w) \to \alpha_o \quad \text{or} \quad \beta_1(w) \to \alpha_o w. \tag{4.8}$$

Thus, we have the following relations for the integrals in (4.6) involving an ALA $\alpha(w)$,[c]

$$\int \gamma A(w) dw = \gamma \beta \frac{1}{\alpha(w)\gamma_0^2} + a_o, \tag{4.9}$$

$$\int \gamma \beta B(w) dw = \gamma \frac{1}{\alpha(w)\gamma_0^2} + b_o. \tag{4.10}$$

By differentiation of (4.9) and (4.10), we can determine the two unknown functions $A(w)$ and $B(w)$,

$$A(w) = \frac{\gamma^2}{\gamma_0^2} - \frac{\beta J_e}{\alpha^2(w)\gamma_0^2}, \quad B(w) = \frac{\gamma^2}{\gamma_0^2} - \frac{J_e}{\beta\alpha^2(w)\gamma_0^2}, \quad J_e(w) = \frac{d\alpha}{dw}, \tag{4.11}$$

where $J_e = J_e(w)$ is the "jerk," which is the third-order time derivative of the coordinates.

It thus follows from (4.6), (4.7) and (4.11) that the most general space-time coordinate transformations for the case where the acceleration and velocity of the non-inertial frame are both along parallel x- and x_I-axes are

$$w_I = \gamma\beta\left(x + \frac{1}{\alpha(w)\gamma_0^2}\right) - \frac{\beta_o}{\alpha_o\gamma_o} + w_o,$$

$$x_I = \gamma\left(x + \frac{1}{\alpha(w)\gamma_0^2}\right) - \frac{1}{\alpha_o\gamma_o} + x_o, \tag{4.12}$$

$$y_I = y + y_o, \quad z_I = z + z_o;$$

$$\gamma = \frac{1}{\sqrt{1-\beta^2}}, \quad \gamma_o = \frac{1}{\sqrt{1-\beta_o^2}}, \quad \beta = \beta_1(w) + \beta_o < 1.$$

A constant space-time shift x_o^μ has been included so that both the Lorentz and Poincaré transformations are special limiting cases of (4.12). We call (4.12) the "taiji transformations" for a frame F with an acceleration $\alpha(w)$ that is an arbitrary function of w.

---

[c]The generalization with the replacement $\alpha_o \to \alpha(w)$ (for the accelerations in the denominators of (4.9) and (4.10)) is crucial. Mathematically, this replacement implies that the two variables w and x in $W(w, x)$ cannot be separated, in general. In contrast, a general transformation for ALA frames was discussed in a previous paper based on the separation of w and x in $W(w, x)$. (See Ref. 4.) However, the generality in this conference paper turns out to be restricted and hence, not completely satisfactory because an additional technical assumption of the separation of variables, $W(w, x) = W_1(w)W_2(x)$, prevents the ALA transformations from being fully realized.

The inverse of the taiji transformation (4.12) is found to be

$$\beta(w) = \frac{w_I + \beta_o/(\alpha_o\gamma_o) - w_o}{x_I + 1/(\alpha_o\gamma_o) - x_o} \equiv \beta_I(w_I, x_I),$$

$$x + \frac{1}{\alpha(w)\gamma_o^2} = \sqrt{\left(x_I + \frac{1}{\alpha_o\gamma_o} - x_o\right)^2 - \left(w_I + \frac{\beta_o}{\alpha_o\gamma_o} - w_o\right)^2}, \qquad (4.13)$$

$$y = y_I - y_o, \quad z = z_I - z_o.$$

Although the arbitrary velocity function $\beta(w) = \beta_I(w_I, x_I)$ can always be expressed in terms of w_I and x_I, the arbitrary acceleration function $\alpha(w)$ on the left-hand side of (4.13) cannot, in the most general case. Thus, the inverse taiji transformations (4.13) cannot necessarily be written explicitly in the usual form $x^\mu = x^\mu(w_I, x_I)$.

However, if a specific function for $\beta(w)$ is given and one can solve for the time w in terms of $\beta(w)$, then it will be possible for the first equation in (4.13) to be written in the usual form $w = w(w_I, x_I)$. For example, if the jerk J_{eo} is constant, so that $\beta(w) = J_{eo}w^2/2 + \alpha_o w + \beta_o$, then when w is positive,

$$w = \frac{1}{J_{eo}}\left[-\alpha_o + \sqrt{\alpha_o^2 + 2J_{eo}[\beta_I(w_I, x_I) - \beta_o]}\right] = w(w_I, x_I), \qquad (4.14)$$

where $\beta(w) = \beta_I(w_I, x_I)$, is a function of x_I and w_I and is given by (4.13). Moreover, the result (4.14) enables $\alpha(w)$ to be expressed in terms of w_I and x_I

$$\alpha(w) = \alpha(w_I, x_I),$$

so that in this case, we can write the inverse taiji transformations (4.13) in the usual form, $w = w(w_I, x_I)$ and $x = x(w_I, x_I)$.

The taiji transformations can also be generalized to the case in which the velocity $\boldsymbol{\beta}(w)$ and the linear acceleration $\boldsymbol{\alpha}(w)$ are both in the same arbitrary direction, and not necessarily along the x- and x_I-axes. In this case, $\boldsymbol{\beta}(w)$, $\boldsymbol{\beta}_o$ and $\boldsymbol{\alpha}(w)$ are in the same fixed direction and the resultant taiji transformations take the form

$$w_I = \gamma\left(\frac{\beta}{\alpha(w)\gamma_o^2} + \boldsymbol{\beta}\cdot\mathbf{r}\right) - \frac{\beta_o}{\alpha_o\gamma_o} + w_o,$$

$$\mathbf{r}_I = \mathbf{r} + (\gamma - 1)(\boldsymbol{\beta}\cdot\mathbf{r})\frac{\boldsymbol{\beta}}{\beta^2} + \left(\frac{\gamma}{\alpha(w)} - \frac{\gamma_o}{\alpha_o}\right)\frac{\boldsymbol{\beta}}{\beta\gamma_o^2} + \mathbf{r}_o. \qquad (4.15)$$

The corresponding "inverse" taiji transformation is

$$\beta(w) = \left[\frac{w_I - w_o + \beta_o/(\alpha_o\gamma_o)}{\mathbf{n} \cdot (\mathbf{r}_I - \mathbf{r}_o) + 1/(\alpha_o\gamma_o)} \right] \equiv \beta_I(w_I, \mathbf{r}_I),$$

$$\mathbf{r} + \frac{\mathbf{n}}{\alpha(w)\gamma_o^2} = \mathbf{r}_I - \mathbf{r}_o + \left(\frac{\gamma_o}{\alpha_o} \right) \frac{\mathbf{n}}{\gamma_o^2} - \frac{(\gamma - 1)\mathbf{n}}{\gamma\beta} \left(w_I - w_o + \frac{\beta_o}{\alpha_o\gamma_o} \right), \quad (4.16)$$

$$\mathbf{n} = \frac{\boldsymbol{\beta}}{\beta} = \frac{\boldsymbol{\beta}_o}{\beta_o}.$$

Since $\gamma = 1/\sqrt{1 - \beta^2(w)} = 1/\sqrt{1 - \beta_I^2(w_I, \mathbf{r}_I)}$ and $\beta(w) = \beta_I(w_I, \mathbf{r}_I)$ can be expressed in terms of w_I and $\mathbf{r}_I$, the right-hand side of the "inverse" taiji transformations (4.16) consists solely of quantities measured from inertial frame F_I. However, the left-hand side of (4.16) can only be expressed in terms of $\beta(w)$ and $\mathbf{r} + \mathbf{n}/\alpha(w)\gamma_o^2$, because $\beta(w)$ and $\alpha(w) = d\beta(w)/dw$ are arbitrary (and unspecified) functions. Thus, in general, one cannot write the inverse transformations in analytical form. However, mathematically, (4.15) defines a function from one neighborhood of the origin in space-time to another. The implicit function theorem guarantees the existence of an inverse in the neighborhood of any point where the Jacobian is non-zero. For example, a simple calculation says this holds at the origin, and indeed in the open set defined by $x > -1/[\alpha(w)\gamma_o^2]$ for the special case (4.12). In practical terms, methods going back to Newton give a numerical approximation scheme which converges (quadratically) to the inverse of any given point where the Jacobian is non-zero.[d]

The taiji transformations appear to be the most generalized form of the Lorentz transformations from a constant velocity to the case of a frame with an ALA, as defined previously. Throughout the rest of this chapter, we shall set $x_o^\mu = (w_o, x_o, y_o, z_o) = 0$ and $\boldsymbol{\beta} = (\beta, 0, 0)$ in the taiji transformations (4.12) and (4.16) for simplicity.

4.2. Poincaré Metric Tensors for ALA Frames

The invariant interval ds and Poincaré metric tensors in ALA frames can be obtained from (4.2), (4.5), and (4.11). We find

$$ds^2 = \eta_{\mu\nu}dx_I^\mu dx_I^\nu = P_{\mu\nu}dx^\mu dx^\nu, \quad (4.17)$$

---

[d]We would like to thank Dana Fine for helpful discussions.

$$P_{00} = W_\alpha^2, \quad P_{01} = P_{10} = U, \quad P_{11} = P_{22} = P_{33} = -1;$$

$$W_\alpha^2 = \left[\gamma^2(\gamma_o^{-2} + \alpha(w)x)\right]^2 - \left[\frac{J_e(w)}{\alpha^2(w)\gamma_o^2}\right]^2 > 0, \quad U = \frac{J_e(w)}{\alpha^2(w)\gamma_o^2}, \quad (4.18)$$

where $J_e(w)$ is defined in (4.11).[e]

Mathematically, transformation (4.2) can be obtained by making the replacement $(dw, dx) \rightarrow (dw^*, dx^*) = (W_o dw, dx - U dw)$, where $W_o = \gamma^2(\alpha x + 1/\gamma_o^2)$, and then performing a four-dimensional rotation in the w–x plane with y- and z-axes fixed:

$$dw_I = \gamma(dw^* + \beta dx^*), \quad dx_I = \gamma(dx^* + \beta dw^*), \quad dy_I = dy, \quad dz_I = dz;$$

$$dw^* \equiv W_o dw, \quad dx^* \equiv dx - U dw. \quad (4.19)$$

This is analogous to the four-dimensional rotation of coordinates in special relativity.

The contravariant metric tensors for a general non-inertial frame can be obtained from (4.18):

$$P^{00} = \frac{1}{W_o^2}, \quad P^{01} = P^{10} = \frac{U}{W_o^2},$$

$$P^{11} = \frac{-W_\alpha^2}{W_o^2}, \quad P^{22} = P^{33} = -1, \quad (4.20)$$

$$W_o^2 = W_\alpha^2 + U^2 = \left[\gamma^2(\gamma_o^{-2} + \alpha(w)x)\right]^2.$$

Using (4.18) and (4.20), one can verify that $P^{\alpha\gamma} P_{\gamma\beta} = \delta_\beta^\alpha$.

4.3.   New Properties of the Taiji Transformations

As might be expected, the taiji transformations have many of the same mathematical properties as the Wu transformations. Similar to the Wu transformations, the taiji transformations (4.12) also imply a singular wall in the non-inertial frame at $x = -1/\alpha(w)\gamma_o \equiv x_s$, the location of which depends on the acceleration function $\alpha(w)$. This singular wall separates physical space ($x > x_s$ for $\alpha(w) > 0$) from unphysical space ($x < x_s$), which does not correspond to any physical part

---

[e]For $\beta(w) = \mathbf{n}\beta$ in an arbitrary direction, the invariant interval $ds^2 = P_{\mu\nu}dx^\mu dx^\nu = W^2 dw^2 + 2U dw \mathbf{n} \cdot \mathbf{dr} - \mathbf{dr}^2$ with $W = W(w, \mathbf{r})$ is given by

$$W_\alpha^2(w, \mathbf{r}) = \left[\gamma^2\left(\alpha(w)(\mathbf{n} \cdot \mathbf{r}) + \frac{1}{\gamma_o^2}\right)\right]^2 - \left[\frac{J_e(w)}{\alpha^2(w)\gamma_o^2}\right]^2.$$

of inertial frames. Thus, the taiji transformations map the entire space-time of an inertial frame to just a portion of space-time in an ALA frame. The taiji transformations between two ALA frames, say F and F', would map a portion of the space-time of the F frame to a portion of the space-time in the F' frame.

Also like the Wu transformations, the taiji transformations form a pseudo-group rather than a usual group.[3] As in the previous chapter, we first demonstrate that the subset of taiji transformations for all ALA frames moving along a particular direction (such as parallel x/x'-axes) satisfies all the properties of a normal group.

In the limit $\alpha(w) = \alpha_o \to 0$ and $\beta_o \to 0$, one can verify that the taiji transformation (4.12) reduces to the identity transformation (identity property) and that the taiji transformation (4.12) has an inverse (as given by (4.16)) (inverse property). To demonstrate the other group properties of the taiji transformations for an arbitrary acceleration along the x-axis, let us consider two other ALA frames F' and F'', characterized by arbitrary velocities $\beta'(w')$, $\beta(w'')$, initial accelerations α'_o, α''_o, and initial velocities β'_o, β''_o, respectively. Using (4.12), the taiji transformations among F_I, F, F' and F'' are

$$w_I = \gamma\beta\left(x + \frac{1}{\alpha(w)\gamma_o^2}\right) - \frac{\beta_o}{\alpha_o\gamma_o} = \gamma'\beta'\left(x' + \frac{1}{\alpha'(w')\gamma_o'^2}\right) - \frac{\beta'_o}{\alpha'_o\gamma'_o}$$

$$= \gamma''\beta''\left(x'' + \frac{1}{\alpha''(w'')\gamma_o''^2}\right) - \frac{\beta''_o}{\alpha''_o\gamma''_o},$$

$$x_I = \gamma\left(x + \frac{1}{\alpha(w)\gamma_o^2}\right) - \frac{1}{\alpha_o\gamma_o} = \gamma'\left(x' + \frac{1}{\alpha'(w')\gamma_o'^2}\right) - \frac{1}{\alpha'_o\gamma'_o}$$

$$= \gamma''\left(x'' + \frac{1}{\alpha''(w'')\gamma_o''^2}\right) - \frac{1}{\alpha''_o\gamma''_o}, \tag{4.21}$$

where we have ignored the y and z transformations for simplicity.

The taiji transformations between F and F' can be obtained from (4.21):

$$\beta(w) = \frac{\gamma'\beta'Q' - \beta'_o/(\alpha'_o\gamma'_o) + \beta_o/(\alpha_o\gamma_o)}{\gamma'Q' - 1/(\alpha'_o\gamma'_o) + 1/(\alpha_o\gamma_o)},$$

$$x + \frac{1}{\alpha(w)\gamma_o^2} = \sqrt{\left(\gamma'Q' - \frac{1}{\alpha'_o\gamma'_o} + \frac{1}{\alpha_o\gamma_o}\right)^2 - \left(\gamma'\beta'Q' - \frac{\beta'_o}{\alpha'_o\gamma'_o} + \frac{\beta_o}{\alpha_o\gamma_o}\right)^2},$$

$$\tag{4.22}$$

$$y = y', \quad z = z'; \quad Q' = \left(x' + \frac{1}{\alpha'(w')\gamma_o'^2}\right).$$

From (4.21), one can also obtain the inverse of (4.22)

$$\beta'(w') = \frac{\gamma\beta Q - \beta_o/(\alpha_o\gamma_o) + \beta_o'/(\alpha_o'\gamma_o')}{\gamma Q - 1/(\alpha_o\gamma_o) + 1/(\alpha_o'\gamma_o')},$$

$$x' + \frac{1}{\alpha'(w')\gamma_o^2} = \sqrt{\left(\gamma Q - \frac{1}{\alpha_o\gamma_o} + \frac{1}{\alpha_o'\gamma_o'}\right)^2 - \left(\gamma\beta Q - \frac{\beta_o}{\alpha_o\gamma_o} + \frac{\beta_o'}{\alpha_o'\gamma_o'}\right)^2},$$

$$(4.23)$$

$$y' = y, \quad z' = z; \quad Q = \left(x + \frac{1}{\alpha(w)\gamma_o^2}\right).$$

Similarly, the taiji transformations between F and F'' can be obtained from (4.21):

$$\beta(w) = \frac{\gamma''\beta''Q'' - \beta_o''/(\alpha_o''\gamma_o'') + \beta_o/(\alpha_o\gamma_o)}{\gamma''Q'' - 1/(\alpha_o''\gamma_o'') + 1/(\alpha_o\gamma_o)},$$

$$x + \frac{1}{\alpha(w)\gamma_o^2} = \sqrt{\left(\gamma''Q'' - \frac{1}{\alpha_o''\gamma_o''} + \frac{1}{\alpha_o\gamma_o}\right)^2 - \left(\gamma''\beta''Q'' - \frac{\beta_o''}{\alpha_o''\gamma_o''} + \frac{\beta_o}{\alpha_o\gamma_o}\right)^2},$$

$$(4.24)$$

$$y = y'', \quad z = z''; \quad Q'' = \left(x'' + \frac{1}{\alpha''(w'')\gamma_o''^2}\right).$$

From (4.23) and (4.24), one can show that the taiji transformation between $F'(x')$ and $F''(x'')$ is given by

$$\beta'(w') = \frac{\gamma''\beta''Q'' - \beta_o''/(\alpha_o''\gamma_o'') + \beta_o'/(\alpha_o'\gamma_o')}{\gamma''Q'' - 1/(\alpha_o''\gamma_o'') + 1/(\alpha_o'\gamma_o')},$$

$$x' + \frac{1}{\alpha'\gamma_o'^2} = \sqrt{\left(\gamma''Q'' - \frac{1}{\alpha_o''\gamma_o''} + \frac{1}{\alpha_o'\gamma_o'}\right)^2 - \left(\gamma''\beta''Q'' - \frac{\beta_o''}{\alpha_o''\gamma_o''} + \frac{\beta_o'}{\alpha_o'\gamma_o'}\right)^2},$$

$$(4.25)$$

$$y' = y'', \quad z' = z'', \quad \alpha' = \alpha'(w').$$

Thus the transformation (4.25) between two ALA frames F' and F'' has the same form as (4.22) for F and F' frames. Substituting (4.23) into (4.22), we obtain the "identity transformation"

$$\beta(w) = \beta(w), \quad x + \frac{1}{\alpha(w)\gamma_0^2} = x + \frac{1}{\alpha(w)\gamma_0^2}, \quad y = y, \quad z = z. \quad (4.26)$$

Thus, just as with the Wu transformations, all the usual group properties are satisfied by this subset of taiji transformations. Also as was the case with the Wu transformations, when the full set of taiji transformations is considered, including those for non-inertial frames with accelerations in different directions, the combination of two taiji transformations with accelerations in different directions cannot be expressed as a combination of a single such transformation combined with a rotation. The set obtained from generators of the taiji transformations using the Lie bracket, the brackets of brackets, and so on, never closes. Therefore, the taiji transformations define a group with infinitely many generators.[5]

Because the group properties of the taiji transformations appear to be somewhat more general than those of traditional pseudo-groups, it may be appropriate to denote such a pseudo-group as a "taiji pseudo-group." The taiji pseudo-group includes the Wu pseudo-groups (with constant accelerations) and the Lorentz and Poincaré groups (with constant velocities) as special cases. We observe that the fundamental metric tensors (4.18) for ALA frames appear to depend on both the coordinates x^μ and the differentials dx^μ.[f]

As a final note regarding the mathematical properties of the taiji transformations, we see that it is, in general, not possible to write the ALA transformations in the usual form

$$x^\mu = x^\mu(w', x', y', z') \quad \text{and} \quad x'^\mu = x'^\mu(w, x, y, z). \quad (4.27)$$

Instead, the best that can be done is to express them in the symmetric form,

$$f^\mu(w, x, y, z, \alpha, \alpha_o, \beta, \beta_o) = f^\mu(w', x', y', z', \alpha', \alpha'_o, \beta', \beta'_o), \quad (4.28)$$

---

[f] Although the space-time of all ALA frames has a zero Riemann-Christoffel curvature tensor, the metric tensors $P_{\mu\nu}$ in (4.17) and (4.18), etc. have unusual structures. The quantity $P_{00} = W^2$, for example, depends on x^μ, $\alpha(w) = d\beta(w)/dw$, and $J_e(w) = d\alpha(w)/dw$. This property of P_{00} is interesting because in Riemannian geometry the metric tensors $g_{\mu\nu}(x)$ depend only on the position coordinates x^μ. However, the metric tensors of space-time in ALA frames appear to be more general than that of Riemannian geometry. In the Finsler geometry, the metric tensor, $g_{\mu\nu}(x^\rho, dx^\sigma)$ depends on the coordinates and their differentials (or the velocities). In this sense, the geometry of space-time in ALA frames appears to be a generalized Riemannian geometry. Possible applications of the Finsler geometry in physics have been suggested by the mathematician S. S. Chern and merit further investigation.

as shown in (4.21)–(4.26). However, using this symmetric form, one can still discuss the space-time transformations for the differentials dx^μ and dx'^ν ((4.2)), the fundamental space-time metric tensors $P_{\mu\nu}$ ((4.17) and (4.18)), and the physics of particle dynamics and field theory in ALA frames.

If F is an ALA frame and F' is a different type of non-inertial frame, say, a rotating frame (to be discussed in the next chapter), then even the symmetric form (4.28) does not hold.

4.4.   Physical Implications

Some physical implications of the taiji transformations for ALA frames worthy of note are:

(A) *Existence of a "preferred" coordinate system:* Similar to the way in which Cartesian coordinates are the preferred coordinates for writing down the Lorentz transformation, the coordinates x^μ with the Poincaré metric tensors given in (4.17) and (4.18) are the preferred coordinates for expressing the taiji transformations. No other choice to coordinates satisfies the principle of limiting continuation of physical laws. Thus, the taiji symmetry framework based on a flat space-time for both inertial and non-inertial frames does not have the property of general coordinate invariance, in contrast to general relativity.[8]

(B) *Space-time-dependent speed of light in ALA frames:* In an inertial frame, the law for the propagation of light is $ds = 0$. Analogously, the propagation of light in a non-inertial frame is described by the same invariant law (4.17) with $ds = 0$. Let us consider some specific and simple cases. Suppose a light signal travels along the x-axis, i.e., $dy = dz = 0$. The speed of light β_{Lx} is then

$$\beta_{Lx} = \frac{dx}{dw} = \gamma^2 \left(\alpha x + \frac{1}{\gamma_o^2} \right) + \frac{J_e(w)}{\alpha^2(w)\gamma_o^2}, \quad \frac{dy}{dw} = \frac{dz}{dw} = 0, \qquad (4.29)$$

which is certainly different from the speed of light $\beta_L = 1$ (derived from $ds^2 = dw_I^2 - dx_I^2 = 0$) in an inertial frame. If the light signal moves in the y-direction,

---

[8]The generalization of the Lorentz transformations to the taiji transformations (4.12) does not change the curvature of space-time. Thus, the curvature of space-time in ALA frames is the same as that of inertial frames, i.e., all these frames have vanishing Riemann-Christoffel curvature tensor (cf. Ref. 6).

i.e., $dx = dz = 0$, equation (4.14) with $ds = 0$ leads to the speed of light β_{Ly},

$$\beta_{Ly} = \frac{dy}{dw} = \sqrt{\left[\gamma^2\left(\alpha x + \frac{1}{\gamma_o^2}\right)\right]^2 - \left[\frac{J_e(w)}{\alpha^2(w)\gamma_o^2}\right]^2}, \qquad \frac{dx}{dw} = \frac{dz}{dw} = 0. \quad (4.30)$$

(C) *Operational definition of the space-time coordinates in ALA frames:* Since the speed of light in ALA frames is not constant, as shown in (4.29), it is very complicated to use light signals to synchronize clocks in an ALA frame $F(w, x, y, z)$. However, as discussed in Chapter 3 in reference to the Wu transformations, one can synchronize a set of clocks in the ALA frame F by using a grid of computerized "space-time clocks" that are programmed to accept information concerning their positions x_I relative to the F_I frame, obtain w_I from the nearest F_I clock, and then compute and display w and x using the inverse transformation (4.16) with given parameters α_o, β_o and x_o^μ and functions $\beta(w)$ and $\alpha(w)$.[h] If this cannot be done for a given $\beta(w)$ in the taiji transformations, then the space-time coordinates have no operational meaning and we would consider such an ALA frame to be unphysical.

Thus, the operational meaning of space and time coordinates is completely determined by such a grid of "space-time clocks." In the limit of constant velocity, this grid of "space-time clocks" will reduce to the grid of clocks in special relativity.

In an ALA frame, the values the physical time w can take on are restricted by the condition $\beta^2(w) < 1$ and the coordinates of physical space are limited by $x > -1/(\alpha(w)\gamma_o^2)$, as shown in (4.12).

(D) *The velocity-addition laws in ALA frames:* In general, the law for velocity addition can be obtained from (4.2),

$$\frac{dx_I}{dw_I} = \frac{dx/dw + \beta W_b}{W_a + \beta dx/dw},$$

$$\frac{dy_I}{dw_I} = \frac{dy/dw}{\gamma(W_a + \beta dx/dw)}, \qquad \frac{dz_I}{dw_I} = \frac{dz/dw}{\gamma(W_a + \beta dx/dw)}. \quad (4.31)$$

The condition $dx = 0$ implies that an object is at rest in F and its velocity as measured by observers in F_I is then dx_I/dw_I. Similarly, the condition $dx_I = 0$ in (4.31) corresponds to the case where an object is at rest in F_I. Its velocity as measured by observers in the ALA frame F is dx/dw.

---

[h]See the operational meaning of space-time coordinates, discussed in Section 3.3.

If $\beta_L = dx_I/dw_I = 1$, the velocity-addition law (4.31) leads to the same result for the space-time-dependent speed of light in an ALA frame as equation (4.29). Similarly, (4.31) with $dx = dz = 0$ is consistent with (4.30).

(E) *Classical electrodynamics in inertial and non-inertial frames*: For a continuous charge distribution in space, the invariant action for electromagnetic fields and their interactions is postulated to be

$$S_{em} = -\int \left[a_\mu j^\mu + \frac{1}{4} f_{\mu\nu} f^{\mu\nu} \right] \sqrt{-P} d^4 x,$$

$$\sqrt{-P} = \sqrt{-\det P_{\mu\nu}} = \sqrt{W_\alpha^2 + U^2} = \gamma^2 \left(\alpha(w) x + \frac{1}{\gamma_o^2} \right) > 0, \qquad (4.32)$$

$$f_{\mu\nu} = D_\mu a_\nu - D_\nu a_\mu = \partial_\mu a_\nu - \partial_\nu a_\mu,$$

where D_μ denotes the partial covariant derivative with respect to the metric tensor $P_{\mu\nu}$. The most general Maxwell equations for both inertial and non-inertial frames can be derived from the action S_{em}. The Lagrange equation of a_μ in a general frame characterized by the Poincaré tensor $P_{\mu\nu}$ leads to

$$\frac{1}{\sqrt{-P}} \partial_\mu (\sqrt{-P} f^{\mu\nu}) + j^\nu = 0. \qquad (4.33)$$

This implies the continuity equation of charge,

$$\partial_\mu (\sqrt{-P} j^\mu) = 0, \qquad (4.34)$$

which implies charge conservation. This general continuity equation (4.34) also implies that the electric charge is constant in any frame of reference based on taiji symmetry framework.

(F) *Generalized Klein–Gordon and Dirac equations for non-inertial frames*: The general Klein–Gordon and Dirac equations for both inertial and non-inertial frames are

$$\left[P^{\mu\nu} (D_\mu - iea_\mu)(D_\nu - iea_\nu) + m^2 \right] \phi = 0, \quad e > 0, \qquad (4.35)$$

$$i\Gamma^\mu (\partial_\mu - iea_\mu)\psi + \frac{1}{2} i (\partial_\mu \Gamma^\mu)\psi + \frac{1}{2} (\partial_\mu \ln \sqrt{-P}) i\Gamma^\mu \psi - m\psi = 0, \qquad (4.36)$$

where $\{\Gamma^\mu, \Gamma^\nu\} = 2P^{\mu\nu}(x)$. The Lagrangians for these equations are

$$L_\phi = \sqrt{-P}(P^{\mu\nu}[(D_\mu - iea_\mu)\phi]^*(D_\nu - iea_\nu)\phi - m^2 \phi^* \phi), \qquad (4.37)$$

$$L_\psi = \sqrt{-P} \left[\frac{i}{2} \overline{\psi} \Gamma^\mu (D_\mu - iea_\mu) - \frac{i}{2}[(D_\mu - iea_\mu)^* \overline{\psi}] \Gamma^\mu - m\overline{\psi} \right] \psi, \quad (4.38)$$

$$D_\mu \phi = \partial_\mu \phi, \quad D_\mu \psi = \partial_\mu \psi, \quad D_\mu U_\nu = \partial_\mu U_\nu - \Gamma^\lambda_{\mu\nu} U_\lambda,$$

$$\Gamma^\lambda_{\mu\nu} = \frac{1}{2} P^{\lambda\sigma} (\partial_\mu P_{\sigma\nu} + \partial_\nu P_{\sigma\mu} - \partial_\sigma P_{\mu\nu}),$$

where the asterisk * denotes complex conjugate. The invariance of the action involving the Lagrangians (4.37) and (4.38) implies that the quantity corresponding to the electric charge in the Klein–Gordon and Dirac equations has the same value in an arbitrary frame of reference and hence, is a true universal constant, as opposed to the speed of a light signal.

4.5.　Experimental Tests of the Taiji Transformations

Because it is extremely difficult to construct an experimental set up in which objects moving at relativistic speeds also have a non-constant or arbitrary acceleration, obtaining experimental support for the taiji transformations will be difficult. However, we can still use the taiji transformations to predict the shift in the frequency of radiation from a source undergoing an ALA.

　　With the help of the Poincaré metric tensors (4.18) and (4.20), the taiji transformations for covariant vectors can be obtained from the transformations for the contravariant coordinate vectors dx^μ. Thus, we can derive the transformations for the covariant wave vectors k_μ between an inertial frame F_I and an ALA frame F

$$k_{I0} = \gamma \left(\frac{k_0}{W_o} - \left[\beta - \frac{U}{W_o} \right] k_1 \right), \quad k_{I1} = \gamma \left(k_1 \left[1 - \frac{\beta U}{W_o} \right] - \frac{\beta k_0}{W_o} \right), \quad (4.39)$$

and

$$k_{I0}^2 - k_{I1}^2 = \left(\frac{k_0}{W_o} \right)^2 - (k_1)^2 \left(\frac{W_\alpha}{W_o} \right)^2 + 2k_1 k_0 \frac{U}{W_o^2} = 0, \quad (4.40)$$

where $k_{I2} = k_2 = 0$ and $k_{I3} = k_3 = 0$. The quantities k_{I0} and k_0 are the frequencies (or equivalently, energies) of the same wave measured by observers in F_I and F, respectively.

　　Consider a radiation source at rest in an ALA frame F. Suppose k_0(rest) denotes the frequency of the waves emitted from this source as measured by observer in the F frame and k_{I0}(rest) denotes the frequency of the radiation from the same source at rest in an inertial frame F_I as measured by observers in F_I.

In practice, it is difficult to determine the relationship between $k_0(\text{rest})$ and $k_{I0}(\text{rest})$ experimentally because of the difficulty of putting observers in the ALA frame. However, the postulate of the weak equivalence of non-inertial frames (3.37) gives the relationship between $k_0(\text{rest})$ and $k_{I0}(\text{rest})$ as

$$k_0(\text{rest})\sqrt{P^{00}} = \frac{k_0(\text{rest})}{W_o} = k_{I0}(\text{rest}), \quad W_o = \gamma^2(\gamma_o^{-2} + \alpha(w)x). \tag{4.41}$$

Based on (4.39), (4.40), and (4.41), we obtain the shift in the frequency of the radiation emitted from a source undergoing an ALA as

$$k_{I0} = k_{I0}(\text{rest})\frac{\gamma(1-\beta)}{[1 - U/W_o]}, \tag{4.42}$$

$$\frac{1}{\lambda_I} = \frac{1}{\lambda_I(\text{rest})}\frac{\gamma(1-\beta)}{[1 - U/W_o]}, \tag{4.43}$$

where we have used (4.40) to find the relation $k_0 = k_1(W_o - U)$ in the ALA frame F, which holds for the case when the radiation source is at rest in F. Results (4.42) and (4.43) may be called the taiji-Doppler effect. The taiji-Doppler shift includes the Wu–Doppler shift (3.43) and (3.44) in the limit where the jerk vanishes $J_e = 0$ or $U = 0$, or equivalently, when the acceleration is constant, $\alpha(w) = \alpha_o$.

References

1. A. Einstein, *Jahrb. Rad. Elektr.* **4**, 411 (1907); L. Page, *Phys. Rev.* **49**, 254 (1936); C. Møller, *Danske Vid. Sel. Mat. Fys.* **20**, (1943); T. Fulton, F. Rohrlich and L. Witten, *Nuovo Cimento* **24**, 652 (1962); *Rev. Mod. Phys.* **34**, 442 (1962); E. A. Desloge and R. J. Philpott, *Am. J. Phys.* **55**, 252 (1987).

2. J.-P. Hsu and L. Hsu, *Nuovo Cimento B* **112**, 575 (1997); *Chin. J. Phys.* **35**, 407 (1997); **40**, 265 (2002). J.-P. Hsu, *Einstein's Relativity and Beyond — New Symmetry Approaches* (World Scientific, Singapore, 2000), Chapters 21–23. (Available online at "google books, Jong-Ping Hsu.")

3. E. Cartan, *Ann. Sci. École Norm. Sup. (3)* **21**, 153 (1904); O. Veblen and J. Whitehead, *Proc. Natl. Acad. Sci. USA* **17**, 551 (1931); O. Veblen and J. H. C. Whitehead, *The Foundations of Differential Geometry* (Cambridge University Press, 1932), pp. 37–38.

4. J. P. Hsu, in *Frontiers of Physics at the Millennium Symposium* eds. Y. L. Wu and J. P. Hsu (World Scientific, 2001). (Available online at "google books, Jong-Ping Hsu.")

5. D. Fine, Private correspondence (2012).

6. C. Møller, in *The Theory of Relativity* (Oxford University Press, London, 1952), p. 285.

Chapter 5

Coordinate Transformations for Rotating Frames
and Experimental Tests

5.1. Rotational Taiji Transformations

In Chapters 3 and 4, we applied the principle of limiting continuation to derive coordinate transformations between an inertial frame and a frame with a velocity and acceleration pointing along the same line (either parallel or antiparallel). We now use the same approach to derive coordinate transformations between an inertial frame and a frame that rotates with a constant angular velocity.[1] As we shall see, our results are not only consistent with the results of high energy experiments involving unstable particles in a circular storage ring, but also support Pellegrini and Swift's analysis of the Wilson experiment,[2] in which they point out that rotational transformations cannot be locally replaced by Lorentz transformations.

Suppose $F_I(w_I, x_I, y_I, z_I)$ is an inertial frame and $F(w, x, y, z)$ (which we subsequently refer to as $F(\Omega)$) is a frame that rotates with a constant angular velocity Ω (to be defined more precisely below). The origins of both frames coincide at all times and we use a Cartesian coordinate system in both frames. The usual classical transformation equations between F_I and F are

$$w_I = w, \quad x_I = x \cos(\Omega w) - y \sin(\Omega w),$$
$$y_I = x \sin(\Omega w) + y \cos(\Omega w), \quad z_I = z. \tag{5.1}$$

In order to derive a set of coordinate transformations between F_I and F that satisfy the requirements of limiting Lorentz and Poincaré invariance (i.e., limiting four-dimensional symmetry),[a] we first consider a slightly more general case. In this

---

[a]This is a special case of the principle of limiting continuation of physical laws when the accelerated frame becomes an inertial frame in the limit of zero acceleration. There is no additional constant velocity term here.

more general case, there is an inertial reference frame F_I and a non-inertial frame $F_R(\Omega)$ whose origin orbits the origin of the inertial frame at a constant distance R, with a constant angular velocity Ω. A Cartesian coordinate system is used in both frames, set up in such a way that the positive portion of the y-axis of the $F_R(\Omega)$ frame always extends through the origin of F_I. This is useful because in the limit $R \to \infty$ and $\Omega \to 0$ such that the product $R\Omega = \beta_o$ is a finite non-zero constant velocity, the two frames become inertial frames with coordinates related by the familiar Lorentz transformations. With that in mind, the classical coordinate transformations between F_I and $F_R(\Omega)$ (the orbiting frame) are

$$w_I = w, \quad x_I = x \cos(\Omega w) - (y - R) \sin(\Omega w),$$
$$y_I = x \sin(\Omega w) + (y - R) \cos(\Omega w), \quad z_I = z. \tag{5.2}$$

According to the principle of limiting continuation, we postulate that the transformations between F_I and $F_R(\Omega)$ that satisfy the requirements of limiting Lorentz and Poincaré invariance (i.e., the transformations satisfy Lorentz and Poincaré invariance in the limit of zero acceleration) should have the form

$$w_I = Aw + B\rho \cdot \beta, \quad x_I = Gx \cos(\Omega w) + E(y - R) \sin(\Omega w),$$
$$y_I = Ix \sin(\Omega w) + H(y - R) \cos(\Omega w), \quad z_I = z, \tag{5.3}$$

where $\rho = (x, y)$, $S = (x, y - R)$, $\beta = \Omega \times S$, $\Omega = (0, 0, \Omega)$, and the functions A, B, E, G, H, and I may, in general, depend on the coordinates x^μ.

One unusual feature of the transformation equations (5.3) is that we will treat both the constant angular velocity Ω (defined as $\Omega = d\phi/dw$) and the orbital radius R as quantities that are measured by observers in the non-inertial frame $F_R(\Omega)$. This is counter to the usual procedure of measuring such parameters from the "lab" or "inertial" reference frame, but will simplify the following discussion since all quantities on the right side of the transformation equations (5.3) are measured with respect to $F_R(\Omega)$ observers. Thus, when $w = w_I = 0$, the y- and y_I-axes overlap and the origin of F_I, i.e., $x_I = y_I = 0$, is found at the coordinates $(x, y) = (0, +R)$. Equivalently, the origin of the rotating frame $F_R(\Omega)$, i.e., $x = y = 0$, is found at the inertial coordinates $(x_I, y_I) = (0, -HR)$, where H is an unknown function to be determined.

To determine the unknown functions A, B, E, G, H, and I, we consider the following limiting cases. First, when $R = 0$, transformations (5.3) must have x/y

symmetry (i.e., be symmetric under an exchange of x and y). This implies that

$$-E = G = H = I \quad \text{when } R = 0. \tag{5.4}$$

Second, in the limit of small Ω (or $|\mathbf{\Omega} \times \mathbf{S}| \ll 1$) with $R \to 0$, transformations (5.3) should reduce to the classical rotational transformations (5.1). Thus,

$$-E \approx G \approx H \approx I \approx 1 \quad \text{for small } \Omega \text{ with } R = 0. \tag{5.5}$$

Finally, and perhaps most importantly, in the limit where $R \to \infty$ and $\Omega \to 0$ such that their product $R\Omega = \beta_o$ is a non-zero constant velocity parallel to the x_I-axis, the finite and differential forms of (5.3) must reduce to the Lorentz transformations, in both its finite form

$$w_I = \gamma_o(w + \beta_o x), \quad x_I = \gamma_o(x + \beta_o w), \quad y_I = y, \quad z_I = z;$$
$$\gamma_o = \frac{1}{\sqrt{1 - \beta_o^2}}, \tag{5.6}$$

and its differential form,

$$dw_I = \gamma_o(dw + \beta_o dx), \quad dx_I = \gamma_o(dx + \beta_o dw), \quad dy_I = dy, \quad dz_I = dz. \tag{5.7}$$

(Strictly speaking, $y_I = y$ should be replaced by $y_I = -\infty$ in (5.6), because $R \to \infty$ in this limit.) This is necessary for (5.3) to satisfy the limiting Lorentz and Poincaré invariance, as specified by the principle of limiting continuation of physical laws. The taiji time w in (5.2) reduces to the usual relativistic time in this limit. Thus, we have

$$A = B = G = -E = \gamma_o, \quad H = 1,$$
$$\text{when } R \to \infty \text{ and } \Omega \to 0 \text{ such that } \beta_o = R\Omega. \tag{5.8}$$

The requirements put on A, B, E, G, H, and I by considering these three limiting cases do not lead to a unique solution for the unknown functions. This is analogous to the case in which gauge symmetry does not uniquely determine the electromagnetic action[3] and one must also postulate a minimal electromagnetic coupling. Here, as we did in Chapters 3 and 4 in deriving the Wu and taiji transformations, we postulate a minimal generalization of the classical rotational transformations (5.2).

Based on the limiting cases considered above, it is not unreasonable for transformations (5.3) to have the following two properties: (i) for non-zero Ω and finite R, the functions A, B, G, I and $-E$ are simply generalized from $\gamma_o = (1 - \beta_o^2)^{-1/2}$ to $\gamma = (1 - \beta^2)^{-1/2}$, where $\beta = S\Omega$, and (ii) only H depends on R and it is required to be the simplest function involving only the first power of γ. Combining (5.4), (5.5), and (5.8) with the two properties just named leads to the following solutions:

$$A = B = G = I = \gamma = (1 - \beta^2)^{-1/2}, \quad E = -\gamma,$$
$$H = (\gamma + R/R_o)(1 + R/R_o), \quad \gamma = (1 - \beta^2)^{-1/2}. \tag{5.9}$$

The quantity R_o in (5.9) is an undetermined length parameter that seems to be necessary for the transformations to satisfy the limiting criteria specified in (5.4), (5.5), and (5.8), as well as the limiting Lorentz and Poincaré invariance. From a comparison of rotational transformations and their classical approximation, R_o should be very large value. However, since it is unsatisfactory to have such undetermined parameters in a theory and there are no known experimental results that can help to determine a value for R_o, we shall take the limit $R_o \to \infty$ in the following discussions for simplicity.

Thus, a simple rotational taiji transformation corresponding to the classical transformations (5.2) is

$$w_I = \gamma(w + \boldsymbol{\rho} \cdot \boldsymbol{\beta}), \quad x_I = \gamma[x \cos(\Omega w) - (y - R)\sin(\Omega w)],$$
$$y_I = \gamma[x \sin(\Omega w) + (y - R)\cos(\Omega w)], \quad z_I = z; \tag{5.10}$$
$$\beta = |\boldsymbol{\Omega} \times \mathbf{S}| = \Omega\sqrt{x^2 + (y - R)^2} = \Omega S < 1, \quad \boldsymbol{\rho} \cdot \boldsymbol{\beta} = xR\Omega.$$

Because all known experiments can be analyzed in the $R = 0$ case, for the rest of this chapter, we set $R = 0$ and concentrate solely on the implications of the transformations under that condition. With $R = 0$ the coordinate transformations between an inertial frame F_I and a non-inertial frame $F(\Omega)$ whose origin coincides with that of F_I and that rotates with a constant angular velocity Ω about the origin are

$$w_I = \gamma(w + \boldsymbol{\rho} \cdot \boldsymbol{\beta}) = \gamma w, \quad x_I = \gamma[x \cos(\Omega w) - y \sin(\Omega w)],$$
$$y_I = \gamma[x \sin(\Omega w) + y \cos(\Omega w)], \quad z_I = z, \tag{5.11}$$

where $\gamma = 1/\sqrt{1 - \rho^2\Omega^2}$ and $\boldsymbol{\rho} \cdot \boldsymbol{\beta} = \Omega Rx = 0$ for $R = 0$. Again, we take the constant angular velocity Ω (defined as $\Omega = |d\phi/dw|$) as a quantity that is measured by observers in the non-inertial frame $F(\Omega)$.

To derive the inverse transformations of (5.11), we must first find a way to express Ωw and $\gamma = 1/\sqrt{1 - \rho^2\Omega^2}$ in terms of quantities measured in the inertial frame F_I. While the coordinate transformations (5.10) for the non-zero R case can only be written in Cartesian coordinates (so that they satisfy the limiting Lorentz and Poincaré invariance), the transformations in the $R = 0$ case (5.11) can be written in terms of cylindrical coordinates. We introduce the relations $x_I = \rho_I \cos\phi_I, y_I = \rho_I \sin\phi_I, x = \rho\cos\phi$, and $y = \rho\sin\phi$ so that (5.11) can be written as

$$w_I = \gamma w, \quad \rho_I = \sqrt{x_I^2 + y_I^2} = \gamma\rho, \quad \phi_I = \phi + \Omega w, \quad z_I = z$$

with

$$\Omega_I \equiv \frac{d\phi_I}{dw_I} = \frac{d\phi_I}{dw}\frac{dw}{dw_I} = \Omega\frac{1}{\gamma}, \quad (\rho \text{ and } \phi \text{ fixed}).$$

The conditions that ρ and ϕ are fixed mean that an object is at rest in the rotating frame, so that an observer in F_I can measure its angular velocity Ω_I and identify it with the angular velocity of the rotating frame. These equations give the operational definitions of Ω_I and Ω. Thus, we have

$$w_I\Omega_I = w\Omega, \quad \rho_I\Omega_I = \rho\Omega, \quad \gamma = \frac{1}{\sqrt{1 - \rho^2\Omega^2}} = \frac{1}{\sqrt{1 - \rho_I^2\Omega_I^2}}. \tag{5.12}$$

From (5.11) and (5.12), we can then derive the inverse rotational transformations

$$w = \frac{w_I}{\gamma}, \quad x = \frac{1}{\gamma}[x_I\cos(\Omega_I w_I) + y_I\sin(\Omega_I w_I)],$$

$$y = \frac{1}{\gamma}[-x_I\sin(\Omega_I w_I) + y_I\cos(\Omega_I w_I)], \quad z = z_I, \tag{5.13}$$

where

$$\gamma = \frac{1}{\sqrt{1 - \Omega_I^2(x_I^2 + y_I^2)}}.$$

If one imagines the frame $F(\Omega)$ as a carousel, then for any given angular velocity, objects at rest relative to $F(\Omega)$ that are sufficiently far from the origin would have a classical linear velocity that exceeds the speed of light in the inertial frame F_I. As expected, the taiji rotational transformations display some unusual behavior near that region. From the relationship between Ω and Ω_I, i.e., $\Omega_I = \Omega/\gamma = \Omega\sqrt{1 - \rho^2\Omega^2}$, one can see that for a given constant angular velocity

Ω, the corresponding angular velocity Ω_I depends on ρ as well as Ω. In particular, when $\rho = 1/\Omega$, the value of Ω_I becomes zero. Furthermore, the reading of $F(\Omega)$ clocks at that radius, as viewed by F_I observers, stops changing, much like the case in special relativity where the rate of ticking of clocks slows down and stops as the speed of those clocks approaches the speed of light.

These effects are not unexpected because the rotational transformations (5.11) map only a portion of the space in the rotating frame $F(\Omega)$ ($\rho < 1/\Omega$) to the entire inertial frame F_I. Similar to the singular walls found in frames with a linear acceleration, there is a cylindrical singular wall at $\rho = 1/\Omega$ in the rotating frame $F(\Omega)$.

The rotational transformations (5.11) have the inverse transformations (5.13), but when $R \neq 0$ in (5.10), it appears that there is no algebraic expression for the inverse of the transformations (5.10). However, mathematically, (5.10) defines a function from one neighborhood of the origin in space-time to another. The implicit function theorem then guarantees the existence of an inverse in a neighborhood of any point where the Jacobian is non-zero.[b] The situation is similar to the inverse of the space-time transformations (4.15) involving arbitrary linear accelerations in Chapter 4.

5.2. Metric Tensors for the Space-Time of Rotating Frames

We note that the transformations of the contravariant 4-vectors $dx_I^\mu = (dw_I, dx_I, dy_I, dz_I)$ and $dx^\mu = (dw, dx, dy, dz)$ can be derived from (5.11). We have

$$dw_I = \gamma[dw + (\gamma^2\Omega^2 wx)dx + (\gamma^2\Omega^2 wy)dy],$$

$$\begin{aligned}
dx_I = \gamma\,\{&[\cos(\Omega w) + \gamma^2\Omega^2 x^2\cos(\Omega w) - \gamma^2\Omega^2 xy\sin(\Omega w)]dx \\
&- [\sin(\Omega w) + \gamma^2\Omega^2 y^2\sin(\Omega w) - \gamma^2\Omega^2 xy\cos(\Omega w)]dy \\
&- [\Omega x\sin(\Omega w) + \Omega y\cos(\Omega w)]dw\},
\end{aligned}$$

$$\begin{aligned}
dy_I = \gamma\,\{&[\sin(\Omega w) + \gamma^2\Omega^2 x^2\sin(\Omega w) + \gamma^2\Omega^2 xy\cos(\Omega w)]dx \\
&+ [\cos(\Omega w) + \gamma^2\Omega^2 y^2\cos(\Omega w) + \gamma^2\Omega^2 xy\sin(\Omega w)]dy \\
&+ [\Omega x\cos(\Omega w) - \Omega y\sin(\Omega w)]dw\},
\end{aligned}$$

$$dz_I = dz. \tag{5.14}$$

---

[b]We would like to thank Dana Fine for helpful discussions.

To find the Poincaré metric tensor $P_{\mu\nu}$ for the rotating frame $F(\Omega)$, it is convenient to use (5.11) to first write $ds^2 = dw_I^2 - dx_I^2 - dy_I^2 - dz_I^2$ as

$$ds^2 = d(\gamma w)^2 - (x^2 + y^2)\gamma^2\Omega^2 dw^2 - d(\gamma x)^2 - d(\gamma y)^2 - dz^2$$
$$+ 2\gamma\Omega y\, d(\gamma x)dw - 2\Omega\gamma x\, d(\gamma y)dw. \tag{5.15}$$

Then, with the help of the relation $d\gamma = \gamma^3\Omega^2(xdx + ydy)$, (5.15) can be written as

$$ds^2 = P_{\mu\nu}dx^\mu dx^\nu, \quad \mu, \nu = 0, 1, 2, 3, \tag{5.16}$$

where the non-vanishing components of $P_{\mu\nu}$ are given by

$$P_{00} = 1, \quad P_{11} = -\gamma^2\left[1 + 2\gamma^2\Omega^2 x^2 - \gamma^4\Omega^4 x^2(w^2 - x^2 - y^2)\right],$$
$$P_{22} = -\gamma^2\left[1 + 2\gamma^2\Omega^2 y^2 - \gamma^4\Omega^4 y^2(w^2 - x^2 - y^2)\right], \quad P_{33} = -1,$$
$$P_{01} = \gamma^2\left[\Omega y + \gamma^2\Omega^2 wx\right], \quad P_{02} = \gamma^2[-\Omega x + \gamma^2\Omega^2 wy], \tag{5.17}$$
$$P_{12} = -\gamma^4\Omega^2 xy\left[2 - \gamma^2\Omega^2(w^2 - x^2 - y^2)\right].$$

The contravariant metric tensor $P^{\mu\nu}$ is found to be[c]

$$P^{00} = \gamma^{-2}\left[1 - \Omega^4 w^2(x^2 + y^2)\right], \quad P^{33} = -1,$$
$$P^{11} = -\gamma^{-2}\left[\gamma^{-2}(1 - \Omega^2 x^2) - 2\gamma^{-2}\Omega^3 wxy + \Omega^6 w^2 y^2(x^2 + y^2)\right],$$
$$P^{22} = -\gamma^{-2}\left[\gamma^{-2}(1 - \Omega^2 y^2) + 2\gamma^{-2}\Omega^3 wxy + \Omega^6 w^2 x^2(x^2 + y^2)\right],$$
$$P^{01} = -\gamma^{-2}\left[-\Omega y - \gamma^{-2}\Omega^2 wx + \Omega^5 w^2 y(x^2 + y^2)\right], \tag{5.18}$$
$$P^{02} = -\gamma^{-2}\left[\Omega x - \gamma^{-2}\Omega^2 wy - \Omega^5 w^2 x(x^2 + y^2)\right],$$
$$P^{12} = \gamma^{-2}\left[\gamma^{-2}\Omega^2 xy - \gamma^{-2}\Omega^3 w(x^2 - y^2) + \Omega^6 w^2 xy(x^2 + y^2)\right].$$

All other components in (5.17) and (5.18) are zero. Indeed, one can verify $P_{\mu\lambda}P^{\lambda\nu} = \delta_\mu^\nu$.

5.3. The Rotational Pseudo-Group

As mentioned previously, the rotational transformations (5.11) imply the existence of a cylindrical singular wall at $\rho = \sqrt{x^2 + y^2} = 1/\Omega \equiv \rho_s$, which depends on the angular velocity Ω. Like the Wu and taiji transformations, the rotational

---

[c]The components of this contravariant tensor could also be obtained using the momentum transformations (5.30) below and the invariant relation, $p_{I\nu}p_I^\nu = P^{\mu\nu}p_\mu p_\nu$.

transformations map only the portion of space-time of the rotating frame within the singular wall to the entire space-time of an inertial frame and the "group" of rotating transformations is a pseudo-group.

Two of the group properties of the rotational transformations are straightforward to verify. One can easily see that in the limit $\Omega \to 0$, the rotational transformations (5.11) reduce to the identity transformation. Also, the inverse transformations of (5.11) are given by (5.13).

In order to see other group properties of the rotational transformations, let us consider two other rotating frames, $F'(\Omega')$ and $F''(\Omega'')$, which are characterized by two different constant angular velocities Ω' and Ω'', respectively. With the help of (5.11), we can derive the rotational transformations between $F(0) = F_I, F(\Omega), F'(\Omega')$ and $F''(\Omega'')$,

$$w_I = \gamma w = \gamma' w' = \gamma'' w'',$$

$$x_I = \gamma[x \cos(\Omega w) - y \sin(\Omega w)] = \gamma'[x' \cos(\Omega' w') - y' \sin(\Omega' w')]$$
$$= \gamma''[x'' \cos(\Omega'' w'') - y'' \sin(\Omega'' w'')],$$

$$y_I = \gamma[x \sin(\Omega w) + y \cos(\Omega w)] = \gamma'[x' \sin(\Omega' w') + y' \cos(\Omega' w')]$$
$$= \gamma''[x'' \sin(\Omega'' w'') + y'' \cos(\Omega'' w'')],$$

$$\gamma' = \frac{1}{\sqrt{1 - \beta'^2}}, \quad \beta' = \Omega' \rho', \quad \gamma'' = \frac{1}{\sqrt{1 - \beta''^2}}, \quad \beta'' = \Omega'' \rho'', \tag{5.19}$$

where we have neglected to write the trivial transformation equations for the z-direction. The rotational transformations between $F(\Omega)$ and $F'(\Omega')$ can be obtained from (5.19):

$$w = \frac{\gamma'}{\gamma} w',$$

$$x = \gamma' \left\{ [x' \cos(\Omega' w') - y' \sin(\Omega' w')] \frac{\cos(\Omega w)}{\gamma} \right.$$
$$\left. + [x' \sin(\Omega' w') + y' \cos(\Omega' w')] \frac{\sin(\Omega w)}{\gamma} \right\},$$

$$y = \gamma' \left\{ [x' \sin(\Omega' w') + y' \cos(\Omega' w')] \frac{\cos(\Omega w)}{\gamma} \right.$$
$$\left. - [x' \cos(\Omega' w') - y' \sin(\Omega' w')] \frac{\sin(\Omega w)}{\gamma} \right\}. \tag{5.20}$$

If one compares these transformations with those in equations (5.11)–(5.13), one can see that the transformations between two rotating frames are more algebraically complicated than those between an inertial frame and a rotating frame. In general, it does not seem possible to write the inverse of the transformations (5.20) in analytic form. However, as was the case for the taiji transformations (4.15) and (4.16), the implicit function theorem guarantees the existence of an inverse in the neighborhood of any point where the Jacobian is non-zero.

5.4. Physical Implications

Some physical implications of the rotational taiji transformations for rotating frames worthy of note are as follows:

(A) *The invariant action for electrodynamics in rotating frames*: We are now able to write the invariant action S_{em} in natural units in a rotating frame for a charged particle with mass m and charge $e > 0$ moving in the 4-potential A_μ:

$$S_{em} = \int \left[-mds + eA_\mu dx^\mu - \frac{1}{4} F_{\mu\nu} F^{\mu\nu} \sqrt{-P} d^4 x \right], \qquad (5.21)$$

$$F_{\mu\nu} = D_\mu A_\nu - D_\nu A_\mu = \partial_\mu A_\nu - \partial_\nu A_\mu, \quad P = \det P_{\mu\nu}, \qquad (5.22)$$

where ds is given by (5.16).

The Lagrange equation of motion of a charged particle can be derived from (5.21). We obtain

$$m \frac{Du_\mu}{ds} = -e F_{\mu\nu} u^\nu, \qquad (5.23)$$

$$Du_\mu = D_\nu u_\mu dx^\nu, \quad u^\nu = \frac{dx^\nu}{ds}, \quad u_\mu = P_{\mu\nu} u^\nu, \qquad (5.24)$$

where D_ν denotes the partial covariant derivative associated with the rotational metric tensors $P_{\mu\nu}$ in (5.17).

Starting with the invariant action (5.21) and replacing the second term with

$$\int A_\mu j^\mu \sqrt{-P} d^4 x \qquad (5.25)$$

for a continuous charge distribution in space, we obtain the invariant Maxwell's equations in a rotating frame

$$D_\nu F^{\mu\nu} = -j^\mu, \quad \partial_\lambda F_{\mu\nu} + \partial_\mu F_{\nu\lambda} + \partial_\nu F_{\lambda\mu} = 0. \qquad (5.26)$$

Based on gauge invariance and the rotational invariance of the action (5.21), the electromagnetic potential must be a covariant vector A_μ in non-inertial frames. Since the force F and the fields E and B are naturally related to a change of the potential A_μ with respect to a change of coordinates (i.e., x^μ, by definition), the fields E and B are naturally identified with components of the covariant tensor $F_{\mu\nu}$ as given by (5.22) in non-inertial frames. The metric tensor $P_{\mu\nu}$ behaves like a constant under covariant differentiation, $D_\alpha P_{\mu\nu} = 0$.

(B) *Absolute contraction of a rotating radius and absolute slow-down of a rotating clock:* Comparing the taiji rotational transformations (5.11) and the classical rotational transformations (5.1), one can see that the rotational taiji transformations (5.11) predict that the length of a rotating radius $\sqrt{x^2 + y^2}$ is contracted by a factor of γ

$$\sqrt{x_I^2 + y_I^2} = \gamma\sqrt{x^2 + y^2}. \tag{5.27}$$

This contraction is absolute, meaning that both observers in the inertial frame F_I and in the rotating frame $F(\Omega)$ agree that the radius, as measured in the rotating frame $F(\Omega)$ is shorter, because there is no relativity between an inertial frame and a rotating (non-inertial) frame.

In some discussions of phenomena involving circular motion at high speeds, such as the lifetime dilation of unstable particles traveling in a circular storage ring, the argument is made that during a very short time interval, one can approximate the true rotational transformations using the Lorentz transformations.[2] However, making this approximation leads to a completely different conclusion regarding the radius, namely, that it does not contract because it is always perpendicular to the direction of motion.

Furthermore, for a given ρ, (5.11) implies $\Delta w_I = \gamma\Delta w$, which is independent of the spatial distance between two events.[d] In other words, clocks at rest relative to a rotating frame and located at a distance $\rho = \sqrt{x^2 + y^2}$ from the center of rotation slow down by a factor of $\gamma = \sqrt{(1 - \beta^2)}$ in comparison with clocks in the inertial frame F_I. Analogous to the absolute contraction of radial lengths as shown in (5.27), this time dilation is also an absolute effect in that observers in both F_I and $F(\Omega)$ agree that it is the accelerated clocks that are slowed.

---

[d]For comparison with the usual time t, if one were to define $w = ct$ and $w_I = ct_I$, then one would have the usual relation $\Delta t_I = \gamma\Delta t$. However, the constant 'speed of light' c in a non-inertial frame is not well-defined because the speed of a light signal is no longer isotropic or constant.

Both the contraction of radial distances and the slowing down of clocks are consequences of requiring the rotational transformations to satisfy the limiting Lorentz and Poincaré invariance.

(C) *The covariant four-momentum in a rotating frame.* Although the speed of light is not a universal constant in a rotating frame F, one can still formulate a covariant four-momentum $p_\mu = (p_0, p_i)$, $i = 1, 2, 3$ using the time w as the evolution variable in the Lagrangian formalism. From the invariant "free" action $S_f = -\int m\,ds = \int L\,dw$ for a "non-interacting particle" with mass m in the rotating frame $F(\Omega)$, the spatial components of the covariant four-momentum are

$$p_i = -\frac{\partial L}{\partial v^i} = mP_{vi}\frac{dx^v}{ds} = P_{vi}p^v, \quad L = -m\sqrt{P_{\mu v}v^\mu v^v}, \tag{5.28}$$

where $i = 1, 2, 3$. Both L and p_i have the dimension of mass and $v^\mu \equiv dx^\mu/dw = (1, v^i)$. The zeroth component p_0 (or the Hamiltonian) with the dimension of mass is defined as usual

$$p_0 = v^i \frac{\partial L}{\partial v^i} - L = mP_{v0}\frac{dx^v}{ds} = P_{v0}p^v. \tag{5.29}$$

The rotational taiji transformations of the differential operators $\partial/\partial x_I^\mu$ and $\partial/\partial x^\mu$ can be calculated from (5.13). Just as in quantum mechanics, the covariant momentum p_μ has the same transformation properties as the covariant differential operator $\partial/\partial x^\mu$. From (5.13), we obtain

$$p_{I0} = \gamma^{-1}(p_0 + \Omega y p_1 - \Omega x p_2),$$

$$p_{I1} = \left[-\gamma^{-2}\Omega^2 wx_I\right]p_0 + \gamma^{-2}\left[\gamma\cos(\Omega w) - \Omega^2 x_I x - \Omega^3 wx_I y\right]p_1$$

$$+ \gamma^{-2}\left[-\gamma\sin(\Omega w) - \Omega^2 x_I y + \Omega^3 wx_I x\right]p_2,$$

$$p_{I2} = \left[-\gamma^{-2}\Omega^2 wy_I\right]p_0 + \gamma^{-2}\left[\gamma\sin(\Omega w) - \Omega^2 y_I x - \Omega^3 wy_I y\right]p_1$$

$$+ \gamma^{-2}\left[\gamma\cos(\Omega w) - \Omega^2 y_I y + \Omega^3 wy_I x\right]p_2,$$

$$p_{I3} = p_3, \quad \gamma = \frac{1}{\sqrt{1 - \rho^2\Omega^2}}, \tag{5.30}$$

where x_I and y_I can be expressed in terms of x, y, and w of the rotating frame $F(\Omega)$ using (5.11).

Consider the case of a particle traveling at high speed in a circular storage ring. Such a particle can be considered to be at rest in a rotating frame F, so that $dx^i = 0$ and hence, $ds = dw$. Based on $p^v = mdx^v/ds$ in (5.28), the contravariant momenta are $p^i = 0$, $i = 1, 2, 3$, and $p^0 = m$. In this case the covariant momenta of this particle in F are

$$p_0 = m, \quad p_1 = m\gamma^2[\Omega y + \gamma^2\Omega^2 wx],$$
$$p_2 = m\gamma^2[-\Omega x + \gamma^2\Omega^2 wy], \quad p_3 = 0. \tag{5.31}$$

This difference between p_μ and p^μ is due to the presence of the metric tensor components P_{01} and P_{02}. The covariant momenta of the particle, as measured in an inertial frame F_I, are given by (5.30) and (5.31),

$$p_{I0} = \gamma m, \quad p_{I1} = m\gamma[\Omega x \sin(\Omega w) + \Omega y \cos(\Omega w)],$$
$$p_{I2} = -m\gamma[\Omega x \cos(\Omega w) - \Omega y \sin(\Omega w)], \quad p_{I3} = 0. \tag{5.32}$$

Thus, the expression for the energy of a rotating particle $p_{I0} = \gamma m$ agrees with the well-established results of high energy experiments performed in an inertial laboratory frame F_I.

5.5. Experimental Tests of the Rotational Taiji Transformations

Because particles moving in a straight line at relativistic speeds travel large distances in very short times, designing experiments to test relativistic effects with linear motion, where the Lorentz transformations are most directly applicable, can be challenging. Experiments in which objects move in a circular path at relativistic speeds can be performed with much more compact apparatus. However, as was discussed briefly in the previous section, applying the Lorentz transformations to analyzing such experiments is problematic because of the non-relativity between inertial frames and rotating frames. In this section, we discuss three experiments that can test the usefulness of the rotational taiji transformations.

(A) *Absolute dilation of decay-length for particle decay in circular motion:* Let us first consider the lifetime dilation of unstable particles traveling in a circular storage ring.[4] If the particle's rest lifetime in $F(\Omega)$ is denoted by $\Delta w(\text{rest})$, then the rotational taiji transformation (5.11) gives

$$\Delta w_I = \gamma\Delta w(\text{rest}), \quad \gamma = \frac{1}{\sqrt{1 - \beta^2}}, \quad \beta = \Omega\rho. \tag{5.33}$$

As mentioned in Chapters 5 and 6, the rest lifetime of the particle $\Delta w(\text{rest})$ in the rotating frame F cannot be directly measured. However, according to the weak equivalence of non-inertial frames (3.36),

$$\Delta w(\text{rest})\sqrt{P_{00}} = \Delta w_I(\text{rest})\sqrt{\eta_{00}}, \tag{5.34}$$

where we have taken the limit (5.8) for the non-inertial frame F' in (3.36), so that F' becomes F_I in the limit and $P'_{00} = \eta_{00}$. Since the lifetime is typically very short, $\Delta w(\text{rest})$ can be considered as the zeroth component of the contravariant differential coordinate vector V^μ:

$$\Delta w = dw = dx^0 = V^0.$$

It follows from (5.33) and (5.34) that

$$\Delta w_I = \gamma \Delta w_I(\text{rest}), \tag{5.35}$$

where we have used $P_{00} = 1$ in (5.17).

Since Δw_I has the dimension of length, (5.35) can be understood as a dilation of the decay length, i.e., the distance traveled by the particle before it decays. Result (5.35) is consistent with well-established experimental results of the decay lifetime dilation of muons in flight in a circular storage ring.[5] The lifetime dilation (5.35) is an absolute effect in that observers in both F_I and $F(\Omega)$ agree that it is the accelerated muons whose mean decay length is dilated.

(B) *Davies–Jennison experiment:* A second experiment that can test the usefulness of the taiji rotational transformations is the Davies–Jennison experiment,[5,e] in which the shift in frequency of transversely emitted radiation from an orbiting source was measured. The arrangement of the apparatus for this experiment is roughly as follows: A laser beam is directed straight downward onto a mirror at the center of a horizontal rotating table. The mirror sends the beam radially outward, parallel to the surface of the table, where it strikes a second mirror mounted at the edge of the table. This second mirror reflects the beam in a direction perpendicular to its velocity, thus acting as a source emitting radiation in the transverse direction. This resultant beam is then mixed with light from the original laser in order to produce interference fringes and measure any shift in its frequency compared to the original laser light. In the experiments of Davies and Jennison, no such shift was observed. Let us analyze this experiment using the rotational taiji transformations.

---

[e]The reflected radiation from the moving mirror is considered to be transversely-emitted radiation from an orbiting radiation source.

Consider a radiation source at rest in a rotating frame $F(\Omega)$, located at $\vec{\rho}_s = (x_s, y_s)$, that emits a wave with a frequency $k_0 = k_0(\text{rest})$, as measured in $F(\Omega)$. If this wave, travels along the $-\vec{\rho}_s$ direction to the origin, then $k_1/k_2 = k_x/k_y = x/y$, because the two vectors $-\vec{\rho}$ and $\vec{k}$ are on the x–y plane and parallel to each other. Because the rotational taiji transformations for the covariant wave vector k_μ are the same as those for the covariant momentum $p_\mu = \hbar k_\mu$, the frequency shift of transversely emitted radiation as measured by observers in an inertial laboratory frame F_I is given by (5.30). To detect the transverse effect, the observer is assumed to be at rest at the origin $x_I = y_I = x = y = 0$. Equation (5.30) with $p_\mu = \hbar k_\mu$ gives

$$k_{I0} = \gamma^{-1}(k_0 + \Omega[yk_1 - xk_2]) = \gamma^{-1}k_0,$$

$$xk_2 = yk_1, \quad \gamma = \frac{1}{\sqrt{1 - \rho_s^2\Omega^2}}, \quad \rho_s^2 = x_s^2 + y_s^2 = constant, \tag{5.36}$$

where $k_0 = k_0(\text{rest})$ is the frequency of the source at rest in the rotating frame $F(\Omega)$, as measured by observers in $F(\Omega)$.

Just as in our discussion of the lifetime dilation of unstable particles traveling in a circle, it is extremely difficult to measure the frequency $k_0(\text{rest})$ directly because we cannot move our measuring apparatus to the rotating frame $F(\Omega)$. Furthermore, because of the non-equivalence of inertial and rotating frames, we cannot assume that $k_0(\text{rest})$ is equal to $k_{I0}(\text{rest})$, the frequency of the radiation from the same source at rest in an inertial frame F_I as measured by observers in F_I. However, we can again use the weak equivalence of non-inertial frames (3.36) to write

$$k_0(\text{rest})\sqrt{P^{00}} = k_{I0}(\text{rest})\sqrt{\eta^{00}}, \quad P^{00} = \gamma^{-2}, \quad \eta^{00} = 1. \tag{5.37}$$

Recall that $k_0(\text{rest})$ is the frequency of radiation emitted from the source at rest in the rotating frame $F(\Omega)$, located at the constant radius, $\rho = fixed$ on the x–y plane. When this condition is used in the derivation of (5.14), then $dw_I = \gamma dw$, etc. and the contravariant component P^{00} in (5.18) reduces to $P^{00} = \gamma^{-2} = constant$ in (5.37). It follows from (5.36) and (5.37) that the relationship between the frequency of the transversely emitted radiation emitted from the orbiting source as measured in the inertial laboratory frame k_{I0} and the frequency of the radiation emitted from the same source at rest in the inertial laboratory frame as measured in the inertial laboratory frame $k_{I0}(\text{rest})$ is

$$k_{I0} = \gamma^{-1}k_0(\text{rest}) = k_{I0}(\text{rest}). \tag{5.38}$$

Thus, the rotational taiji transformation predicts that no frequency shift for transverse radiation should be observed for an orbiting radiation source, as measured by observers in inertial frames. This result (5.38) appears to be consistent with the experiments of Davies and Jennison.[6] It is interesting to note that, based on experimental results, *Jennison concluded that rotating radii are contracted and that angular velocities increase, as measured from the rotating frame.* Those conclusions are consistent with those in (5.12), derived from the rotational taiji transformations (5.11). These results thus lend support to the taiji rotational transformations and the concept of limiting Lorentz-Poincaré invariance.

(C) *Thim's experiment with orbiting receivers*: In an experiment by H. W. Thim,[6] a radiation source is located at the center of a rotating disk and eight detectors (monopole antennas, in this case) are mounted at the edges of the disk so that they move in a circle with a radius ρ. This experiment is similar to that of Davies and Jennison, but with the source and receiver interchanged. Like Davies and Jennison, Thim found no shift in the radiation frequency. However, because of the absolute difference between an inertial and a rotating frames, the two null results cannot be explained by a simple relativity argument.

In Thim's experiment, the source at rest at the center of the rotating frame can also be considered to be at rest in the inertial laboratory frame F_I. Thus, the frequency of its radiation k_{I0} measured in F_I is the same as $k_0(\text{rest})$,

$$k_{I0} = k_{I0}(\text{rest}) = k_0(\text{rest}). \tag{5.39}$$

Consider one specific detector located at the rim of the rotating disk. The wave with the wave vector $\vec{k} = (k_1, k_2)$ propagates from the center of the disk to this detector along the radius $\vec{\rho} = (x, y)$, so that $k_1/k_2 = x/y$. Thus, similar to (5.36), we have

$$k_{I0} = \gamma^{-1}k_0, \quad xk_2 = yk_1, \quad \gamma = \frac{1}{\sqrt{1 - \rho_a^2\Omega^2}}. \tag{5.40}$$

From (5.39) and (5.40), we derive the transverse frequency shift,

$$k_0 = \gamma k_{I0} = \gamma k_0(\text{rest}), \tag{5.41}$$

as measured by an observer at rest in the rotating frame $F(\Omega)$. Result (5.41), derived from the rotational space-time transformations (5.11) holds for each of the eight detectors and implies that if the frequency k_0 is measured in the rotating frame $F(\Omega)$, there is a shift in frequency by a factor of γ as a result of the orbiting motion of the detectors.

Although Thim measured no shift in the radiation frequency, result (5.41) actually does not contradict Thim's null result. The reason is as follows: In Thim's experiment,[6] the frequency of the radiation k_0 received by the orbiting detector is not directly measured in the rotating frame $F(\Omega)$. Instead, the signal received by the detector on the rotating disk is transferred through a second rotating disk into a stationary detector and its frequency is measured in the inertial laboratory F_I.

According to the rotational transformations (5.11), clocks in the rotating frame located at a radius ρ are slowed by a factor of γ, resulting in an increase in the frequency k_0 by a factor of γ. This is shown in the first equation in (5.36) with $xk_2 = yk_1$, because the wave vector points in the radial direction in the rotating frame. Conversely, the frequency of the radiation $k_{I0}(\text{detector})$ received by the detector, as measured in the laboratory, will be smaller by a factor of γ because the clocks in the inertial frame run faster by the factor of γ, $k_{I0}(\text{detector}) = k_0/\gamma$. This relation and equation (5.41) lead to the result

$$k_{I0}(\text{detector}) = k_o(\text{rest}) = k_{I0}, \qquad (5.42)$$

in the inertial frame F_I. Thus, the rotational transformations (5.11) imply that when the frequency of the radiation $k_{I0}(\text{detector})$ received by the detector is measured by standard mixer and interferometer techniques in the inertial laboratory frame and compared with the frequency $k_0(\text{rest})$ of the source in (5.39), no shift should be measured.

Furthermore, according to the rotational transformations (5.11) and (5.36), conclusion (5.42) should be independent of the angular velocity Ω' of the second disk. This property can be seen by regarding the second disk as a second rotating frame $F(\Omega')$. The rotational transformations for F_I, $F(\Omega)$ and $F(\Omega')$ relevant to the experiment are given by

$$w_I = \gamma w = \gamma' w' = w_I, \quad etc., \qquad (5.43)$$

$$k_{I0} = \gamma^{-1} k_0 = \gamma'^{-1} k_0' = k_{I0}, \quad etc., \qquad (5.44)$$

$$\gamma = \frac{1}{\sqrt{1 - \rho_a^2 \Omega^2}}, \quad \gamma' = \frac{1}{\sqrt{1 - \rho'^2 \Omega'^2}};$$

where ρ_a and ρ' are constant. Thus, if the frequency of the radiation were to be measured by an apparatus in either the $F(\Omega)$ (disk 1) or the $F(\Omega')$ (disk 2) frame, it would differ from its original value (measured in the inertial laboratory frame). However, because this radiation is eventually transferred back and measured by a detector that is at rest in the same inertial laboratory frame as the source from which it was emitted, there will be no overall frequency shift.

We note that the transverse frequency shift given by (5.41) should be detectable if one were to measure the frequency received by the orbiting detector directly, in the rotating frame, rather than transferring the signal back to a stationary detector to be measured in the inertial laboratory frame. It would be an interesting challenge to redesign the experiment to measure directly the frequency of the signal received by the orbiting antenna by mounting a standard mixer and interferometer on the rotating disk. Assuming that the centrifugal effects on the apparatus are small and negligible, one can test the transverse frequency shift predicted in (5.41). In fact, one could test for centrifugal effects by repeating the experiment with different angular velocities.

In conclusion, the three experiments discussed here are important because they suggest new principles in physics and novel space-time properties of non-inertial frames, which are beyond the realm of special relativity. They can help to develop a deeper and broader understanding of physics in non-inertial frames based on the general principle of limiting continuation of physical laws and, in particular, limiting Lorentz–Poincaré invariance.

References

1. L. Hsu and J.-P. Hsu, *Nuovo Cimento B* 112, 1147 (1997); J.-P. Hsu and L. Hsu, in *JingShin Theoretical Physics Symposium in Honor of Professor Ta-You Wu*, eds. J. P. Hsu and L. Hsu (World Scientific, Singapore; New Jersey, 1998), pp. 393–412; *Nuovo Cimento B* 112, 575 (1997).
2. G. N. Pellegrini and A. R. Swift, *Am. J. Phys.* 63, 694–705 (1995).
3. J. J. Sakurai, in *Invariance Principles and Elementary Particles* (Princeton University Press, 1964), p. v, pp. 3–5, and p. 10.
4. F. J. M. Farley, J. Bailey and E. Picasso, *Nature* 217, 17–18 (1968).
5. P. A. Davies and R. C. Jennison, *Nature* 248 (1974); *J. Phys. A. Math.* 8, 1390 (1975).
6. H. W. Thim, *IEEE Trans. Instrum. Meas.* 52, 1660 (2003).

Chapter 6

Conservation Laws and Symmetric
Energy–Momentum Tensors

6.1.   Conservation Laws in the Taiji Symmetry Framework

Traditionally, theories of physics have been formulated and discussed in inertial frames based on flat space-time and are consistent with the principle of Lorentz and Poincaré invariance. In particular, conservation laws are usually derived from global symmetries in inertial frames through Noether's well-known Theorem I.[1] Einstein's theory of gravity, however, is based on the principle of general coordinate invariance in curved space-time. Despite its success, this theory of gravity fails to preserve certain fundamental elements of physics such as the law of conservation of energy because the symmetry group of general coordinate transformations involves a continuously infinite number of generators, as shown by Noether's Theorem II.[1] Although the covariant continuity equation for the energy–momentum tensor and the usual continuity equation for the energy–momentum pseudo-tensor can be found within general relativity, they are not conservation laws in the usual sense and are thus not ideal. Furthermore, since non-inertial frames in general relativity are assumed to involve curved space-time, it is not obvious that energy and momentum are conserved in such reference frames. Although there is a well-known general result that states that any equations in inertial frames can be expressed in a covariant form, so that it holds in arbitrary coordinate system,[2] the framework of general relativity does not allow for inertial frames.[3] In this chapter, we show how the taiji symmetry framework, which is based on a flat space-time for both inertial and non-inertial frames of reference, avoids all the complications caused by the curved space-time of general relativity and naturally accommodates all known conservation laws.

6.2. Symmetric Energy–Momentum Tensors and Variations of Metric Tensors in Taiji Space-Time

We assume that the space-time of all frames of reference, both inertial and non-inertial, is described by coordinates x^μ and characterized by symmetric Poincaré metric tensors $P_{\mu\nu} = P_{\nu\mu}$. As we have discussed in previous chapters, we apply the principle of limiting continuation and postulate that in the limit of zero acceleration, all non-inertial frames reduce to inertial frames and the Poincaré metric tensors $P_{\mu\nu}$ reduce to the Minkowski metric tensor $\eta_{\mu\nu} = (1, -1, -1, -1)$. In general, in order for x^0 and $x^i = (x, y, z)$ to have the character of temporal and spatial coordinates respectively, it is necessary and sufficient that

$$P_{00} > 0, \quad P_{ik}dx^i dx^k < 0, \tag{6.1}$$

where i and k are summed over 1, 2 and 3 and the components P_{10}, P_{20} and P_{30} may be positive or negative.

Noether's theorem allows one to obtain the conserved energy–momentum tensor $t_{\mu\nu}$ for a physical system in a flat space-time with global translational symmetry. Such a tensor $t_{\mu\nu}$ is, in general, not symmetric. Although, it is possible to symmetrize the energy–momentum tensor, the procedure can be quite complicated. A simple and direct way to obtain a symmetric energy–momentum tensor was first discussed by Hilbert in 1915 when he combined Einstein's idea of general covariance and the variation of an invariant action involving scalar curvature to derive the gravitational field equation. The relationship between the symmetric energy–momentum tensor and the variation of the metric tensor have been discussed by Hilbert, Fock, and others.[4,5]

Hilbert's symmetric energy–momentum tensor $T_{HF}^{\mu\nu}$ in arbitrary coordinates is given by[4,5]

$$T_{HF}^{\mu\nu} = \frac{-2}{\sqrt{-P}} \left(\frac{\partial L\sqrt{-P}}{\partial P_{\mu\nu}} - \partial_\lambda \frac{\partial L\sqrt{-P}}{\partial(\partial_\lambda P_{\mu\nu})} \right). \tag{6.2}$$

This definition can be applied in a general frame with the Poincaré metric tensor $P_{\mu\nu}$. Because the ten elements of $\delta P_{\mu\nu}$ are constrained by the condition that the Riemann–Christoffel curvature tensor must vanish, they are not independent.

To obtain a symmetric energy–momentum tensor $U_{\alpha\beta}$ for an electromagnetic field within the taiji space-time framework, let us consider the U(1) gauge invariant action for electromagnetic fields,

$$S_f = \int L\sqrt{-P}\, d^4x, \quad L = -\frac{1}{4}F_{\mu\nu}F^{\mu\nu}. \tag{6.3}$$

Under the variation δ_p of the metric tensor $P_{\mu\nu}$, we have

$$\delta_p S_f = -\frac{1}{4} \int [F_{\mu\nu} \delta_p (F^{\mu\nu} \sqrt{-P})] \, d^4x = -\frac{1}{2} \int U^{\alpha\beta} \delta P_{\alpha\beta} \sqrt{-P} \, d^4x,$$

$$(6.4)$$

$$U^{\alpha\beta} = \frac{1}{4} F_{\mu\nu} F^{\mu\nu} P^{\alpha\beta} - P_{\mu\nu} F^{\mu\alpha} F^{\nu\beta}, \qquad (6.5)$$

where we have used the relations,

$$F_{\mu\nu} \delta_p F^{\mu\nu} = -2 F_{\mu\nu} F^{\mu\alpha} P^{\nu\beta} \delta P_{\alpha\beta}, \quad \delta_p P_{\alpha\beta} = \delta P_{\alpha\beta}, \qquad (6.6)$$

$$\delta_p \sqrt{-P} = \frac{1}{2} \sqrt{-P} P^{\alpha\beta} \delta P_{\alpha\beta}. \qquad (6.7)$$

Note that A_μ and $F_{\mu\nu}$ are not varied because A_μ is originally introduced through the U(1) gauge covariant derivative, $\partial_\mu - ieA_\mu$, $e > 0$. Only the contravariant potential $A^\mu = P^{\mu\nu} A_\nu$ depends on the metric tensors.

In the Fock–Hilbert approach to the energy–momentum tensor, the law of conservation of energy–momentum is derived on the basis of a local transformation in flat space-time.[2] Specifically, that the action (6.3) is invariant under local translations in flat space-time,

$$x'^\mu = x^\mu + \Lambda^\mu(x), \qquad (6.8)$$

where $\Lambda^\mu(x)$ are arbitrary infinitesimal functions. Transformation (6.8) can also be viewed as a general coordinate transformation. The invariance of actions under this local translation is the mathematical expression of the space-time "taiji symmetry."[a] In this case, Noether's theorem assumes a global space-time translational-symmetry of the action to derive the conservation of momentum.

According to the general transformation formula for tensors, the transformation of $P_{\mu\nu}$ is given by

$$P'^{\mu\nu}(x') = P^{\alpha\beta}(x) \frac{\partial x'^\mu}{\partial x^\alpha} \frac{\partial x'^\nu}{\partial x^\beta}, \qquad (6.9)$$

which can be calculated to first order using (6.8) and

$$P'^{\mu\nu}(x') = P'^{\mu\nu}(x) + \Lambda^\lambda (\partial_\lambda P'^{\mu\nu}) = P'^{\mu\nu}(x) + \Lambda^\lambda (\partial_\lambda P^{\mu\nu}).$$

---

[a] This taiji symmetry is essential for Yang–Mills gravity. It enables one to introduce the gauge covariant derivative involving the generators of the space-time translation group. All the properties of the gravitational interaction appear to be manifestations of this taiji symmetry, as we shall see in Part II of this book.

The difference between $\Lambda^\lambda(\partial_\lambda P'^{\mu\nu})$ and $\Lambda^\lambda(\partial_\lambda P^{\mu\nu})$ is of the second order and negligible. We obtain expressions for $\delta P^{\mu\nu}(x)$ and $\delta P_{\mu\nu}$

$$\delta P^{\mu\nu}(x) = P'^{\mu\nu}(x) - P^{\mu\nu}(x) = -(\partial_\lambda P^{\mu\nu})\Lambda^\lambda$$

$$+ P^{\mu\alpha}\partial_\alpha\Lambda^\nu + P^{\nu\alpha}\partial_\alpha\Lambda^\nu = D^\mu\Lambda^\nu + D^\nu\Lambda^\mu, \qquad (6.10)$$

$$\delta P_{\mu\nu} = -D_\mu\Lambda_\nu - D_\nu\Lambda_\mu, \qquad (6.11)$$

where

$$D_\mu\Lambda_\nu = \partial_\mu\Lambda_\nu - \Gamma^\lambda_{\mu\nu}\Lambda_\lambda, \quad \Gamma^\lambda_{\mu\nu} = \frac{1}{2}P^{\lambda\alpha}(\partial_\nu P_{\mu\alpha} + \partial_\mu P_{\nu\alpha} - \partial_\alpha P_{\mu\nu}).$$

$$(6.12)$$

Since the action S_f is invariant under an infinitesimal local space-time translation (6.8),

$$\delta_p S_f = -\frac{1}{2}\int U^{\alpha\beta}\delta P_{\alpha\beta}\sqrt{-P}d^4x = -\frac{1}{2}\int U^{\mu\nu}(-D_\mu\Lambda_\nu - D_\nu\Lambda_\mu)$$

$$= -\int (D_\mu U^{\mu\nu})\Lambda_\nu\sqrt{-P}d^4x = 0, \qquad (6.13)$$

where we have used (6.10) and (6.11). From (6.13) with the arbitrary infinitesimal functions Λ_ν, we obtain the following equation for the symmetric energy–momentum tensor $U^{\alpha\beta}$ in (6.5),

$$D_\alpha U^{\alpha\beta} = 0, \qquad (6.14)$$

which is the covariant continuity equation for all frames and arbitrary coordinates. The form of equation (6.14) holds for all physical systems with space-time translational gauge symmetry. Combining the continuity equation (6.14) with the metric tensors derived in the preceding chapters, one can obtain explicit transformation equations for energy and momentum between inertial frames, frames with an arbitrary linear acceleration, and rotating frames.

In Einstein's theory of gravity, there is a covariant equation analogous to (6.14) for the energy–momentum tensor, but the field equation does not necessarily imply that the energy–momentum tensor is conserved. In the Section 6.3, we discuss the conditions under which the covariant equation (6.14) implies the conservation of energy–momentum in the usual sense (see Ref. 6).[b]

---

[b]Mathematically, any space-time with a constant curvature K possesses maximal symmetry. Fock showed that the system of differential equations (6.14) is completely integrable if the Riemann–Christoffel curvature takes the form $R_{\mu\nu,\alpha\beta} = K(g_{\nu\alpha}g_{\mu\beta} - g_{\mu\alpha}g_{\nu\beta})$, where K is the constant of

6.3. Integral Form of Conservation Laws in Non-Inertial Frames

Understanding the conservation laws in non-inertial frames is not a simple thing. For example, consider a simple two-particle scattering process. In an inertial frame, it is easy to see that energy, momentum, and angular momentum are conserved because the energy–momentum 4-vector of each particle is constant. However, in a non-inertial frame, the energy–momentum 4-vector of each particle is a vector function of time because of the acceleration of the reference frame.

To see how the conservation laws are instantiated in the taiji symmetry framework, we first note the following three mathematical results.

(i) In taiji space-time, the symmetric energy–momentum tensor $T^{\mu\nu} = T^{\nu\mu}$ satisfies the covariant continuity condition given by (6.14),

$$D_\nu T^{\mu\nu} = \partial_\nu T^{\mu\nu} + \Gamma^\nu_{\rho\nu} T^{\mu\rho} + \Gamma^\mu_{\rho\nu} T^{\rho\nu} = 0, \qquad (6.15)$$

for any physical system whose action is invariant under a space-time translation. The first two terms can then be combined, so that

$$\frac{1}{\sqrt{-P}} \partial_\nu(\sqrt{-P} T^{\mu\nu}) + \Gamma^\mu_{\rho\nu} T^{\rho\nu} = 0. \qquad (6.16)$$

(ii) It can be shown that vanishing curvature, $R^\rho_{\mu,\nu\alpha} = 0$, is a necessary and sufficient condition to ensure that the following equation has a solution,[4,5]

$$\partial_\mu \partial_\nu \phi - \Gamma^\rho_{\mu\nu} \partial_\rho \phi = \partial_\mu \phi_\nu - \Gamma^\rho_{\mu\nu} \phi_\rho = 0, \quad \phi_\rho = \partial_\rho \phi. \qquad (6.17)$$

(iii) Let us introduce a vector ϕ_μ and integrate the equation

$$D_\nu(T^{\mu\nu} \phi_\mu) = \frac{1}{\sqrt{-P}} \partial_\nu(\sqrt{-P} T^{\mu\nu} \phi_\mu) = 0 \qquad (6.18)$$

over a three-dimensional volume. As usual, we assume the energy–momentum tensor $T^{\mu\nu}$ vanishes on the surface of this volume, obtaining

$$\int D_\nu(T^{\mu\nu} \phi_\mu)\sqrt{-P} d^3r = \frac{d}{dx^0} \int T^{\mu 0} \phi_\mu \sqrt{-P} d^3r$$

$$= \frac{1}{2} \int T^{\mu\nu}(D_\nu \phi_\mu + D_\mu \phi_\nu)\sqrt{-P} d^3r, \qquad (6.19)$$

curvature. Thus, only in this case do the covariant continuity equations (6.14) imply the conservation laws (see Ref. 4). However, if $K \neq 0$, we have neither inertial frames nor the principle of relativity for physical laws in such a space-time framework. This is the physical reason that the taiji symmetry framework is based on flat space-time, i.e., zero Riemann–Christoffel curvature, $R_{\mu\nu,\alpha\beta} = 0$.

where we have used $T^{\mu\nu} = T^{\nu\mu}$ and equation (6.15). If the vector ϕ_μ satisfies

$$D_\nu\phi_\mu + D_\mu\phi_\nu = 0, \tag{6.20}$$

then equation (6.19) leads to

$$K \equiv \int T^{\mu 0}\phi_\mu\sqrt{-P}d^3r = \text{constant}, \tag{6.21}$$

which is independent of time x^0.

To demonstrate the conservation laws in the taiji symmetry framework in an inertial frame F_I with coordinates $x_I^\mu = (x_I^0, x_I^1, x_I^2, x_I^3)$, we introduce the vector

$$\phi_\mu^{(0)} = \frac{\partial x_I^0}{\partial x^\mu}, \quad \phi_\mu^{(1)} = \frac{\partial x_I^1}{\partial x^\mu}, \quad \phi_\mu^{(2)} = \frac{\partial x_I^2}{\partial x^\mu}, \quad \phi_\mu^{(3)} = \frac{\partial x_I^3}{\partial x^\mu}. \tag{6.22}$$

These quantities constitute a 4-vector with respect to the unprimed coordinates (x^0, x^1, x^2, x^3).[4] Using (6.17), i.e., $\partial_\mu\phi_\nu - \Gamma^\rho_{\mu\nu}\phi_\rho = D_\mu\phi_\nu = 0$, and (6.22) we have

$$D_\nu\phi_\mu^{(\alpha)} = 0, \quad (\alpha = 0, 1, 2, 3). \tag{6.23}$$

Thus, each of the elements of the 4-vector in (6.22) also satisfies (6.23). One can also introduce the 6-vector

$$\phi_\mu^{\alpha\beta} = x_I^\alpha\partial_\mu x_I^\beta - x_I^\beta\partial_\mu x_I^\alpha, \tag{6.24}$$

which satisfies

$$D_\nu\phi_\mu^{\alpha\beta} + D_\mu\phi_\nu^{\alpha\beta} = 0. \tag{6.25}$$

This is the same as the relation $D_\nu\phi_\mu + D_\mu\phi_\nu = 0$ in (6.20). Substituting (6.22) and (6.24) into (6.21), we obtain two sets of constant integrals p_I^α and $M_I^{\alpha\beta}$,

$$p_I^\alpha = \int T^{\mu 0}\frac{\partial x_I^\alpha}{\partial x^\mu}\sqrt{-P}d^3r \tag{6.26}$$

and

$$M_I^{\alpha\beta} = \int T^{\mu 0}(x_I^\alpha\partial_\mu x_I^\beta - x_I^\beta\partial_\mu x_I^\alpha)\sqrt{-P}d^3r, \tag{6.27}$$

where p_I^α is the constant energy–momentum 4-vector and $M_I^{\alpha\beta}$ is the constant angular momentum tensor in an inertial frame $F_I(x_I)$. Physically, the components $M_I^{0k}(k = 1, 2, 3)$ are the constants of the motion of the center of mass, and M_I^{ik} is the constant angular momentum in the inertial frame F_I. Comparison of (6.26) and (6.27) with (6.21) shows both of these quantities to be constant, as expected, demonstrating the conservation of the energy–momentum 4-vector and the angular momentum.[6]

To see the analogous result in non-inertial frames explicitly, consider a frame with an arbitrary linear acceleration (ALA) $F(x) \equiv F(x^\mu)$ with coordinates $x^\mu = (w, x, y, z)$. According to (4.12), the coordinate transformation between the inertial frame $F_I(x_I)$ and an ALA frame $F(x)$ is

$$x_I{}^0 = \gamma\beta\left(x^1 + \frac{1}{\alpha(x^0)\gamma_o^2}\right) - \frac{\beta_o}{\alpha_o\gamma_o} + x_o^0,$$

$$x_I{}^1 = \gamma\left(x^1 + \frac{1}{\alpha(x^0)\gamma_o^2}\right) - \frac{1}{\alpha_o\gamma_o} + x_o^1,$$

$$x_I{}^2 = x^2 + x_o^2, \quad x_I{}^3 = x^3 + x_o^3; \tag{6.28}$$

$$\gamma = \frac{1}{\sqrt{1-\beta^2}}, \quad \gamma_o = \frac{1}{\sqrt{1-\beta_o^2}},$$

$$\beta = \beta_1(x^0) + \beta_o, \quad \alpha(x^0) = \frac{d\beta}{dx^0}.$$

It follows from (6.28) that

$$\frac{\partial x_I^0}{\partial x^\nu} = (A, \gamma\beta, 0, 0), \quad A \equiv \gamma[W_o - \beta U],$$

$$\frac{\partial x_I^1}{\partial x^\nu} = (B, \gamma, 0, 0), \quad B \equiv \gamma\beta\left[W_o - \frac{U}{\beta}\right],$$

$$\frac{\partial x_I^2}{\partial x^\nu} = (0, 0, 1, 0), \quad \frac{\partial x_I^3}{\partial x^\nu} = (0, 0, 0, 1); \quad \alpha = \alpha(x^0); \tag{6.29}$$

$$W_o = \gamma^2\left(\alpha x^1 + \frac{1}{\gamma_o^2}\right), \quad U = \frac{d\alpha/dx^0}{\alpha^2\gamma_o^2}.$$

Thus, in an ALA frame, we have the energy–momentum vector

$$p_\nu = p_{I\alpha}\frac{\partial x_I^\alpha}{\partial x^\nu} = \eta_{\alpha\lambda}p_I^\lambda\frac{\partial x_I^\alpha}{\partial x^\nu}, \tag{6.30}$$

where $p_{I\alpha}$ is the constant covariant momentum 4-vector as measured in the inertial frame $F_I(x_I)$. Similarly, the angular momentum tensor in an ALA frame is

$$M_{\mu\nu} = \eta_{\alpha\lambda}\eta_{\beta\rho}M_I^{\lambda\rho}\frac{\partial x_I^\alpha}{\partial x^\mu}\frac{\partial x_I^\beta}{\partial x^\nu}. \tag{6.31}$$

The ten elements of p_ν and $M_{\mu\nu}$ are not constant. However, they satisfy the covariant continuity equations,

$$D_\lambda p_\nu = 0, \qquad D_\lambda M_{\mu\nu} = 0. \tag{6.32}$$

The momentum p_ν and the antisymmetric angular momentum $M_{\mu\nu}$ are functions of space-time and depend on ten constants, which are just the ten constants of motion $p_{I\alpha}$ and $M_I^{\lambda\rho}$ in the inertial frame F_I. Therefore, the covariant continuity equations (6.32) in an ALA frame with a flat space-time imply the conservation of momentum and angular momentum. This is a specific demonstration of the general case in equations (6.15), (6.21), (6.26) and (6.27) using the space-time transformations between an ALA frame and an inertial frame.

To see explicitly the relationship between the 4-momentum vectors p_ν and $p_{I\alpha}$ in (6.30), let us consider an ALA frame $F(x)$ and an inertial frame F_I. Equations (6.29) and (6.30) lead to the transformation between p_ν and $p_{I\alpha}$,

$$p_0 = \gamma \left[(W_o - \beta U)p_{I0} + \beta \left(W_o - \frac{U}{\beta} \right) p_{I1} \right],$$

$$p_1 = \gamma(p_{I1} + \beta p_{I0}), \qquad p_2 = p_{I2}, \qquad p_3 = p_{I3}, \tag{6.33}$$

where the functions W_o and U are given in (6.29) and p_ν ($\nu = 0, 1, 2, 3$) is the time-dependent energy and momentum vector as measured in the ALA frame $F(x)$. It can be verified that the transformation of the covariant 4-vector in (6.33) can also be derived from the covariant coordinate differentials dx_μ. These, in turn, can be obtained using the transformations (6.28) and the relation $dx^\mu = P^{\mu\nu} dx_\nu$, where $P^{\mu\nu}$ is given in equation (4.20).

The inverse transformations of (6.33) are found to be

$$p_{I0} = \frac{\gamma}{W_o} \left[p_0 - \beta \left(W_o - \frac{U}{\beta} \right) p_1 \right],$$

$$p_{I1} = \frac{\gamma}{W_o} [(W_o - \beta U)p_1 - \beta p_0], \qquad p_{I2} = p_2, \qquad p_{I3} = p_3. \tag{6.34}$$

It is clear that the momentum p_1 and energy p_0, as measured in the accelerated frame are not separately constant because β and γ are time-dependent, as shown in (6.28). However, their combinations to form the energy and momentum (p_{I0} and p_{I1}, respectively) in an inertial frame are time-independent constants, as shown in (6.26).

The results (6.34) are the same as those in (3.17) obtained on the basis of Wu transformations since p'^μ is the 4-momentum vector in an inertial frame, $p'_\mu = p_{I\mu}$.

6.4. Symmetry Implications of Global and Local Space-Time Translations

The previous discussions may be puzzling. How, one might ask, can the invariance of an action under one a single transformation (6.8) lead to two different conservation laws for energy–momentum and angular momentum? Conventionally, it is the translational invariance that leads to the conservation of energy–momentum and rotational invariance that leads to the conservation of angular momentum.

In physics, one usually discusses the relations between symmetry and conservation laws based on global transformations and Noether's theorem in inertial frames. For example, the most general space-time transformations for two inertial frames $F_I(x_I)$ and $F'_I(x'_I)$ are the Poincaré transformations,

$$x_I'^\mu = x_I{}^\mu + \epsilon^\mu{}_\nu x_I{}^\nu + \epsilon^\mu, \tag{6.35}$$

where we have written the infinitesimal transformation for simplicity. The quantities $\epsilon^\mu{}_\nu$ and ϵ^μ denote the constant (or global) rotations and translations in a flat four-dimensional space-time, respectively. Noether's theorem implies that if the Lagrangian of a physical system is invariant under an infinitesimal translation,

$$x_I'^\mu = x_I{}^\mu + \epsilon^\mu, \tag{6.36}$$

then one has the conserved energy–momentum 4-vector p_ν,

$$p_\nu = \int d^x T_{0\nu}, \quad \partial^\mu T_{\mu\nu} = 0, \tag{6.37}$$

where $T_{\mu\nu}$ may not be symmetric in general. If the Lagrangian is invariant under an infinitesimal four-dimensional rotation (i.e., the Lorentz transformation),

$$x_I'^\mu = x_I{}^\mu + \epsilon^\mu{}_\nu x_I{}^\nu, \tag{6.38}$$

then one has the conserved angular momentum tensor,

$$M^{\nu\lambda} = \int d^3 x M^{0\nu\lambda}, \quad \partial_\mu M^{\mu\nu\lambda} = 0. \tag{6.39}$$

Let us compare these two transformations (6.36) and (6.38), with the single local translation (6.8). Since $\Lambda^\mu(x)$ in (6.8) are arbitrary infinitesimal functions, they include both the infinitesimal translation ϵ^μ in (6.36) and the infinitesimal Lorentz transformation $\epsilon^\mu{}_\nu x_I{}^\nu$ in (6.38) as two special cases. Thus, if the action of a physical system is invariant under the local space-time translations (6.8), the system actually has both the symmetries of global translations and infinitesimal Lorentz transformations.

Therefore, in using the Fock–Hilbert approach to considering the energy–momentum tensor (6.2) based on the local translation (6.8) and the variation of the metric tensor $P_{\mu\nu}$ (6.10), we can obtain both the conservation laws of the energy–momentum vector (6.26) and the angular momentum tensor (6.27). The Fock–Hilbert results have two interesting differences from Noether's theorem, as usually presented in physics:

(a) The Fock–Hilbert approach based on local symmetries leads directly to a symmetric energy–momentum tensor (6.2) in arbitrary coordinates in the taiji symmetry framework.

(b) The resultant conservation laws hold not only in inertial frames with arbitrary coordinates, but also in non-inertial frames based on the taiji symmetry framework.

Noether's well-known theorem shows the deep connection between symmetries and conservation laws. In her 1918 paper,[1] Noether discussed her Theorem II, which showed that general relativity does not include the usual law for the conservation of energy. The reason is that the group of general coordinate transformations in Einstein's gravity is a Lie group with a continuously infinite number of generators (an infinite continuous group).[7,c,d] The law of conservation of energy in classical mechanics and in special relativity holds because the groups have a finite number of generators. Thus, from the viewpoint of physics, general coordinate invariance in curved space-time is not a "proper symmetry" because it is not associated with the law of conservation of energy.

---

[c] Noether's results were consistent with Hilbert's assertion (1918) that the failure of the law of conservation of energy was a characteristic of Einstein's theory of gravity.

[d] If one compares Noether's theorem and Fock's results in Section 6.2, they seem to suggest some sort of equivalence between invariance under a finite (or countably infinite) number of infinitesimal generators of external space-time groups, and symmetry in a space-time with a constant curvature (cf. footnote b).

References

1. E. Noether, *Goett. Nachr.* **235** (1918). English translation of Noether's paper by M.A. Tavel is online. Google search: M.A. Tavel, Noether's paper.
2. V. A. Fock, in *The Theory of Space Time and Gravitation,* Trans. by N. Kemmer (Pergamon Press, 1959), p. 149.
3. V. Fock, in *The Theory of Space Time and Gravitation,* Trans. by N. Kemmer (Pergamon Press, 1959), p. xvii; E. P. Wigner, *Symmetries and Reflections, Scientific Essays* (The MIT Press, 1967), pp. 52–53; L. Landau and E. Lifshitz, *The Classical Theory of Fields,* Trans. by M. Hamermesh (Addison-Wesley, Cambridge, MA. 1951), pp. 248–289; C. Møller, *The Theory of Relativity* (Oxford University Press, London, 1969), p. 226.
4. V. Fock, in *The Theory of Space Time and Gravitation,* Trans. by N. Kemmer (Pergamon Press, 1959), pp. 158–161; L. Landau and E. Lifshitz, in *The Classical Theory of Fields,* Trans. by M. Hamermesh (Addison-Wesley, Cambridge, MA. 1951), pp. 293–296, pp. 316–318.
5. D. Hilbert, Translated by D. Fine in *100 Years of Gravity and Accelerated Frames: The deepest Insights of Einstein and Yang–Mills,* (eds. J. P. Hsu and D. Fine (World Scientific, 2005), p. 120. (Available online: google books, 100 years of gravity).
6. V. Fock, in *The Theory of Space Time and Gravitation,* Trans. by N. Kemmer (Pergamon Press, 1959), pp. 163–165.
7. N. Byers, E. Noether's discovery of the deep connection between symmetries and conservation laws, arXiv:physics/9807044 [physics.hist.ph]. Google search: N. Byers, Noether's discovery.

Part II

Quantum Yang–Mills Gravity

Jong-Ping Hsu and Leonardo Hsu

Chapter 7

Internal and External Gauge Symmetries and Lie Derivative
Application to Gauge Fields

7.1. The Yang–Mills–Utiyama–Weyl Framework

Since the creation of the Yang–Mills theory in 1954, quantum field theories with gauge symmetries based on a flat space-time have been successfully applied to all fundamental interactions in nature except for the gravitational interaction. This problem is not necessarily surprising since it has been noted that one of "the most glaring incompatibility of concepts in contemporary physics" is that between the principle of general coordinate invariance and a quantum-mechanical description of all of nature.[1]

In contrast to the field-theoretic approach, Einstein's approach involving the geometrization of physics using the Riemannian geometry appears to be successful only in the formulation of classical gravity. Einstein's principle of general coordinate invariance implies that space-time points can be given arbitrary labels. Because the presence of gravity implies a change in the space-time metric, there is no way to define a space-like separation between two regions.[1] Thus, it is difficult to construct quantum gravity because local commutativity is meaningless. Another difficulty with Einstein's principle of general coordinate invariance is the failure of the conservation law for energy–momentum, as was discussed in the Overview.

It is fair to say that general coordinate invariance, which is one of the most powerful ideas in Einstein's theory of the gravitational interaction, is also the origin of its incompatibility with quantum field theory and its non-conservation of the energy–momentum tensor. If a theory of gravity could be formulated in flat space-time, both the difficulties of the quantization of gravity and of conserving the energy–momentum tensor would automatically be resolved. In addition, the aforementioned conceptual difficulty related to the operational meaning of coordinates and momentum would also disappear.

From the viewpoint of established field theories such as quantum electrodynamics and unified electroweak theory, a geometrization of physics should be based on the geometry of fiber bundles with a flat space-time as the base manifold, rather than something like the Riemannian or Finsler geometries.[2,a] In such a Yang–Mills approach, the central roles are played by the connections (i.e., gauge potential fields) on a principal fiber bundle and the gauge curvatures on the connections. These gauge curvatures are closely related to the symmetry of groups. However, despite extensive investigation over several decades, it seems unlikely that one can develop a satisfactory formulation of gravity within the usual Yang–Mills framework. Previous such studies based on the Yang–Mills idea and curved space-time[3,4] have not shed light on the problems of quantization and energy–momentum conservation.

One reason is that the usual gauge fields in the Yang–Mills framework are "phase fields" associated with internal symmetry groups, whose generators have constant matrix representations. Consequently, all gauge potential fields are vector fields in space-time with dimensionless coupling constants. However, tensor fields and a coupling constant with the dimension of length appear to be necessary for a description of gravity, and so a new framework to accommodate these different properties of the gravitational gauge field must be found. One effective and simple way that we describe here is to generalize the traditional Yang–Mills framework based on flat space-time so as to accommodate an external symmetry such as the space-time translational symmetry. This symmetry is what leads to the conserved energy–momentum tensor.

The generalized Yang–Mills framework we will describe in this chapter can accommodate both internal symmetry groups and external symmetry groups, enabling us to discuss all the interactions of nature, including gravity, within the same gauge symmetry framework. Because one can trace the idea of the unification of interactions with gauge symmetry to Weyl,[5] and the insight of gravity as a Yang–Mills gauge field to Utiyama,[3] we call this generalized Yang–Mills framework the Yang–Mills–Utiyama–Weyl (YMUW) framework.

Wigner divided symmetry in physics into two classes: (i) geometric symmetries and (ii) dynamical symmetries. It now appears clear that dynamical

---

[a]If a velocity-dependent interaction, such as the electromagnetic interaction, were to be treated using Einstein's geometric approach, one might reasonably employ the Finsler geometry, rather than the Riemannian geometry. The result of such a geometrization of electrodynamics was not satisfactory. See also discussion in Section 0.3.

Table 7.1.   YMUW framework for gauge symmetry.

Symmetry	
(a) geometric symmetry	(b) dynamical symmetry
↓	↓
(a′) conservation law only	(b′) conservation law and interaction
	↓ ↘
	(b′1) internal symmetry (b′2) external symmetry
	↓ ↓
	(b″1) (electroweak force, (b″2) (gravitational force,
	strong force, charge energy–momentum
	conservation, etc.) conservation, etc.)

symmetries are the gauge symmetries. The generators of the dynamical groups, such as $SU_2 \times U_1$ and SU_3 for the electroweak theory and quantum chromodynamics, respectively, have constant matrix representations.[6] The gauge transformations of these dynamical groups and parallel transport correspond to phase changes. These are key properties of the usual gauge theories with internal symmetry groups. Analogously, Yang–Mills gravity suggests that an "external" space-time translational group is also a dynamical group (see Table 7.1).[7] Parallel transport related to translational group corresponds to a scale change.

In Yang–Mills gravity, the flat space-time translational gauge symmetry insures that the wave equations of quarks, leptons, photons, and all other particles reduce to Hamilton–Jacobi type equations that explicitly display "effective Riemannian metric tensors" in the geometric-optics (or classical) limit. This gives a new and interesting picture of the physical world, namely that in the presence of T_4 gauge fields, classical objects and light rays exhibit motions as if they were in a "curved space-time," even though the real physical space-time for fields, quantum particles, and their interactions is flat. The obvious challenge is to construct a gauge-invariant action in flat space-time that correctly predicts experimental results and in the following chapters, we do just that.

7.2.   The Levi-Civita Connection and Interpretations of Einstein Gravity

In 1917, Levi-Civita introduced a connection on a Riemannian manifold by parallel transport.[8] In Euclidean space, a vector field, whose direction at any point is that of the unit vector **i**, is parallel if the derivative of **i** vanishes at every point and for all directions, but this cannot be true of a vector field in a Riemannian manifold in general. Instead, Levi-Civita defined the "parallel transport" of vectors with respect to a given curve C, where one can express the coordinates of points on the curve C in terms of the arc-length t. A vector field $V^\nu(t)$ of constant magnitude is parallel with respect to the Riemannian manifold along the curve C, if its derived vector in the direction of the curve is zero at all points of C. The differential equation for the parallel transport of the vector field $V^\nu(t)$ is[8]

$$\frac{dV^\nu(t)}{dt} + \Gamma^\nu_{\mu\lambda}\frac{dx^\mu(t)}{dt}V^\lambda(t) = 0, \quad V^\nu(t) = V^\nu(x(t)). \tag{7.1}$$

It can also be written in the form

$$\frac{dx^\mu}{dt}D_\mu V^\nu = 0, \tag{7.2}$$

where

$$D_\mu V^\nu = \partial_\mu V^\nu + \Gamma^\nu_{\mu\lambda}V^\lambda \tag{7.3}$$

is the covariant derivative of V^ν in Riemannian geometry. As expected, equation (7.2) is independent of the choice of coordinates, i.e., it is invariant under general (or arbitrary infinitesimal) coordinate transformations

$$x'^\mu = g^\mu(x^\lambda) \quad \text{or} \quad x'^\mu = x^\mu + \Lambda^\mu(x), \tag{7.4}$$

where g^μ are single-valued functions whose functional determinant does not vanish, and $\Lambda^\mu(x)$ are arbitrary infinitesimal functions. The Levi-Civita connection is the Christoffel symbol $\Gamma^\nu_{\mu\lambda}$, which is expressed in terms of the Riemannian metric tensors. Although the connection is not a tensor, the difference between two connections or any infinitesimal variation of a connection is a tensor. As an aside, the Levi-Civita connection can also be determined by noting that it is the only connection that satisfies $\Gamma^\nu_{\mu\lambda} = \Gamma^\nu_{\lambda\mu}$ (i.e., it is torsionless) and preserves the length of the vector $V^\nu(x)$ in the parallel transport (7.1), i.e., $d(V_\mu V^\mu) = 0$. The Riemann–Christoffel curvature of the Riemannian manifold can be considered as the curvature on the Levi-Civita connection.

The Riemann–Christoffel curvature played a central role in the derivation of the gravitational field equation, as discussed by Hilbert.[9] Utiyama first interpreted Einstein's field equation as a gauge field equation in 1956. C. N. Yang proposed interpreting the Levi-Civita connection as the gravitational gauge potential and the Riemann–Christoffel curvature as the gauge curvature for gravity, constructing an action involving the quadratic gauge curvature and obtaining a new gravitational equation that contained the third-order derivative of the metric tensor.[6] However, these results did not shed light on the basic difficulties of Einstein gravity discussed earlier.

7.3.   Weyl's Parallel Transport of Scale and Electromagnetic Fields

In 1917, G. Hessenberg published a discussion of the vectorial foundations of differential geometry.[10] His idea led to a very broad generalization of Riemannian geometry by dispensing with the Riemannian metric tensor so that in Hessenberg's generalized geometry there is no fundamental metric tensor to define the length of a vector. In place of it, he introduced axiomatically the coefficients of affine connection — a certain function of the coordinates. Thus, Hessenberg and Weyl[11] laid down the mathematical foundations of gauge field theories by liberating the concept of connection from the bondage of metric tensors, so that one can axiomatically introduce connections associated with various internal and external symmetry groups.

Weyl's insight led to the application of parallel transport to any vector field attached to a point of the manifold and even the possible scales at various points as a "non-geometric" vector bundle on space-time, using a prescribed connection. He then used these mathematical properties to investigate a unified field theory involving the electromagnetic and gravitational interactions.[11] The transformation law for the connection is determined uniquely by requiring that the parallel transport be independent of the choice of coordinates. The concepts of connection, parallel transport and covariant differentiation are essentially equivalent. These general properties, which are independent of the Riemannian metric tensors, turned out to be the cornerstones of modern gauge fields and the unification of the interactions.[12,13]

Weyl defined the parallel transport of "scale" $s(t)$ by introducing the following corresponding relation to (7.1):

$$A_\mu \, dx^\mu \leftrightarrow \Gamma^\nu_{\mu\lambda} \, dx^\mu. \tag{7.5}$$

In analogy to (7.1), the differential equation that describes the parallel transport for scale $s(t)$ is given by

$$\frac{ds(t)}{dt} + \frac{dx^\mu(t)}{dt} A_\mu(t) s(t) = 0, \quad s(t) = s(x(t)), \tag{7.6}$$

which leads to the solution

$$s(t) = T \exp\left[-\int_0^t A_\mu \frac{dx^\mu}{dt} dt \right] s(0). \tag{7.7}$$

Here, T denotes that $s(t)$ is to be evaluated by path-ordering each term in the series expansion for the exponential,[14] and the line integral is along the path C from 0 to t. According to Weyl's view, the scales at various points can be considered as a vector bundle on space-time, connections can be introduced on this bundle, and their curvatures as 2-forms can be defined on space-time.[b] This curvature turned out to be the electromagnetic field strength, which plays a key role in the physics of electromagnetism.

After the birth of quantum mechanics, it was realized that the vector potential A_μ should be related to the phase change of a wave function rather than to the scale change. In this case, the parallel transport of phase and the solution are formally the same as (7.6) and (7.7), except that the A_μ in (7.6) and (7.7) is replaced by $-ieA_\mu$, where e is the electric charge associated with the coupling of the electromagnetic field A_μ. One can also generalize (7.6) and (7.7) to Yang–Mills fields by the replacement, $A_\mu \rightarrow ifb_\mu^a \tau^a/2$, where the SU_2 generators satisfy the relations, $[\tau^a/2, \tau^b/2] = i\epsilon^{abc}\tau^c/2$. Thus, the differential equation that describes the parallel transport for $s(t)$ is given by

$$\frac{ds(t)}{dt} = \frac{dx^\mu(t)}{dt} Z_\mu(x(t))s(t), \tag{7.8}$$

$$Z_\mu = ieA_\mu \quad \text{for } U_1, \quad \text{and} \quad Z_\mu = -ifb_\mu^a \frac{\tau^a}{2} \quad \text{for } SU_2.$$

This leads to the solution

$$s(t) = T\exp\left[\int_0^t Z_\mu(t')\frac{dx^\mu(t')}{dt'}dt'\right]s(0). \tag{7.9}$$

The equation of the parallel transport (7.8) corresponds to the gauge covariant derivative of the form,

$$\Delta_\mu = \partial_\mu - ieA_\mu \quad \text{for } U_1; \quad \Delta_\mu = \partial_\mu + ifb_\mu^a \frac{\tau^a}{2} \quad \text{for } SU_2. \tag{7.10}$$

This can be seen by expressing ds/dt as $(\partial s/\partial x^\mu)(dx^\mu/dt)$ in (7.8), so that we have

$$\frac{dx^\mu}{dt}\Delta_\mu s(t) = 0, \quad \Delta_\mu = \partial_\mu - Z_\mu. \tag{7.11}$$

The parallel transport (7.9) in this case is the gauge-dependent or non-integrable phase factor.

---

[b]The expression $(\Gamma_{\mu\lambda}^\nu dx^\mu)$ in (7.1) can be considered as the matrix A of 1-form. The matrix A and its transformation law define the connection on the manifold. The components of F in the matrix 2-form $dA = F$ are $F_{\mu\nu} = \partial_\mu A_\nu - \partial_\nu A_\mu$, which is the electromagnetic field strength.

Thus, the definitions of the gauge covariant derivatives in (7.10) and (7.11) resemble the covariant derivative (7.3) in Riemannian geometry.

The U_1 and SU_2 gauge curvatures $F_{\mu\nu}$ and $B^a_{\mu\nu}$ are defined by

$$[\Delta_\mu, \Delta_\nu] = -ieF_{\mu\nu}, \quad F_{\mu\nu} = \partial_\mu A_\nu - \partial_\nu A_\mu \quad \text{for } U_1, \tag{7.12}$$

$$[\Delta_\mu, \Delta_\nu] = +ifB^a_{\mu\nu}\frac{\tau^a}{2} \quad \text{for } SU_2, \tag{7.13}$$

$$B^a_{\mu\nu} = \partial_\mu b^a_\nu - \partial_\nu b^a_\mu + f\epsilon^{abc}b^b_\mu b^c_\nu. \tag{7.14}$$

In physics, the generalization from the Abelian group U_1 for electromagnetic fields to the non-Abelian group SU_2 for Yang–Mills fields is highly non-trivial, especially with respect to their quantum properties. Yang and Mills attempted to make this generalization, motivated by the question of whether one could understand the conservation of isospin quantum number in a way similar to the conservation of electric charge. Although their original formulation of a gauge theory for the strong interaction was incorrect, their idea turned out to be very significant, gradually becoming the mainstream of particle physics and paving the way for the unification of the electromagnetic and weak interactions. In hindsight, the generalization of U_1 to SU_2 is mathematically simple or even trivial if one knows their relations to fiber bundles. Nevertheless, physicists initially struggled to understand non-Abelian gauge fields, without the help of fiber bundles.

7.4. Curvatures on the Connections

Gauge curvatures associated with various internal (dynamical) gauge groups are essential for our understanding of fundamental interactions in physics. To see the wonderful analogy of gauge curvatures and the Riemann–Christoffel curvature, let us calculate the relation between the gauge curvatures (e.g., given by (7.12), (7.13), and (7.14)), and the parallel transport along a small square with side c in the x^1–x^2 plane. For convenience of treating U_1 and SU_2 simultaneously, we use the following notations:

$$\Delta_\mu = \partial_\mu - Z_\mu, \quad Z_\mu = -ifb^a_\mu(x)\frac{\tau^a}{2} \quad \text{for } SU_2, \tag{7.15}$$

$$\Delta_\mu = \partial_\mu - Z_\mu, \quad Z_\mu = ieA_\mu(x) \quad \text{for } U_1. \tag{7.16}$$

To calculate the parallel transport (7.9) along the x-direction on one side of the square from $(0, x, 0, 0)$ to $(0, x', 00)$, let us use the following parametrization for $x^\mu(t)$:

$$f^\mu(t) = x^\mu + ct\delta_1^\mu, \quad df^\mu = c\delta_1^\mu \, dt. \tag{7.17}$$

We have

$$\exp\left[\int_x^{x'} Z_\lambda(f^\mu(t)) df^\lambda\right] = \exp\left[\int_0^1 cZ_\lambda(x^\mu + ct\delta_1^\mu) dt \delta_1^\lambda\right]$$

$$= \exp\left[-c \int_0^1 [Z_1(x^\mu) + \partial_\mu Z_1 ct\delta_1^\mu] dt\right]. \tag{7.18}$$

For the parallel transport along the $+y$ direction on the side of the square from $(0, 0, y, 0)$ to $(0, 0, y', 0)$, we use the parameterization

$$f^\mu(t) = x^\mu + c\delta_1^\mu + ct\delta_2^\mu, \quad df^\mu = c\delta_2^\mu \, dt. \tag{7.19}$$

We have

$$\exp\left[\int_y^{y'} Z_\lambda(f^\mu(t)) df^\lambda\right] = \exp\left[\int_0^1 cZ_\lambda(x^\mu + c\delta_1^\mu + ct\delta_2^\mu) dt \delta_2^\lambda\right]$$

$$= \exp\left[-c \int_0^1 (Z_2(x^\mu) + (\partial_\mu Z_2)[c\delta_1^\mu + ct\delta_2^\mu]) dt\right]$$

$$= \exp\left[-c\left(Z_2(x^\mu) + (\partial_\mu Z_2)\left[c\delta_1^\mu + \frac{c}{2}\delta_2^\mu\right]\right)\right]$$

$$= \exp\left[-cZ_2\left(x^\mu + c\delta_1^\mu + \frac{c}{2}\delta_2^\mu\right)\right]. \tag{7.20}$$

Using (7.18) and (7.20), the parallel transport along the perimeter of a small square with side c is then

$$\Phi_{\text{square}} = \exp\left[cZ_2\left(x^\mu + \frac{c}{2}\delta_2^\mu\right)\right] \exp\left[cZ_1\left(x^\mu + \frac{c}{2}\delta_1^\mu + c\delta_2^\mu\right)\right]$$

$$\times \exp\left[-cZ_2\left(x^\mu + c\delta_1^\mu + \frac{c}{2}\delta_2^\mu\right)\right] \exp\left[-cZ_1\left(x^\mu + \frac{c}{2}\delta_1^\mu\right)\right]$$

$$= \exp\left[cZ_2(x^\mu) + c^2\partial_\mu Z_2 \frac{1}{2}\delta_2^\mu\right] \exp\left[cZ_1(x^\mu) + c^2\partial_\mu Z_1 \left[\frac{1}{2}\delta_1^\mu + \delta_2^\mu\right]\right]$$

$$\times \exp\left[cZ_2(x^\mu) - c^2\partial_\mu Z_2 \left[\delta_1^\mu + \frac{1}{2}\delta_2^\mu\right]\right]$$

$$\times \exp\left[cZ_1(x^\mu) - c^2\partial_\mu Z_1 \frac{1}{2}\delta_1^\mu\right]. \tag{7.21}$$

Let us use the Baker–Campbell–Hausdorff formula for operators A and B,

$$e^A e^B = e^{A+B} e^{\frac{1}{2}[A,B]}, \qquad (7.22)$$

which suffices for our purpose of calculating terms of order $O(c^2)$. The phase factor Φ_{square} for a small square with side c in the x^1–x^2 plane can be expressed as

$$\Phi_{\text{square}} = e^A e^B e^C e^D = e^{A+B+C+D+[A,B]/2+[C,D]/2} e^{[A+B,C+D]/2},$$

$$A = cZ_2(x^\mu) + c^2 \partial_\mu Z_2 \frac{1}{2} \delta_2^\mu, \quad B = cZ_1(x^\mu) + c^2 \partial_\mu Z_1 \left[\frac{1}{2} \delta_1^\mu + \delta_2^\mu \right],$$

$$(7.23)$$

$$C = cZ_2(x^\mu) - c^2 \partial_\mu Z_2 \left[\delta_1^\mu + \frac{1}{2} \delta_2^\mu \right], \quad D = cZ_1(x^\mu) - c^2 \partial_\mu Z_1 \frac{1}{2} \delta_1^\mu,$$

to the second order in c. We thus obtain the curvature $C_{\mu\nu}$ with $\mu = 2, \nu = 1$,

$$\Phi_{\text{square}} = \exp[c^2 C_{21} + O(c^3)],$$
$$C_{21} \equiv (\partial_2 Z_1 - \partial_1 Z_2 + Z_2 Z_1 - Z_1 Z_2), \qquad (7.24)$$

where $\Delta_\mu = \partial_\mu - Z_\mu$.

For gauge theories with U_1 and SU_2 groups, we have the usual results

$$C_{21} = ie(\partial_2 A_1 - \partial_1 A_2) \equiv ieF_{21} \quad \text{for } U_1, \qquad (7.25)$$

$$C_{21} = -if(\partial_2 b_1^a - \partial_1 b_2^a + f\epsilon^{abc} b_2^b b_1^c) \frac{\tau^a}{2} \equiv -ifB_{21}^a \frac{\tau^a}{2}, \quad \text{for } SU_2. \quad (7.26)$$

Other components of $C_{\mu\nu}$ can also be obtained. We note again that the derivations of gauge curvatures in (7.24), (7.25) and (7.26) also resemble the Riemann–Christoffel curvatures $R_{\alpha,\mu\nu}^\beta$ given by

$$[D_\mu, D_\nu]V_\alpha = R_{\alpha,\mu\nu}^\beta V_\beta,$$
$$R_{\alpha,\mu\nu}^\beta = \partial_\nu \Gamma_{\alpha\mu}^\beta - \partial_\mu \Gamma_{\alpha\nu}^\beta + \Gamma_{\alpha\mu}^\rho \Gamma_{\rho\nu}^\beta - \Gamma_{\alpha\nu}^\rho \Gamma_{\rho\mu}^\beta. \qquad (7.27)$$

7.5. Taiji Space-Time Symmetry and the Translational Symmetry Group T_4

7.5.1. *Taiji space-time symmetry*

Yang–Mills gravity[15] is based on the external space-time translational gauge group T_4 in flat space-time with arbitrary coordinates, which is applicable to both inertial and non-inertial frames.[7] Translational gauge symmetry can be expressed as a local space-time translation with an arbitrary infinitesimal vector gauge function $\Lambda^\mu(x)$,

$$x^\mu \to x'^\mu = x^\mu + \Lambda^\mu(x), \quad x \equiv x^\lambda = (w, x, y, z), \quad (7.28)$$

where $\Lambda^\mu(x)$ must not be affected by the transformations at a fixed point because they are gauge functions (i.e., group parameters). The Jacobian is assumed to be non-zero.

   However, we observe that the mathematical forms of the local space-time translations (7.28) are also the arbitrary infinitesimal transformations of coordinates in flat space-time. Therefore, we postulate to interpret that (7.28) is both a local translation and a coordinate transformation with the same arbitrary function $\Lambda^\mu(x)$ in Yang–Mills gravity. Such a framework for physics in general frames of reference may be called taiji space-time symmetry framework.

   The invariance of the gravitational action for a physical system under this *local translational group in flat space-time* is the physical essence and the mathematical expression of taiji symmetry. Furthermore, such a gauge symmetry is intimately related to the gravitational field and its universal interaction with all other fields in nature, as we shall see below.

7.5.2. T_4 *gauge transformations and external groups*

The dual transformation property embedded in (7.28) makes the formulation of the gauge invariant gravity highly non-trivial. It appears to be difficult to treat the external translation group in space-time with the generator ∂_μ in the fiber bundles, which describe internal gauge symmetry. Nevertheless, we can accommodate these two mathematical implications (i.e., local space-time translations and arbitrary coordinate transformations) of the local transformations (7.28) by defining suitable gauge transformations of space-time translations for the various functions that appear in a Lagrangian.[7] For example, the infinitesimal T_4 gauge transformations for a scalar $Q(x) \to Q^\$(x)$ and a vector $\Gamma^\mu(x) \to (\Gamma^\mu(x))^\$$ etc. are given by

$$Q^\$(x) = Q(x) - \Lambda^\lambda(x)\partial_\lambda Q(x), \quad (7.29)$$

$$(\Gamma^\mu(x))^\$ = \Gamma^\mu(x) - \Lambda^\lambda(x)\partial_\lambda\Gamma^\mu(x) + \Gamma^\lambda(x)\partial_\lambda\Lambda^\mu(x),$$

$$(\Gamma_\mu(x))^\$ = \Gamma_\mu(x) - \Lambda^\lambda(x)\partial_\lambda\Gamma_\mu(x) - \Gamma_\lambda(x)\partial_\mu\Lambda^\lambda(x),$$

$$(T_v^\mu(x))^\$ = T_v^\mu(x) - \Lambda^\lambda(x)\partial_\lambda T_v^\mu(x) - T_v^\lambda(x)\partial_\lambda\Lambda^\mu(x)$$
$$+ T^\mu_{\ \lambda}(x)\partial_v\Lambda^\lambda(x).$$

$$(7.30)$$

We may remark that if, for example, $T_v^\mu(x)$ is a tensor density with weight $w \neq 0$, there is an additional term $-wT_v^\mu(x)\partial_\lambda\Lambda^\lambda(x)$ in its T_4 gauge transformation, i.e., after the term $+T^\mu_{\ \lambda}(x)\partial_v\Lambda^\lambda(x)$. (Cf. Appendix A.)

Mathematically, only the second terms, involving $\Lambda^\lambda\partial_\lambda$, on the right-hand side of the gauge transformations (7.29) and (7.30) correspond to a local space-time translation (i.e., the external T_4 group with generators $p_\lambda = i\partial_\lambda$). All other terms reflect a much more complicated group of arbitrary infinitesimal transformations of coordinates in flat space-time. This complicated group actually has a continuously infinite number of generators which have little to do with the action in Yang–Mills gravity and the physics of the gravitational interaction. (See Section 13.4.)

7.5.3. T_4 and Weyl's parallel transports of "scale" $s(t)$ and T_4 gauge curvature

The following displays of the mathematical properties of the translational gauge symmetry (in Yang–Mills gravity) and the corresponding fiber bundle are based purely on algebraic analogy, rather than a rigorous derivation. The geometric picture of the external translational gauge symmetry in Yang–Mills gravity seems to be much more complicated than that of gauge symmetries of internal groups, the reason being that the local space-time translations (7.28) are also the general coordinate transformations. Thus, the T_4 gauge transformations (7.29) and (7.30) and the associated group generators related to arbitrary coordinate transformations are complicated and again, will be discussed only briefly in Section 13.4.

The tensor potential field $\phi_{\mu v}$ in Yang–Mills gravity is associated with the space-time translational group T_4 with the generators $p_\mu = i\partial/\partial x^\mu$ in inertial frames with the Minkowski metric tensor. Thus, the results for T_4 can be obtained from (7.8)–(7.14) with the replacement

$$ifb_\mu^a\frac{\tau^a}{2} \rightarrow ig\phi_\mu^v p_v \equiv Y_\mu, \quad p_v = i\partial/\partial x^v. \tag{7.31}$$

In Yang–Mills gravity, we have the following corresponding results:

$$\frac{ds(t)}{dt} - ig\frac{dx^\mu(t)}{dt}\phi^\nu_\mu p_\nu s(t) = 0, \quad c = \hbar = 1, \tag{7.32}$$

$$s(t) = T\exp\left[\int_0^t dt'\frac{dx^\mu(t')}{dt'}Y_\mu(t')\right]s(0)$$

$$= \exp\left[-g\int_{C(t)} dx^\mu\,\phi^\nu_\mu\partial_\nu\right]s(0), \quad s(t) = s(x^\mu(t)), \tag{7.33}$$

where $C(t)$ is the path of integration. It is important to note that the generators of the T_4 group are $p_\mu = i\partial/\partial x^\mu$, so that $s(t)$ is a scale factor rather than a phase factor. These properties have important physical implications, namely that the gauge fields associated with scale factor lead to only one kind of force (e.g., attractive force), while those associated with phase factors lead to both attractive and repulsive forces.

Equations (7.32) and (7.33) for T_4 resemble Weyl's parallel transport in (7.6) and (7.7) for a scale. The equation for parallel transport (7.32) can be written as

$$\frac{dx^\mu(t)}{dt}\Delta_\mu s(t) = \frac{dx^\mu}{dt}[\partial_\mu - ig\phi^\nu_\mu p_\nu]s(x(t)) = 0. \tag{7.34}$$

We note that the derivative $\partial_\mu - ig\phi^\nu_\mu p_\nu = \partial_\mu + g\phi^\nu_\mu\partial_\nu$ is not independent of the choice of the coordinates. In order for the gravitational action in the Yang–Mills gravity to hold for both inertial and non-inertial frames (i.e., to be independent of the choice of coordinates), we must replace the ordinary derivative ∂_μ in $\partial_\mu + g\phi^\nu_\mu\partial_\nu$ by the covariant derivative D_μ associated with the Poincaré metric tensors $P_{\mu\nu}$ in general frames based on flat space-time. Thus, the T_4 gauge covariant derivatives are

$$\Delta_\mu = D_\mu + g\phi^\nu_\mu D_\nu, \tag{7.35}$$

which can be written in the form

$$\Delta_\mu = J^\nu_\mu D_\nu, \quad J^\nu_\mu = \delta^\nu_\mu + g\phi^\nu_\mu. \tag{7.36}$$

Thus, the T_4 gauge curvature $C_{\mu\nu\alpha}$ is given by

$$[\Delta_\mu, \Delta_\nu] = gC_{\mu\nu\alpha}D^\alpha, \tag{7.37}$$

$$C_{\mu\nu\alpha} = J^\sigma_\mu D_\sigma J_{\nu\alpha} - J^\sigma_\nu D_\sigma J_{\mu\alpha} = -C_{\nu\mu\alpha}, \tag{7.38}$$

in general frames of reference.

There is a difference between the invariance under the gauge transformations of internal symmetry groups and that of external symmetry groups. Equation (7.11) with $\Delta_\mu = \partial_\mu - ieA_\mu$ for the parallel transport with the U_1 group or the Lagrangian $L = [(\partial_\mu + ieA_\mu)\Phi^*][(\partial^\mu - ieA^\mu)\Phi]$ can be verified to be invariant under the internal U_1 gauge transformations

$$A'_\mu = A_\mu + \partial_\mu \Lambda(x), \quad s'(x) = e^{[ie\Lambda(x)]}s(x). \tag{7.39}$$

However, the corresponding equation (7.34) with $\Delta_\mu = \partial_\mu + g\phi_\mu^\nu \partial_\nu$ for the parallel transport with the external T_4 group or the Lagrangian $L = [(\partial_\mu + g\phi_\mu^\nu \partial_\nu)\Phi^*][(\partial^\mu + g\phi_\lambda^\mu \partial^\lambda)\Phi]$ will be changed by a term due to an infinitesimal space-time shift $\Lambda^\lambda(x)$,

$$\left[\frac{dx^\mu(t)}{dt}\Delta_\mu s(t)\right]^{\$} = \frac{dx^\mu(t)}{dt}\Delta_\mu s(t) - \Lambda^\lambda(x)\partial_\lambda\left[\frac{dx^\mu(t)}{dt}\Delta_\mu s(t)\right], \tag{7.40}$$

$$L^{\$} = L - \Lambda^\lambda(x)\partial_\lambda L, \quad L = [(\partial_\mu + g\phi_\mu^\nu \partial_\nu)\Phi^*][(\partial^\mu + g\phi_\nu^\mu \partial^\nu)\Phi]. \tag{7.41}$$

For a gauge invariant theory, the Lagrangian itself may not be invariant under gauge transformations, i.e., $L^{\$} \neq L$. However, the action S,

$$S = \int L\sqrt{-P}\, d^4x, \tag{7.42}$$

must be gauge invariant. We shall discuss the space-time gauge transformations and the invariance of the action (7.42) in Chapter 8.

The YMUW framework in flat space-time can accommodate both gauge fields associated with SU_N bundle and T_4 bundle. It suggests that all interactions are dictated by the dynamical gauge symmetries in this framework.

The internal (compact) symmetry groups such as SU_N lead to vector gauge fields, while the external (non-compact) space-time translation group T_4 leads to tensor gauge fields $\phi_{\mu\nu}$.[7] The T_4 gauge curvature $C_{\mu\nu\alpha}$ in (7.38) will play a central role in Yang–Mills gravity, in which the gravitational interaction is dictated by the T_4 gauge symmetry. In this sense, gravity is a manifestation of the space-time translation gauge symmetry or, equivalently, the taiji space-time symmetry. As a result, the difficulties of Einstein gravity discussed earlier in Section 0.1 in the Overview are resolved because Yang–Mills gravity is based on flat space-time with inertial frames, in which space and time coordinates have operational meaning.

Table 7.2.   Gauge fields in YMUW framework and fiber bundle terminology.*

"Phase" and "scale" gauge fields	Fiber bundle terminology
Vector "phase" gauge field b_μ^k	Connection on an SU_N bundle
Tensor "scale" gauge field $\phi_{\mu\nu}$	Connection on a "T_4 bundle"
Phase factor	Parallel transport (with SU_N)
Scale factor	Parallel transport (with T_4)
SU_N gauge curvature $f_{\mu\nu}^k$	Curvature on SU_N connections
T_4 gauge curvature $C_{\mu\nu\lambda}$	Curvature on "T_4 connection"
Electromagnetism	Connection on a U_1 bundle
Chromodynamics	Connection on $(SU_3)_{\text{color}}$ bundle
Gravity	Connection on a "T_4 bundle"
$(\partial_\mu - ieA_\mu)\Phi,\ \ (\partial_\mu + ieA_\mu)\Phi^*$	U_1 covariant differentiation
"Attrac. and repul. electromagnetic force" with dimensionless coupling constants $\pm e$	
$(\partial_\mu + g\phi_\mu^\nu\partial_\nu)\Phi,\ \ (\partial_\mu + g\phi_\mu^\nu\partial_\nu)\Phi^*$	T_4 covariant differentiation
"Only attractive gravitational force" where g has the dimension of length	

*Note. It seems to be non-trivial to treat the external space-time gauge group in the language of fiber bundles as it describes internal gauge group. Nevertheless, based on the algebraic analogy, the correspondences between the physical gauge fields in the YMUW framework and mathematical fiber bundle are summarized in Table 7.2.[c,d]

7.6.   Lie Derivative Application to Gauge Fields and Quantum Yang–Mills Gravity

It is intriguing that the T_4 gauge transformations (7.29)–(7.30) with infinitesimal vector gauge function $\Lambda^\mu(x)$ in quantum Yang–Mills gravity turn out to be identical to the Lie derivatives (in the coordinate expression in flat space-time) and Pauli's variation. Furthermore, the invariance of the action S in (7.42) under T_4 gauge transformations is equivalent to the vanishing of the Lie derivative of the action S with respect to the vector function $\Lambda^\mu(x)$, i.e., $\mathcal{L}_\Lambda S = 0$ and the invariance under Pauli's variation. (See Appendix A.)

---

[c] For fiber bundle terminology related to internal gauge symmetry groups, see Ref. 16. The Levi-Civita connection does not play a role in Yang–Mills gravity based on flat space-time.

[d] It seems that the phase factor or the parallel transport associated with compact Lie groups is related to gauge transformations and gauge curvatures, while the scale factor or the parallel transport associated with non-compact Lie groups such as T_4 is related only to gauge curvatures.

Historically, Pauli appears to have been the first to discuss a new variation for all tensors in his comprehensive book on relativity published in 1921.[17] For example, in contrast to the usual change $a'^\mu(x') - a^\mu(x)$ under the infinitesimal coordinate transformation with an infinitesimal vector function $\Lambda^\mu(x)$, Pauli introduced of a new variation $\delta^* a^\mu = a'^\mu(x) - a^\mu(x)$. His variation shows the change of the form of the vector function $a^\mu(x)$, where one compares a'^μ and a^μ for the same value of the arguments. His idea was physically motivated by deriving field equations and certain identities from invariant actions such as the electromagnetic action and the Hilbert–Einstein action for gravity. In 1931, Ślebodziński introduced a new differential operator for all tensors in his discussion of Hamilton's equations.[18] Later, it was named the Lie derivative, which became so useful that it was developed into a theory of Lie derivatives with many applications.[19]

Here, we would like to discuss a new application of the Lie derivative (and Pauli's variation) to gauge fields. We demonstrate that such gauge fields in inertial frames are precisely the same as the tensor gauge fields in quantum Yang–Mills gravity. Note that we have three apparently different ideas for the changes of tensor fields:

(I)   The gauge transformations of tensor fields are dictated by the space-time translation group with the generators $p_\mu = i\partial_\mu$ in an inertial frame with coordinates x^μ.

(II) Pauli's variation of tensor fields, e.g., $\delta^* a^\mu = a'^\mu(x) - a^\mu(x)$, are defined at the same point x^μ under a coordinate transformation with an infinitesimal vector function.

(III) Lie derivatives (in the coordinate expression) define the directional derivative of tensor fields determined by an arbitrary vector function. Ślebodziński originally started with operators X and $\overline{X}$ and gave an infinitesimal transformation $X(f) = X^r \partial_r f$, where $X^r = X^r(x^1, \dots, x^m)$.[18] He then extended it to a general tensor $X(Q^{ab\dots}_{ef\dots})$.

Idea (I) is physically motivated by the gauge symmetry principle that any basic physical interaction is associated with a dynamical symmetry group. In particular, gauge invariant gravity is postulated to be based on the space-time translation (T_4) gauge group with the generators $p_\mu = i\partial_\mu$. The physical reason is that this displacement generator implies the existence of a tensor gauge field $\phi_{\mu\nu}$

through the T_4 gauge covariant derivative,

$$\partial_\mu \to \partial_\mu - ig\phi_{\mu\nu}p^\nu = (\eta_{\mu\nu} + g\phi_{\mu\nu})\partial^\nu \equiv \Delta_\mu. \tag{7.43}$$

Furthermore, the space-time translation group implies that the interaction has only one kind of force for particles and antiparticles, in sharp contrast to the electromagnetic forces. The reason is as follows: The electromagnetic U_1 gauge covariant derivative takes the form

$$\partial_\mu \to \partial_\mu - ieA_\mu,$$

where the imaginary part changes sign under the complex conjugate. As a result, the presence of i in the U_1 gauge covariant derivatives in, say, the Dirac equation implies two kinds of forces, namely, attractive and the repulsive forces. In sharp contrast, the T_4 gauge covariant derivative in (7.43) does not involve i.

At the first glance, ideas (II) and (III) seem to be unrelated to gauge symmetry group. Lie derivatives may be defined in several equivalent ways. It is not easy to see without explicitly using a coordinate system such as inertial frames in the formulations of gauge theories. We shall show, however, that there is a "hidden" external gauge symmetry group in the Lie derivative, in Ślebodziński's infinitesimal transformation, as well as in Pauli's variation.

First, let us show the relationship between the space-time gauge transformations (7.29)–(7.30) and the Lie derivative with the coordinate expressions. We use the usual coordinate system for inertial frames in gauge theories. Let us consider a vector function $\Gamma^\mu(x)$ under the infinitesimal transformation (7.28) in an inertial frame, for simplicity,

$$\Gamma'^\mu(x') = \Gamma^\lambda(x)\frac{\partial x'^\mu}{\partial x^\lambda} = \Gamma^\mu(x) + \Gamma^\lambda(x)\partial_\lambda\Lambda^\mu(x),$$
$$\Gamma^\mu(x) \to \Gamma'^\mu(x) = \Gamma^\mu(x - \Lambda) + \Gamma^\lambda(x - \Lambda)\partial_\lambda\Lambda^\mu(x). \tag{7.44}$$

Thus, the variation of the form of a function $\Gamma^\mu(x)$ at the space-time point x^μ is given by

$$\Gamma'^\mu(x) - \Gamma^\mu(x) = \Gamma^\mu(x - \Lambda) + \Gamma^\lambda(x - \Lambda)\partial_\lambda\Lambda^\mu - \Gamma^\mu(x)$$
$$= -\Lambda^\lambda(x)\partial_\lambda\Gamma^\mu(x) + \Gamma^\lambda(x - \Lambda)\partial_\lambda\Lambda^\mu$$
$$= -\Lambda^\lambda(x)\partial_\lambda\Gamma^\mu(x) + \Gamma^\lambda(x)\partial_\lambda\Lambda^\mu + O(\Lambda^2). \tag{7.45}$$

This variation is the Lie derivative $-(\mathcal{L}_\Lambda \Gamma)^\mu$,

$$(\mathcal{L}_\Lambda \Gamma)^\mu = \Lambda^\lambda \partial_\lambda \Gamma^\mu - \Gamma^\lambda \partial_\lambda \Lambda^\mu, \tag{7.46}$$

which is also the same as the variation $\delta^* \Gamma^\mu = \Gamma'^\mu(x) - \Gamma^\mu(x)$,

$$(\mathcal{L}_\Lambda \Gamma)^\mu = -\delta^* \Gamma^\mu.$$

One may called $\delta^* \Gamma^\mu$ the Pauli–Ślebodziński's variation.

It follows from (7.30) that the variation of the form of the function at x^μ, $(\Gamma^\mu(x))^\$ - \Gamma^\mu(x)$, is the same as $-(\mathcal{L}_\Lambda \Gamma)^\mu$ and $\delta^* \Gamma^\mu$, i.e.,

$$(\Gamma^\mu(x))^\$ - \Gamma^\mu(x) = -(\mathcal{L}_\Lambda \Gamma)^\mu = \delta^* \Gamma^\mu. \tag{7.47}$$

The relation (7.47) can be generalized to any tensor $Q^{\mu_1 \dots \mu_m}_{\alpha_1 \dots \alpha_n}$, so that the T_4 gauge transformations with infinitesimal vector gauge function $\Lambda^\mu(x)$ are exactly the same as the Lie derivatives in the coordinate expression and the Pauli–Ślebodziński variation,

$$(Q^{\mu_1 \dots \mu_m}_{\alpha_1 \dots \alpha_n})^\$ - Q^{\mu_1 \dots \mu_m}_{\alpha_1 \dots \alpha_n}$$
$$= -(\mathcal{L}_\Lambda Q)^{\mu_1 \dots \mu_m}_{\alpha_1 \dots \alpha_n} = \delta^* Q^{\mu_1 \dots \mu_m}_{\alpha_1 \dots \alpha_n}. \tag{7.48}$$

In particular, for a scalar function $\phi(x)$ and a vector function $\Gamma^\mu(x)$, we have

$$(\mathcal{L}_\Lambda \phi) = -\delta^* \phi = -[\phi^\$(x) - \phi(x)] = \Lambda^\mu \partial_\mu \phi(x), \tag{7.49}$$

$$(\mathcal{L}_\Lambda \Gamma)^\alpha = -\delta^* \Gamma^\alpha = -[(\Gamma^\alpha(x))^\$ - \Gamma^\alpha(x)] = \Lambda^\mu \partial_\mu \Gamma^\alpha - \Gamma^\mu \partial_\mu \Lambda^\alpha, \tag{7.50}$$

under the transformation with an arbitrary infinitesimal vector function $\Lambda^\mu = \Lambda^\mu(x)$,

$$x'^\mu = x^\mu + \Lambda^\mu(x). \tag{7.51}$$

The terms involving $\Lambda^\mu \partial_\mu$ in (7.49) and (7.50) indicate that the Lie derivative and Pauli–Ślebodziński variation are related to a local space-time translation (7.51) with an arbitrary translation $\Lambda^\mu(x)$. (See (7.29).) We note that Pauli's variation in (7.49) is the same as Ślebodziński's infinitesimal transformation $X(f) = X^r \partial_r f$.[18]

Therefore, the Lie derivative and Pauli–Ślebodziński variation are naturally associated with the four-dimensional space-time translation (T_4) group with the generators $p_\mu = i\partial_\mu$.

In the formulation of gauge fields, one must have a gauge symmetry group so that the dynamics of the interactions of the gauge field can be unambiguously determined. The Lie derivative and the Pauli–Ślebodziński variation are intimately associated with the space-time translation group (T_4) with simple generators $p_\mu = i\partial_\mu$ and the T_4 covariant derivative $\Delta_\mu = (\delta_\mu^\nu + g\phi_\mu^\nu)\partial_\nu$. In inertial frames, these results and the T_4 gauge curvature $C_{\mu\nu\alpha}$ are unambiguous and the same as (7.35)–(7.38) in quantum Yang–Mills gravity.

From a physical viewpoint, the vanishing of the action S in (7.42) under the Lie derivative in the coordinate expression or the Pauli–Ślebodziński variation is a manifestation of the invariance of quantum Yang–Mills gravity under simultaneous arbitrary space-time translation and an arbitrary coordinate transformation with the same infinitesimal vector gauge function $\Lambda_\mu(x)$. In this sense, the T_4 gauge invariant quantum Yang–Mills gravity provides a physical interpretation of the Lie derivative in the coordinate expression and the Pauli–Ślebodziński variation.

Furthermore, the Lie derivative in differential geometry and Pauli–Ślebodziński variation substantiate the importance of the metric tensor $G_{\mu\nu(x)}$ that appears in Yang–Mills gravity, i.e., in the relativistic equations of motion for macroscopic objects such as planets and light rays. However, the crucial point is that $G_{\mu\nu}(x)$ appears in Yang–Mills gravity as the effective metric tensor rather than the real metric tensor in Riemannian geometry.[e] Such an effective metric tensor $G_{\mu\nu(x)}$ à la Einstein–Grossmann is just right to help Yang–Mills gravity based on inertial frames to be consistent with experiments, rather than to hinder Yang–Mills gravity to be quantized, as discussed by Dyson in Section 0.1 of the Overview and in Chapter 11. The use of space and time coordinates in inertial frames is also crucial for them to have operational meaning, in contrast to the coordinates in Einstein's gravity, as discussed by Wigner in Section 0.1.

References

1. F. J. Dyson, in *100 Years of Gravity and Accelerated Frames: The Deepest Insights of Einstein and Yang–Mills*, eds. J. P. Hsu and D. Fine (World Scientific, 2005), p. 348.
2. J. P. Hsu, *Nuovo Cimento B* **108**, 183 (1993); **108**, 949 (1993); **109**, 645 (1994).
3. R. Utiyama, *Phys. Rev.* **101**, 1597 (1956).
4. T. W. B. Kibble, *J. Math. Phys.* **2**, 212 (1961); R. Utiyama and T. Fukuyama, *Prog. Theor. Phys.* **45**, 612 (1971); Y. M. Cho, *Phys. Rev. D* **14**, 2521 (1976); Y. M. Cho, *Phys. Rev. D*

---

[e]For a detailed discussion of the emergence of the effective metric in the quantum wave equations in the geometric-optics limit, see Section 8.5 in Chapter 8.

14, 3341 (1976); K. Hayashi and T. Shirafuji, *Phys. Rev. D* **19**, 3524 (1979); T. W. Hehl, P. von der Heyde, G. D. Kerlick and J. M. Nester, *Rev. Mod. Phys.* **48**, 393 (1976).

5. H. Weyl, Preuss. *Akad. Berlin* **465** (1918).

6. C. N. Yang, *Phys. Rev. Lett.* **32**, 445 (1974); "Symmetry and Physics," Talk at the Centenary Symposium of the Peking University (May 5, 1998).

7. J. P. Hsu, *Int. J. Mod. Phys.* **21**, 5119 (2006).

8. T. Levi-Civita, *Rend. Circolo. Mat. Palermo.* **42**, 173 (1917); C. E. Weatherburn, in *An Introduction to Riemannian Geometry and the Tensor Calculus* (Cambridge University Press, 1938), pp. 85–88; V. S. Varadarajan, in *A Tribute to C.S. Seshadri Perspectives in Geometry and Representation Theory*, eds. V. Lakshmibai *et al.* (Hindustan Book Agency, 2003), pp. 502–541.

9. D. Hilbert, in *100 Years of Gravity and Accelerated Frames: The Deepest Insights of Einstein and Yang–Mills*, eds. J. P. Hsu and D. Fine (World Scientific, 2005), p. 120.

10. G. Hessenberg, *Math. Ann.* **78**, 187 (1917).

11. H. Weyl, in *Raum-Zeit-Materie* (Springer, 1918), Section 14.

12. J. P. Hsu, *Mod. Phys. Lett. A* **26**, 1707 (2011).

13. J. P. Hsu, *Chin. Phys. C* **36**, 403 (2012); arXiv, 1108.2423.

14. S. Weinberg, in *The Quantum Theory of Fields*, Vol. 1 (Cambridge University Press, 1995), pp. 142–144.

15. J. P. Hsu and L. Hsu, *A Broader View of Relativity: General Implications of Lorentz and Poincaré Invariance* (World Scientific, 2006), Appendix D.

16. T. T. Wu and C. N. Yang, *Phys. Rev. D* **12**, 3845 (1975).

17. W. Pauli, *Theory of Relativity*, Trans. by G. Field (Pergmon Press, London, 1958), p. 66. Pauli's book was originally published in 1921.

18. W. Ślebodziński, *Bull. Acad. Roy. Belg.* **17**, 864 (1931).

19. K. Yano, in *The Theory of Lie Derivatives and Its Applications* (North-Holland, 1955), p. 14

20. Wikipedia, Lie derivative. See also ion.uwinnipeg.ca, Mathematics Stack Exchange (differential geometry, Lie derivative of volume form).

Chapter 8

Yang–Mills Gravity Based on Flat Space-Time
and Effective Curved Space-Time
for Motions of Classical Objects

8.1. Translational Gauge Transformations in Taiji Space-Time

Let us consider a particle-physics approach to gravity based on the Yang–Mills–Utiyama–Weyl (YMUW) framework with external space-time translational gauge symmetry. The formulation should hold for both inertial and non-inertial frames in taiji space-time. We first briefly review the basic replacement in the formulation of a gauge theory with internal groups. Then we generalize it to a gauge theory with external space-time translation group T_4 and discuss the associated T_4 gauge transformations in flat space-time.

The formulations of electromagnetic and Yang–Mills theories associated with internal gauge groups are all based on the replacement, $\partial_\mu \rightarrow \partial_\mu + igB_\mu$ in flat space-time where the field $B_\mu = B_\mu^a t_a$ involves constant matrix representations of the generators t_a of the gauge groups associated with internal symmetry. However, because the generators of the external space-time translation group T_4 are the displacement operators, $p_\mu = i\partial_\mu$ (in natural units where $c = \hbar = 1$), the replacement takes the form,

$$\partial_\mu \rightarrow \partial_\mu - ig\phi_\mu^\nu p_\nu \equiv J_\mu^\nu \partial_\nu, \quad J_\mu^\nu = \delta_\mu^\nu + g\phi_\mu^\nu, \tag{8.1}$$

in inertial frames, where g is the coupling constant associated with the tensor gauge field $\phi_{\mu\nu}$. As usual, we assume the tensor gauge field to be symmetric,

$$\phi_{\mu\nu} = \phi_{\nu\mu}, \quad \phi_{\mu\nu} = \phi_\mu^\lambda \eta_{\lambda\nu}.$$

We interpret this field as a spin-2 field, which is related to the generators of the space-time translational group T_4 and follow the Yang–Mills approach to formulate

113

a theory of gravity based on T_4 in taiji space-time. It is precisely this unique property (8.1) due to the displacement operators of the T_4 group that leads naturally to an "effective" Riemannian metric tensor,[1,a] and a universal attractive force for all matter and antimatter. The gravitational force is characterized by a coupling constant g with the dimension of length in natural units. This property is in sharp contrast with the dimensionless coupling constants in the usual Yang–Mills theories. For external gauge groups related to space-time, e.g., the de Sitter group or the Poincaré group, the gauge invariant Lagrangian involving fermions turns out to be richer in content (e.g., it also has a new gravitational spin force generated by the fermion spin density as well as genuine scale gauge fields to insure the gauge invariance of the Lagrangian). If the space-time translation group T_4 is replaced by the Poincaré group, there is a new gravitational spin force generated by the fermion spin density.[2] However, we will concentrate on the external gauge group of space-time translations T_4, which is the Abelian subgroup of the Poincaré group and is non-compact. This group T_4 is particularly interesting because it is the minimal group related to the conserved energy–momentum tensor.[3]

Translation gauge symmetry is based on the local space-time translation with an arbitrary infinitesimal vector gauge-function $\Lambda^\mu(x)$,

$$x^\mu \to x'^\mu = x^\mu + \Lambda^\mu(x), \tag{8.2}$$

as given by (7.28), which has two possible interpretations: (i) local translations of the space-time coordinates, and (ii) arbitrary infinitesimal coordinate transformations.[3]

In taiji space-time with arbitrary coordinates, both of these conceptual interpretations can be accommodated by defining an infinitesimal gauge transformation of space-time translations for physical quantities $Q^{\mu_1 \cdots \mu_m}_{\alpha_1 \cdots \alpha_n}(x)$ in the Lagrangian of fields[3]

$$Q^{\mu_1 \cdots \mu_m}_{\alpha_1 \cdots \alpha_n}(x) \to (Q^{\mu_1 \cdots \mu_m}_{\alpha_1 \cdots \alpha_n}(x))^\$ $$

$$= (Q^{\nu_1 \cdots \nu_m}_{\beta_1 \cdots \beta_n}(x) - \Lambda^\lambda(x)\partial_\lambda Q^{\nu_1 \cdots \nu_m}_{\beta_1 \cdots \beta_n}(x)) \frac{\partial x'^{\mu_1}}{\partial x^{\nu_1}} \cdots \frac{\partial x'^{\mu_m}}{\partial x^{\nu_m}} \frac{\partial x^{\beta_1}}{\partial x'^{\alpha_1}} \cdots \frac{\partial x^{\beta_n}}{\partial x'^{\alpha_n}}, \tag{8.3}$$

---

[a]The speculation that Einstein's theory of gravity may be an effective field theory has been around for decades among theorists. The idea of an effective Riemannian space due to the presence of the gravitational field in Minkowski space-time was discussed by Logunov, Mestvirishvili, N. Wu and others. (See Ref. 15.)

where $\mu_1, \nu_1, \alpha_1, \beta_1$, etc. are space-time indices and $\Lambda^\mu(x)$ are infinitesimal functions. As usual, both the (Lorentz) spinor field ψ and the (Lorentz) scalar field Φ are treated as "coordinate scalars" and have the same translational gauge transformation. The gauge transformations for scalar, vector and tensor fields are given by

$$Q(x) \to (Q(x))^\$ = Q(x) - \Lambda^\lambda \partial_\lambda Q(x), \quad Q(x) = \psi, \overline{\psi}, \Phi, \qquad (8.4)$$

$$\Gamma_\mu \to (\Gamma_\mu)^\$ = \Gamma_\mu - \Lambda^\lambda \partial_\lambda \Gamma_\mu - \Gamma_\lambda \partial_\mu \Lambda^\lambda,$$
$$\Gamma^\mu \to (\Gamma^\mu)^\$ = \Gamma^\mu - \Lambda^\lambda \partial_\lambda \Gamma^\mu + \Gamma^\lambda \partial_\lambda \Lambda^\mu, \qquad (8.5)$$

$$S_{\mu\nu} \to (S_{\mu\nu})^\$ = S_{\mu\nu} - \Lambda^\lambda \partial_\lambda S_{\mu\nu} - S_{\mu\alpha} \partial_\nu \Lambda^\alpha - S_{\alpha\nu} \partial_\mu \Lambda^\alpha,$$
$$Q^{\mu\nu} \to (Q^{\mu\nu})^\$ = Q^{\mu\nu} - \Lambda^\lambda \partial_\lambda Q^{\mu\nu} + Q^{\lambda\nu} \partial_\lambda \Lambda^\mu + Q^{\mu\lambda} \partial_\lambda \Lambda^\nu, \qquad (8.6)$$

where $T_{\mu\nu} = J_{\mu\nu}, P_{\mu\nu}$.

Suppose D_μ denotes the partial covariant derivative associated with a Poincaré metric tensor $P_{\mu\nu}(x)$ in a general reference frame (inertial or non-inertial) within flat space-time. We have, for example,

$$D_\mu V^\nu = \partial_\mu V^\nu + \Gamma^\nu_{\rho\mu} V^\rho, \quad D_\mu V_\nu = \partial_\mu V_\nu - \Gamma^\rho_{\nu\mu} V_\rho,$$

where $\Gamma^\nu_{\mu\rho} = \frac{1}{2} P^{\nu\sigma}(\partial_\mu P_{\sigma\rho} + \partial_\rho P_{\sigma\mu} - \partial_\sigma P_{\mu\rho})$. Also, the Riemann–Christoffel curvature tensor in (7.27) vanishes, i.e., $R^\nu_{\mu,\alpha\beta} = 0$. Note that the functions $D_\mu Q$ and $D_\mu D_\nu Q$ transform, by definition, as a covariant vector $\Gamma_\mu(x)$ and a covariant tensor $T_{\mu\nu}(x)$ respectively under the translational gauge transformation. The change of variables in the translational gauge transformations (8.4)–(8.6) in flat space-time is formally similar to the Lie variations in the coordinate transformations in Riemannian geometry.

The coordinate transformation (8.2) indicates that the theory of Yang–Mills gravity is formulated in a general frame of reference (inertial or non-inertial) characterized by an arbitrary coordinate and a Poincaré metric tensor $P_{\mu\nu}$. In general, the ordinary derivative ∂_μ in the basic replacement (8.1) should be expressed in terms of the partial covariant derivative D_μ associated with the Poincaré metric tensor $P_{\mu\nu}$:

$$D_\mu \to D_\mu + g\phi^\nu_\mu D_\nu \equiv J^\nu_\mu D_\nu, \qquad (8.7)$$

$$J_{\mu\nu} = P_{\mu\nu} + g\phi_{\mu\nu}, \quad J_{\mu\nu} = J_{\nu\mu}, \quad J_{\mu\nu} = J^\lambda_\mu P_{\lambda\nu}, \qquad (8.8)$$

for general frames with arbitrary coordinates.

All these equations (8.3)–(8.8) are necessary for the general formulation of Yang–Mills gravity in an arbitrary frame of reference based on taiji space-time with arbitrary coordinates.

8.2. Translational Gauge Symmetry and the Field-Theoretic Origin of Effective Metric Tensors

To see the field-theoretic origin of the effective Riemannian metric tensors, let us consider a fermion field ψ. The kinetic energy term in the Lagrangian for a fermion in a general frame is given by

$$i\overline{\psi}\Gamma_\alpha D^\alpha \psi - m\overline{\psi}\psi = i\overline{\psi}\Gamma_\alpha \partial^\alpha \psi - m\overline{\psi}\psi,$$

$$\{\Gamma_\mu, \Gamma_\nu\} = 2P_{\mu\nu}(x), \quad \Gamma_\mu = \gamma_a e_\mu^a, \tag{8.9}$$

$$\{\gamma_a, \gamma_b\} = 2\eta_{ab}, \quad \eta_{ab} e_\mu^a e_\nu^b = P_{\mu\nu}.$$

Here we have the usual relation $D_\mu \psi = \partial_\mu \psi$ because the "Lorentz spinor" ψ transforms as a "coordinate scalar" and D_μ is the partial covariant derivative defined in terms of the Poincaré metric tensor $P_{\mu\nu}$ in a general frame. In a general frame, the gauge field $\phi_{\mu\nu}$ and the translational gauge symmetry dictate that the replacement be

$$D^\alpha \to D^\alpha + g\phi^{\alpha\beta} D_\beta \equiv J^{\alpha\beta} D_\beta, \quad J^{\alpha\beta} = P^{\alpha\beta} + g\phi^{\alpha\beta},$$

$$i\overline{\psi}\Gamma_\alpha D^\alpha \psi \to i\overline{\psi}\Gamma_\alpha \Delta^\alpha \psi = i\overline{\psi}\Gamma_\alpha J^{\alpha\beta} D_\beta \psi = i\overline{\psi}\gamma_a E^{\alpha\beta} D_\beta \psi, \tag{8.10}$$

$$\Delta^\mu = J^{\mu\alpha} D_\alpha, \quad E^{a\alpha} = e_\mu^a J^{\mu\alpha}.$$

If one considers $E^{a\alpha}$ as an "effective tetrad," then the relation for an "effective metric tensor" is

$$\eta_{ab} E^{a\alpha} E^{b\beta} = \eta_{ab} e_\mu^a J^{\mu\alpha} e_\nu^b J^{\nu\beta} = P_{\mu\nu} J^{\mu\alpha} J^{\nu\beta} = G^{\alpha\beta}. \tag{8.11}$$

Such an "effective metric tensor" $G^{\alpha\beta}$ also shows up if we consider the Lagrangian of a scalar field Φ with the same replacement of D^μ as that in (8.10),

$$\frac{1}{2}[P_{\mu\nu}(D^\mu \Phi)(D^\nu \Phi) - m^2 \Phi^2] \to \frac{1}{2}[G^{\alpha\beta}(D_\alpha \Phi)(D_\beta \Phi) - m^2 \Phi^2]. \tag{8.12}$$

Although it may appear as if the geometry of the space-time has changed from a pseudo-Euclidean space-time to a non-Euclidean space-time due to the presence of the tensor gauge field (or spin-2 field) $\phi_{\mu\nu}$, in the Yang–Mills approach, the presence of $E^{a\alpha}$ in (8.10) and $G^{\alpha\beta}$ is simply the manifestation of the translational gauge symmetry in flat space-time.[3]

In the literature, when one arrives at the crucial step (8.10),[4–6] most authors usually abandon the Yang–Mills approach of a truly gauge invariant theory with a quadratic gauge curvature for a Lagrangian in flat space-time, and instead follow Einstein's approach to gravity by postulating a Riemannian space-time due to the presence of $\phi_{\mu\nu}$ or $J_{\mu\nu}$ in (8.11). In other words, $E^{a\alpha}$ and $G^{\alpha\beta}$ in (8.11) are postulated to be a real tetrad and a real metric tensor of physical space-time, respectively. Such an approach, together with general coordinate invariance, inevitably leads to "the most glaring incompatibility of concepts in contemporary physics" as observed by Dyson.[7]

In the particle-physics approach described here, however, we consistently treat $G^{\mu\nu}$ in (8.11) as merely an "effective metric tensor" for the motion of a classical object in flat space-time and in the presence of the tensor gauge field. This treatment will be substantiated in Section 8.5.

8.3.   Gauge Invariant Action and Quadratic Gauge Curvature

In the YMUW framework with the external translational gauge symmetry, we have the T_4 gauge curvature

$$C^{\mu\nu\alpha} = J^{\mu\lambda}(D_\lambda J^{\nu\alpha}) - J^{\nu\lambda}(D_\lambda J^{\mu\alpha}), \tag{8.13}$$

which is given by the commutation relation of the gauge covariant derivative and satisfies two simple identities,

$$[\Delta^\mu, \Delta^\nu] = C^{\mu\nu\alpha}D_\alpha, \quad \Delta^\mu = J^{\mu\nu}D_\nu,$$
$$C^{\mu\nu\alpha} = -C^{\nu\mu\alpha}, \quad C^{\mu\nu\alpha} + C^{\nu\alpha\mu} + C^{\alpha\mu\nu} = 0. \tag{8.14}$$

The translational gauge curvature $C^{\mu\nu\alpha}$ in (8.13) involves the symmetric tensor gauge field $\phi^{\mu\nu} = \phi^{\nu\mu}$. Thus, it differs from the usual Yang–Mills gauge curvature $f^k_{\mu\nu} = \partial_\nu b^k_\mu - \partial_\mu b^k_\nu - b^i_\mu b^j_\nu c^k_{ij}$, where c^k_{ij} is the structure constant of an internal gauge group whose generators have constant matrix representations. The T_4 gauge covariant derivative, Δ^ν in (8.14) satisfies the Jacobi identity, $[\Delta^\lambda, [\Delta^\mu, \Delta^\nu]] + [\Delta^\mu, [\Delta^\nu, \Delta^\lambda]] + [\Delta^\nu, [\Delta^\lambda, \Delta^\mu]] \equiv 0$, which leads to gauge-Bianchi identity in Yang–Mills gravity:

$$(\delta^\lambda_\alpha \Delta^\rho - (D^\lambda J^\rho_\alpha))C^{\mu\nu}{}_\lambda + (\delta^\lambda_\alpha \Delta^\mu - (D^\lambda J^\mu_\alpha))C^{\nu\rho}{}_\lambda$$
$$+ (\delta^\lambda_\alpha \Delta^\nu - (D^\lambda J^\nu_\alpha))C^{\rho\mu}{}_\lambda \equiv 0. \tag{8.15}$$

It turns out that there are two independent quadratic gauge-curvature scalars, $C_{\mu\alpha\beta}C^{\mu\beta\alpha}$ and $C_{\mu\alpha}{}^{\alpha}C^{\mu\beta}{}_{\beta}$. Other quadratic gauge-curvature scalars can be expressed in terms of them because of the identities in (8.14).

We postulate that, in a general frame with arbitrary coordinates and Poincaré metric tensors, the action $S_{\phi\psi}$ for fermion matter and spin-2 fields involves the linear combination of the two independent quadratic terms of the gauge curvature and the symmetrized fermion Lagrangian:

$$S_{\phi\psi} = \int L_{\phi\psi}\sqrt{-P}\,d^4x, \qquad P = \det P_{\mu\nu}, \tag{8.16}$$

$$\begin{aligned} L_{\phi\psi} &= \frac{1}{2g^2}(C_{\mu\alpha\beta}C^{\mu\beta\alpha} - C_{\mu\alpha}{}^{\alpha}C^{\mu\beta}{}_{\beta}) \\ &\quad + \frac{i}{2}[\overline{\psi}\Gamma_\mu\Delta^\mu\psi - (\Delta^\mu\overline{\psi})\Gamma_\mu\psi] - m\overline{\psi}\psi, \end{aligned} \tag{8.17}$$

$$\Delta^\mu\psi = J^{\mu\nu}D_\nu\psi, \qquad J^{\mu\nu} = P^{\mu\nu} + g\phi^{\mu\nu} = J^{\nu\mu}, \qquad D_\lambda P_{\mu\nu} = 0. \tag{8.18}$$

Note that the quadratic gauge-curvature term in (8.17) can also be expressed as

$$\begin{aligned} L_{\phi\psi} &= \frac{1}{2g^2}\left(\frac{1}{2}C_{\mu\alpha\beta}C^{\mu\alpha\beta} - C_{\mu\alpha}{}^{\alpha}C^{\mu\beta}{}_{\beta}\right) \\ &\quad + \frac{i}{2}[\overline{\psi}\Gamma_\mu\Delta^\mu\psi - (\Delta^\mu\overline{\psi})\Gamma_\mu\psi] - m\overline{\psi}\psi, \end{aligned} \tag{8.19}$$

because $C_{\mu\alpha\beta}C^{\mu\alpha\beta} = 2C_{\mu\alpha\beta}C^{\mu\beta\alpha}$. The different relative sign of the two quadratic gauge curvatures in the Lagrangian (8.17) leads to a simple linearized equation which is mathematically the same as that in Einstein gravity, as we shall see below.

The action (8.16) is a functional rather than a scalar function. Let us show explicitly that the action functional (8.16) of Yang–Mills gravity is invariant under the T_4 gauge transformations (8.3)–(8.6) in taiji symmetry framework. Based on the translational gauge transformations, we have

$$L_{\phi\psi} \to (L_{\phi\psi})^{\$} = L_{\phi\psi} - \Lambda^\lambda(\partial_\lambda L_{\phi\psi}). \tag{8.20}$$

Since Λ^μ is an infinitesimal gauge vector function, the gauge transformation of $P_{\mu\nu}$ can be written in the form,

$$\begin{aligned} (P_{\mu\nu})^{\$} &= P_{\mu\nu} - \Lambda^\lambda\partial_\lambda P_{\mu\nu} - P_{\mu\beta}\partial_\nu\Lambda^\beta - P_{\alpha\nu}\partial_\mu\Lambda^\alpha \\ &= [(1 - \Lambda^\sigma\partial_\sigma)P_{\alpha\beta}](\delta^\alpha_\mu - \partial_\mu\Lambda^\alpha)(\delta^\beta_\nu - \partial_\nu\Lambda^\beta). \end{aligned} \tag{8.21}$$

It follows from (8.21) that

$$\sqrt{-P} \to \sqrt{-P^\$} = [(1 - \Lambda^\sigma \partial_\sigma)\sqrt{-P}](1 - \partial_\lambda \Lambda^\lambda)$$
$$= \sqrt{-P} - \Lambda^\sigma \partial_\sigma \sqrt{-P} - (\partial_\lambda \Lambda^\lambda)\sqrt{-P}, \qquad (8.22)$$

where $P = \det P_{\mu\nu}$ and we have used $\det(AB) = \det A \det B$ and

$$\sqrt{(1 - \Lambda^\sigma \partial_\sigma)(-P)} = [(-P) - 2\sqrt{-P}\Lambda^\lambda \partial_\lambda \sqrt{-P}]^{1/2}$$

$$= \sqrt{-P}\left[\left(1 - \frac{\Lambda^\lambda \partial_\lambda \sqrt{-P}}{\sqrt{-P}}\right)^2\right]^{1/2}$$

$$= \sqrt{-P}\left(1 - \frac{\Lambda^\lambda \partial_\lambda \sqrt{-P}}{\sqrt{-P}}\right),$$

for infinitesimal Λ^λ. Thus, the Lagrangian $\sqrt{-P}L_{\phi\psi}$ changes only by a divergence under the gauge transformation and, hence, $S_{\phi\psi}$ is invariant:

$$S_{\phi\psi} = \int \sqrt{-P}L_{\phi\psi}\, d^4x \to S^\$_{\phi\psi} \equiv \int (\sqrt{-P}L_{\phi\psi}\, d^4x)^\$$$

$$= \int [\sqrt{-P}L_{\phi\psi} - \partial_\lambda(\Lambda^\lambda L_{\phi\psi}\sqrt{-P})]d^4x = S_{\phi\psi}. \qquad (8.23)$$

(See Appendix A.) The divergence term in (8.23) does not contribute to field equations because one can transform an integral over a four-dimensional volume into the integral of a vector over a hypersurface on the boundaries of the volume of integration where fields and their variations vanish. Thus, we have shown that the action functional $S_{\phi\psi}$ is invariant under the T_4 gauge transformations (8.4), although the Lagrangian (8.17) by itself is not gauge invariant. The gauge invariant action is sufficient to ensure that the wave equations in Yang–Mills gravity are T_4 gauge invariant.

8.4. The Gravitational Field Equation and Fermion Equations in General Frames of Reference

In general, gauge field equations with gauge symmetry are not well defined. One usually includes a suitable gauge-fixing term in the Lagrangian to make the solutions of the gauge field equation well-defined. Yang–Mills gravity in a general frame is based on the total Lagrangian $L_{\text{tot}}\sqrt{-P}$, which is the original Lagrangian

$L_{\phi\psi}$ with an additional gauge-fixing term L_{gf} involving the gauge parameter ξ

$$L_{tot}\sqrt{-P} = (L_{\phi\psi} + L_{gf})\sqrt{-P}, \tag{8.24}$$

$$L_{gf} = \frac{\xi}{2g^2}[\partial_\mu J_\alpha^\mu - \frac{1}{2}\partial_\alpha J_\mu^\mu][\partial_\lambda J_\beta^\lambda - \frac{1}{2}\partial_\beta J_\lambda^\lambda]P^{\alpha\beta}. \tag{8.25}$$

Because the gauge-fixing terms are, by definition, not gauge invariant, they involve the ordinary derivatives ∂_μ rather than the covariant derivative D_μ.[b] Here, the gauge-fixing term corresponds to the usual gauge condition $\eta^{\mu\nu}(\partial_\mu J_{\nu\alpha} - \frac{1}{2}\partial_\alpha J_{\mu\nu}) = 0$ for tensor fields. The Poincaré metric tensor $P_{\mu\nu}$ in (8.25) indicates that when one uses spherical coordinates to solve for static solutions, it denotes the metric tensor of the spherical coordinates and that when one uses inertial frames to carry out quantization, it is the Minkowski metric tensor.

The Lagrange equations for the gravitational tensor field $\phi^{\mu\nu}$ in a general frame can be derived from the action $\int L_{tot}\sqrt{-P}d^4x$. We obtain

$$H^{\mu\nu} + \xi A^{\mu\nu} = g^2 S^{\mu\nu}, \tag{8.26}$$

$$H^{\mu\nu} = D_\lambda(J_\rho^\lambda C^{\rho\mu\nu} - J_\alpha^\lambda C^{\alpha\beta}{}_\beta P^{\mu\nu} + C^{\mu\beta}{}_\beta J^{\nu\lambda})$$
$$\quad - C^{\mu\alpha\beta}D^\nu J_{\alpha\beta} + C^{\mu\beta}{}_\beta D^\nu J_\alpha^\alpha - C^{\lambda\beta}{}_\beta D^\nu J_\lambda^\mu, \tag{8.27}$$

$$A^{\mu\nu} = \partial_\alpha \left[\left(\partial_\beta J_\rho{}^\beta - \frac{1}{2}\partial_\rho J_\lambda^\lambda\right) P^{\mu\alpha}P^{\nu\rho} \right]$$
$$\quad + \frac{1}{2}P^{\lambda\rho}(\partial_\alpha P_{\lambda\rho})P^{\mu\alpha}P^{\nu\rho}\left(\partial_\beta J_\rho{}^\beta - \frac{1}{2}\partial_\rho J_\gamma^\gamma\right)$$
$$\quad - \frac{1}{2}\partial_\sigma \left[\left(\partial_\beta J_\rho{}^\beta - \frac{1}{2}\partial_\rho J_\lambda^\lambda\right) P^{\mu\nu}P^{\sigma\rho} \right]$$
$$\quad - \frac{1}{2}P^{\lambda\rho}(\partial_\sigma P_{\lambda\rho})P^{\mu\nu}P^{\sigma\rho}\left(\partial_\beta J_\rho{}^\beta - \frac{1}{2}\partial_\rho J_\gamma^\gamma\right), \tag{8.28}$$

where μ and ν should be made symmetric. We have used the identities in (8.14) for the derivation of (8.26). The "source tensor" $S^{\mu\nu}$ of fermion matter is given by

$$S^{\mu\nu} = \frac{1}{2}\left[\overline{\psi}i\Gamma^\mu D^\nu\psi - i(D^\nu\overline{\psi})\Gamma^\mu\psi\right]. \tag{8.29}$$

---

[b]If one replaces all ∂_μ by D_μ in (8.25), then the static solutions of the field equation (8.26) do not exist beyond the second-order approximation. This property was also verified by symbolic computing. We would like to thank D. W. Yang and J. Westgate for their help.

The Dirac equation for a fermion interacting with the tensor fields $\phi^{\mu\nu}$ in a general frame can also be derived from (8.24):

$$i\Gamma_\mu(P^{\mu\nu} + g\phi^{\mu\nu})D_\nu\psi - m\psi + \frac{i}{2}[D_\nu(J^{\mu\nu}\Gamma_\mu)]\psi = 0,$$

$$i(P^{\mu\nu} + g\phi^{\mu\nu})(D_\nu\overline{\psi})\Gamma_\mu + m\overline{\psi} + \frac{i}{2}\overline{\psi}[D_\nu(J^{\mu\nu}\Gamma_\mu)] = 0,$$

(8.30)

and the partial covariant derivative D_λ associated with the Poincaré metric tensor $P_{\mu\nu}$ is given by

$$D_\lambda J^{\mu\nu} = \partial_\lambda J^{\mu\nu} + \Gamma^\mu_{\lambda\rho}J^{\rho\nu} + \Gamma^\nu_{\lambda\rho}J^{\mu\rho}, \quad \text{etc.,}$$

$$\Gamma^\mu_{\lambda\rho} = \frac{1}{2}P^{\mu\sigma}(\partial_\lambda P_{\sigma\rho} + \partial_\rho P_{\sigma\lambda} - \partial_\sigma P_{\lambda\rho}),$$

(8.31)

where $\Gamma^\mu_{\lambda\rho}$ is the Christoffel symbol and we have used the relation $(1/\sqrt{-P})\partial_\nu[Q^\nu\sqrt{-P}] = D_\nu Q^\nu$.

If one compares the fermion equation (8.30) with the Dirac equation in quantum electrodynamics (i.e., $(i\gamma^\mu\partial_\mu - e\gamma^\mu A_\mu - m)\psi = 0$) in inertial frames, one can see a distinct difference: Namely, if one takes the complex conjugate of the Dirac equations, the kinematical term $i\gamma^\mu\partial_\mu$ and the electromagnetic coupling term $e\gamma^\mu A_\mu$ have different signs. This implies the presence of both repulsive and attractive forces between two charges. However, if one takes the complex conjugate of the fermion equation for ψ in (8.30), the kinematical term and the spin-2 coupling term have the same signs. Thus, the translation gauge symmetry of gravity naturally explains why the gravitational interaction is always attractive between fermion matter and antifermion matter. The physical reason for the electric repulsion and the gravitational attraction between, say, electrons will be explained in terms of the exchange of virtual gauge bosons in Section 14.1. The different nature of these forces originates from the property that the electromagnetic potential transforms like a vector, while the gravitational potential $\phi_{\mu\nu}$ is a tensor.

In inertial frames with $P_{\mu\nu} = \eta_{\mu\nu}$, the gauge-field equation (8.27) can be linearized for weak fields as follows:

$$\partial_\lambda\partial^\lambda\phi^{\mu\nu} - \partial^\mu\partial_\lambda\phi^{\lambda\nu} - \eta^{\mu\nu}\partial_\lambda\partial^\lambda\phi + \eta^{\mu\nu}\partial_\alpha\partial_\beta\phi^{\alpha\beta}$$

$$+ \partial^\mu\partial^\nu\phi - \partial^\nu\partial_\lambda\phi^{\lambda\mu} - gS^{\mu\nu} = 0, \quad \phi = \phi^\lambda_\lambda.$$

(8.32)

This equation can also be written in the form:

$$\partial_\lambda \partial^\lambda \phi^{\mu\nu} - \partial^\mu \partial_\lambda \phi^{\lambda\nu} + \partial^\mu \partial^\nu \phi^\lambda_\lambda - \partial^\nu \partial_\lambda \phi^{\lambda\mu} = g\left(S^{\mu\nu} - \frac{1}{2}\eta^{\mu\nu}S^\lambda_\lambda\right), \qquad (8.33)$$

where we have set $\xi = 0$ and used $J^{\mu\nu} = \eta^{\mu\nu} + g\phi^{\mu\nu}$. It is interesting that the linearized gauge-field equation (8.32) is formally the same as the corresponding equation in general relativity, which may be related to the fact that the transformation (8.2) is formally the same as that in general relativity.

8.5. Derivations of the T_4 Eikonal and Einstein–Grossmann Equations in the Geometric-Optics Limit

In Yang–Mills gravity, the fundamental field equation in the geometric optics limit can be derived as follows: We first postulate the translation gauge invariant Lagrangian L_{em} for the electromagnetic potential A^μ,

$$L_{em} = -\frac{1}{4}P^{\mu\alpha}P^{\nu\beta}F_{\mu\nu}F_{\alpha\beta},$$

$$F_{\mu\nu} = \Delta_\mu A_\nu - \Delta_\nu A_\mu, \quad \Delta_\mu = J_{\mu\nu}D^\nu, \qquad (8.34)$$

where we have used the same replacement as in (8.7). For simplicity, let us consider an inertial frame with $P^{\mu\nu} = \eta^{\mu\nu}$, $\Delta_\mu = J_{\mu\lambda}\partial^\lambda$, and the expression for field A^μ in the geometric-optics limit,

$$A^\mu = a^\mu \exp(i\Psi),$$

where the eikonal Ψ and the wave vector $\partial_\mu \Psi$ are very large.[8]

The generalized electromagnetic wave equations in the geometric-optics limit are

$$0 = \Delta_\mu(\Delta^\mu A^\beta - \Delta^\beta A^\mu) + (\partial_\alpha J^\alpha_\mu)(\Delta^\mu A^\beta - \Delta^\beta A^\mu) \equiv Z^\beta_\mu a^\mu,$$

$$Z^\beta_\mu = \delta^\beta_\mu G^{\alpha\sigma}\partial_\alpha \Psi \partial_\sigma \Psi, \quad G^{\alpha\sigma} = J^\alpha_\lambda J^{\lambda\sigma} \quad \text{(without } \partial_\mu A^\mu = 0), \qquad (8.35)$$

$$Z^\beta_\mu = \delta^\beta_\mu G^{\alpha\sigma}\partial_\alpha \Psi \partial_\sigma \Psi - gJ^{\beta\sigma}\phi^\alpha_\mu \partial_\alpha \Psi \partial_\sigma \Psi \quad \text{(with } \partial_\mu A^\mu = 0),$$

where the terms involving $(\partial_\alpha J^\alpha_\mu)$ are small and the terms involving $\partial_\alpha \partial_\sigma A^\mu$ or $\partial_\alpha \Psi \partial_\sigma \Psi$ are large. If the gauge condition $\partial_\mu A^\mu = 0$ is imposed, we have

$$-\Delta_\mu \Delta^\beta \partial_\nu A^\mu = -J^\sigma_\mu \partial_\sigma (J^{\beta\alpha}\partial_\alpha A^\mu) = -(\delta^\sigma_\mu + g\phi^\sigma_\mu)\partial_\sigma(J^{\beta\alpha}\partial_\alpha A^\mu)$$

$$\approx -g\phi^\sigma_\mu J^{\beta\alpha}\partial_\alpha \partial_\mu A^\mu - g\phi^\sigma_\mu J^{\beta\alpha}\partial_\sigma \partial_\alpha A^\mu = -g\phi^\sigma_\mu J^{\beta\alpha}\partial_\sigma \partial_\alpha A^\mu.$$

Since we are interested in the law for the propagation of light rays and the further simplification of (8.35), we have expressed the amplitude a^μ in terms of the space-like polarization vector $\epsilon^\mu(\lambda)$, i.e., $a^\mu = \epsilon^\mu(\lambda)b(x)$, $b(x) \neq 0$ in the limiting expression for A^μ. As usual, we have $\sum_\lambda \epsilon^\mu(\lambda)\epsilon^\nu(\lambda) \to -\eta^{\mu\nu}$ by summing over all polarizations.[9] Multiplying $Z_\mu^\beta a^\mu$ in (8.35) by $a^\nu \eta_{\nu\beta}$ and summing over all polarizations, we obtain $1/b^2(x) \sum_\lambda Z_\mu^\beta a^\mu a^\nu \eta_{\nu\beta} = -\delta_\beta^\mu Z_\mu^\beta = 0$. After some calculations, we obtain new eikonal equations

$$G^{\mu\nu}\partial_\mu\Psi\partial_\nu\Psi = 0, \quad G^{\alpha\sigma} = J_\lambda^\alpha J^{\lambda\sigma} \quad \text{(without } \partial_\mu A^\mu = 0),$$

$$G_L^{\mu\nu}\partial_\mu\Psi\partial_\nu\Psi = 0, \quad G_L^{\mu\nu} = G^{\mu\nu} - \frac{g}{4}\phi_\lambda^\mu J^{\lambda\nu} \quad \text{(with } \partial_\mu A^\mu = 0). \tag{8.36}$$

We distinguish between the two different effective metric tensors $G^{\mu\nu}$ and $G_L^{\mu\nu}$ in (8.36) because the electromagnetic Lagrangian (8.34) is not U_1 gauge invariant and this non-invariance can be tested experimentally. (See discussions in Chapter 9, Section 9.3 and in Chapter 14, Section 14.4.)

It must be stressed that these simple effective metric tensors in eikonal equations (8.36) emerge only in the geometric-optics limit. Otherwise, equation (8.35) has a very complicated dependence on the wave vector $\partial_\mu \Psi$ and, hence, it is difficult to identify the effective Riemannian metric tensors (and the corresponding effective interval ds_{ei} in (8.41) below) in general.

Next, let us consider the relationship between the classical equation of motion and the massive fermion wave equation. In a general frame with Poincaré metric tensors $P_{\mu\nu}$, the fermion wave equation is given by (8.30),

$$i\Gamma_\mu\Delta^\mu\psi - m\psi + \frac{i}{2}\gamma_a[D_\nu(J^{\mu\nu}e_\mu^a)]\psi = 0. \tag{8.37}$$

Using the limiting expression for the field $\psi = \psi_o \exp(iS)$, we obtain

$$\gamma_a E^{a\mu}\partial_\mu S + m - \frac{i}{2}\gamma_a[D_\nu(J^{\mu\nu}e_\mu^a)] = 0. \tag{8.38}$$

In the classical limit, the momentum $\partial_\mu S$ and mass m are large quantities, and one can neglect the small gravitational interaction term involving e_μ^a. To eliminate the spin variables, we multiply the large terms in (8.38) by a factor of $(\gamma_a E^{a\mu}\partial_\mu S - m)$

so that the resultant equation can be written in the form

$$\frac{1}{2}(\gamma_b \gamma_a + \gamma_a \gamma_b) E^{a\mu} E^{b\nu} (\partial_\mu S)(\partial_\nu S) - m^2 = 0. \tag{8.39}$$

With the help of the anticommutation relation for γ_a in (8.9) and the effective metric tensor (8.11), we can then simplify (8.39), obtaining the following equation for the motion of a classical object in the presence of the gravitational tensor field $\phi^{\mu\nu}$ in flat space-time,

$$G^{\mu\nu}(\partial_\mu S)(\partial_\nu S) - m^2 = 0, \quad G^{\mu\nu} = P_{\alpha\beta} J^{\alpha\mu} J^{\beta\nu}. \tag{8.40}$$

This equation is mathematically the same as the corresponding Hamilton–Jacobi equation in general relativity. Within the framework of Yang–Mills gravity in flat space-time, we shall call (8.40) the Einstein–Grossmann equation of motion.[c] It appears that the classical equation of motion must have the form of (8.40) (or its equivalent form in terms of an "effective geodesic equation" such as (8.44) below) in order to accurately describe the motion of macroscopic objects in a gravitational field, whether the theory is based on a curved or a flat space-time.

    The effective metric tensor $G_L^{\mu\nu}$ in (8.36) with the gauge condition $\partial_\mu A^\mu = 0$ differs from $G^{\mu\nu}$, which is obtained without the requirement of $\partial_\mu A^\mu = 0$. This is due to the violation of the electromagnetic U_1 gauge symmetry in Yang–Mills gravity. The Lagrangian L_{em} and $F_{\mu\nu} = \Delta_\mu A_\nu - \Delta_\nu A_\mu$ in (8.34) are not U_1 gauge invariant because

$$A_\mu \to A'_\mu(x) = A_\mu(x) + \partial_\mu \Lambda(x) \quad \text{and} \quad F_{\mu\nu}(x) \to F'_{\mu\nu}(x) \neq F_{\mu\nu}(x),$$

regardless of how $\phi_{\mu\nu}(x)$ transforms, whether $\phi'_{\mu\nu}(x) = \phi_{\mu\nu}(x) + \partial_\mu \partial_\nu \Lambda(x)$ or $\phi'_{\mu\nu}(x) = \phi_{\mu\nu}(x)$. The appearance of the T_4 gauge covariant derivative $\Delta_\mu = \partial_\mu + g\phi_{\mu\lambda}\partial^\lambda$ (with $P^{\mu\nu} = \eta^{\mu\nu}$) in the electromagnetic Lagrangian L_{em} in (8.34) is due to the universal coupling of the gravitational field to all fields in nature. This universal coupling of the T_4 gravitational gauge field will cause small violations of all internal gauge symmetries, such as U_1 and color SU_3. Such a gravitational violation of the electromagnetic U_1 symmetry can be tested by measurements of the deflection of light. (See discussions in Chapters 9, 14 and 15.)

---

[c] Marcel Grossmann was associated with Einstein in elucidating the mathematical foundation of general relativity.

8.6. Effective Action for Classical Objects and Gauge Invariance

All the observable effects of gravity are directly related to the motion of classical objects and light rays. Thus, it is important to understand the relation between the wave equations of fields and the corresponding classical equations of particles (or objects) and light rays. In Yang–Mills gravity, we have seen that the eikonal equation (8.36) for a light ray and the Einstein–Grossmann equation of motion (8.40) for classical objects are the wave equations in the geometric-optics limit.

We now demonstrate that if one postulates the following effective action S_p for classical particles in general frames with Poincaré metric tensors $P_{\mu\nu}$,

$$S_p = -\int m \, ds_{ei}, \tag{8.41}$$

$$ds_{ei}^2 = I_{\mu\nu} dx^\mu dx^\nu, \quad I_{\mu\nu} G^{\nu\alpha} = \delta^\alpha_\mu,$$

$$G^{\mu\nu} = P_{\alpha\beta} J^{\alpha\mu} J^{\beta\nu} = P^{\mu\nu} + 2g\phi^{\mu\nu} + g^2 \phi^{\mu\lambda} \phi^{\nu\sigma} P_{\lambda\sigma},$$

one can derive the Einstein–Grossmann equation (8.40). We note that the action S_p (8.41) is not invariant under the gauge transformations (8.5) and (8.6). However, this is not surprising or problematic since S_p is only an effective action for classical particles (in the geometric optics limit) rather than the gauge invariant action (8.16) for basic tensor and fermion fields. However, one can show that the effective interval ds_{ei}^2 and hence, the action S_p are invariant under the following tensor transformations

$$Q'^{\mu_1...\mu_m}_{\alpha_1...\alpha_n}(x') = Q^{\nu_1...\nu_m}_{\beta_1...\beta_n}(x) \frac{\partial x'^{\mu_1}}{\partial x^{\nu_1}} \cdots \frac{\partial x'^{\mu_m}}{\partial x^{\nu_m}} \frac{\partial x^{\beta_1}}{\partial x'^{\alpha_1}} \cdots \frac{\partial x^{\beta_n}}{\partial x'^{\alpha_n}}. \tag{8.42}$$

If we consider the variation of the action (8.41), we have

$$\delta S_p = -m \int \left[I_{\lambda\gamma} \frac{d}{ds} \frac{dx^\lambda}{ds} + \Gamma_{\gamma,\alpha\beta} \frac{dx^\alpha}{ds} \frac{dx^\beta}{ds} \right] \delta x^\gamma \, ds - m I_{\mu\nu} \frac{dx^\mu}{ds} \delta x^\nu |, \tag{8.43}$$

$$\Gamma_{\gamma,\alpha\beta} = \tfrac{1}{2}(\partial_\beta I_{\alpha\gamma} + \partial_\alpha I_{\beta\gamma} - \partial_\gamma I_{\alpha\beta}),$$

where $ds = ds_{ei}$. Suppose we consider only the actual path with one of its endpoints left unfixed,[8] i.e.,

$$I_{\lambda\gamma} \frac{d}{ds} \frac{dx^\lambda}{ds} + \Gamma_{\gamma,\alpha\beta} \frac{dx^\alpha}{ds} \frac{dx^\beta}{ds} = 0, \quad ds = ds_{ei}. \tag{8.44}$$

We then have

$$\delta S_p = -m I_{\mu\nu} \left(\frac{dx^\mu}{ds_{ei}} \right) \delta x^\nu. \tag{8.45}$$

The momentum of a classical particle is defined by

$$p_\nu = -\frac{\partial S_p}{\partial x^\nu} = -m\frac{dx^\mu}{ds_{ei}}I_{\mu\nu}, \quad I_{\mu\nu}G^{\nu\alpha} = \delta^\alpha_\mu. \tag{8.46}$$

Since $ds^2_{ei} = I_{\mu\nu}dx^\mu dx^\nu$, we have

$$G^{\mu\nu}p_\mu p_\nu - m^2 = 0, \quad I_{\mu\nu}G^{\nu\lambda} = \delta^\lambda_\mu. \tag{8.47}$$

Equations (8.46) and (8.47) lead to the Einstein–Grossmann equation for a classical particle with mass m,

$$G^{\mu\nu}(\partial_\mu S)(\partial_\nu S) - m^2 = 0, \quad S \equiv S_p. \tag{8.48}$$

This equation is the same as (8.40), which is obtained from the Dirac wave equation (8.37) in the presence of gravity and in the geometric-optics limit. Equations (8.47) and (8.48) are obtained in Yang–Mills gravity based on flat space-time. They are formally the same as the corresponding equations in general relativity. Thus, in the presence of gravity, classical objects in Yang–Mills gravity appear to move in a space-time with an effective metric tensor $G^{\mu\nu}$.

Since the effective action (8.41) and the effective metric tensor $G^{\mu\nu}$ are only approximate quantities in Yang–Mills gravity, it is natural for one to expect that the observable results in the classical limit may not be gauge invariant to all orders (see also discussions at the end of Section 9.1).

8.7.   Classical Objects as a Coherent Collection of Constituent Particles

Equations (8.38)–(8.40) show the effect on the field equations of going to the classical limit. This limiting process is accomplished by making quantities representing the mass and momentum large (taking the limits $m \to \infty$ and $\hbar \to 0$). (In unit systems other than natural units, the phase iS_k in $\psi = \psi_o \exp(iS)$ should be replaced by $iS_k/\hbar$.) However, in the real world, the large mass and large eikonal S of classical objects have their origin in the summation of m_k and S_k over all of the large number of quantum particles k in a macroscopic system, rather than in the existence of any single particle with a large mass and momentum. In this sense, the conventional steps (8.38)–(8.40) do not directly reveal the physical connection between a macroscopic body with a large mass m and the actual physical situation of a macroscopic body consisting of a large number N of constituent particles with masses $m_k, k = 1, 2, 3, \ldots, N,$

$$m_t = \sum_{k=1}^{N} m_k. \tag{8.49}$$

To explore the consequences of applying a more realistic process of taking the classical limit to equations (8.38)–(8.40), we begin with the relation for each particle k,

$$[\gamma_\mu(\eta^{\mu\nu} + g\phi^{\mu\nu})\partial_\nu S]_{(k)} + m_{(k)} = 0, \quad k = 1, 2, \ldots, N. \tag{8.50}$$

After summing over all particles, the expression (8.38) with m replaced by m_t from (8.49) is

$$\gamma_\mu\{\eta^{\mu\nu} + g\phi^{\mu\nu}(x)\}\partial_\nu S(x) + m_t = 0. \tag{8.51}$$

A further natural requirements for the classical limit is that the momentum of particles in a macroscopic object must add up "coherently" in the following sense:

$$\sum_{k=1}^{N}[\partial_\nu S]_{(k)} = \partial_\nu S(x), \quad \sum_{k=1}^{N}[\phi^{\mu\nu}\partial_\nu S]_{(k)} = \phi^{\mu\nu}(x)\partial_\nu S(x). \tag{8.52}$$

Equations (8.51) and (8.52) are reasonable because a classical macroscopic object is, by construction, made of N bounded particles. Although the summation of constant masses in (8.49) is simple, the summation of space-time-dependent momenta in the first equation of (8.52) within the relativistic framework is not, because each particle has its own time and space coordinates in an inertial frame. However, one simplification that might be used to make this equation in (8.52) tractable is to assume that the relative motions of the quantum particles making up a single macroscopic object can be ignored in the classical limit, viewing them as being at rest relative to each other and, hence, describable using a single set of space-time coordinates. As can be seen, the passage from a system of bounded quantum particles to a single macroscopic classical object is not as simple as one might think.

Once one has the result (8.51), one can ignore the spin effect by removing γ_μ from the equation, as done in (8.38)–(8.40). In this way, one obtains the relativistic Einstein–Grossmann equation (8.40) with an effective Riemannian metric tensor $G^{\mu\nu}$, whose presence in the classical limit is crucial for Yang–Mills gravity to be consistent with experiments.

8.8.   Remarks on Yang–Mills Gravity and Teleparallel Gravity

At a first glimpse, the Lagrangian for the electromagnetic field in the presence of a gravitational field $\phi_{\mu\nu}$ appears to be quite different from the Lagrangian (8.17) of the Dirac field. However, if one follows the Yang–Mills approach with the T_4 gauge group, i.e., replacing ∂^μ by the T_4 gauge covariant derivative $\partial^\mu + g\phi^{\mu\nu}\partial_\nu$, one obtains a relativistic Einstein–Grossmann equation (8.40) with $m = 0$ for light rays in the geometric optics limit.

Tensor fields with spin 2 were discussed by Fierz and Pauli in 1939.[10,11] In order to have two independent polarization states for a massless symmetric tensor field "$\psi_{\mu\nu}$," certain subsidiary conditions were imposed. In particular, they discussed tensor field equations which are invariant under a "gauge transformation,"

$$\psi_{\mu\nu} \to \psi_{\mu\nu} + \partial_\mu \Lambda_\nu + \partial_\nu \Lambda_\mu.$$

Without the guidance of modern gauge symmetry, it is non-trivial to find a Lagrangian for tensor fields in such a way that wave equations and subsidiary conditions follow simultaneously from the Hamilton principle.[11] It turns out that the simplest free Lagrangian is equivalent to the free Lagrangian in (8.17), and the free field equation for a massless $\psi_{\mu\nu}$ is exactly the same as the linearized equation (8.32) or (8.33) in Yang–Mills gravity and that in Einstein's theory of gravity.

It is generally believed that general coordinate invariance in Einstein's gravity can be considered a gauge symmetry. In particular, one can use the affine connection to construct covariant derivatives of tensors, just as one uses a gauge field to construct gauge covariant derivatives of matter fields. However, a fundamental difference between Einstein's gravity and Yang–Mills gravity is that Einstein's gravity is based on curved spacetime so that the commutator of two covariant derivatives with respect to x^μ and x^ν leads to the Riemann–Christoffel curvature tensor uniquely and unambiguously. Yang–Mills gravity, however, is based on the T_4 group in flat spacetime, just like the usual Yang–Mills theory and conventional field theories so that there is no Riemann–Christoffel curvature, but only a T_4 gauge curvature $C^{\mu\nu\lambda}$.

The fact that the action (8.16) in Yang–Mills theory involves a quadratic gauge curvature $C_{\mu\nu\alpha}$, while the Hilbert–Einstein action involves a linear curvature of space-time implies a basic and important break down of the analogy between the action (8.16) and the Hilbert–Einstein action.[12] Namely, gauge fields are fundamental and cannot expressed in terms of any more fundamental fields, while the affine connection in Einstein's gravity is itself constructed from first derivatives of the metric tensor. There is no break down in the analogy between Yang–Mills gravity and the usual Yang–Mills theory, however.

There is a formulation of gravity called teleparallel gravity (TG)[13,14] that, like Yang–Mills gravity, is based on a translational gauge symmetry in a flat space-time, but unlike Yang–Mills gravity, has a torsion tensor. In TG, a flat connection with torsion makes translations local gauge symmetries and the curvature tensor is the torsion field. Furthermore, the concept of translational gauge symmetry is realized differently in Yang–Mills gravity and in TG. Although there is a one-to-one correspondence between equations in Yang–Mills gravity and TG, there are conceptual differences between them and the differences show up in the properties of the gauge potentials, gauge curvature and actions as follows.

(a) Gauge covariant derivative.
In Yang–Mills gravity, the T_4 gauge covariant derivative $\Delta_\mu \psi$ is defined in (8.14), where the gauge potential field $\phi_{\mu\nu}$ and $J_{\mu\nu}$ are symmetric tensors in flat space-time. In contrast, the gauge covariant derivative in TG is defined by[13,14]

$$D_\mu \psi = \partial_\mu \psi + B_\mu^a \partial_a \psi = h_\mu^a \partial_a \psi, \qquad h_\mu^a = \partial_\mu x^a + B_\mu^a, \qquad (8.53)$$

where x^μ are the space-time coordinates with the metric tensor $g_{\mu\nu}(x) = \eta_{ab} h_\mu^a(x) h_\nu^b(x)$, and $x^a(x^\mu)$ are the coordinates in a tangent Minkowski space-time with the metric tensor $\eta_{ab} = (+, -, -, -)$. Thus, h_μ^a in TG is treated as a tetrad rather than a tensor.

(b) Gauge curvature.
In Yang–Mills gravity, the gauge curvature $C^{\mu\nu\alpha}$ is a tensor of the third rank and is defined by

$$[\Delta_\mu, \Delta_\nu] = C_{\mu\nu\lambda} \partial^\lambda, \qquad C^{\mu\nu\lambda} = J^{\mu\sigma} \partial_\sigma J^{\nu\lambda} - J^{\nu\sigma} \partial_\sigma J^{\mu\lambda},$$
$$J_{\mu\nu} = \eta_{\mu\nu} + g\phi_{\mu\nu} = J_{\nu\mu}. \qquad (8.54)$$

In contrast, the gauge curvature in TG is a torsion tensor $T_{\mu\nu}^a$:

$$[D_\mu, D_\nu] = T_{\mu\nu}^a \partial_a, \qquad T_{\mu\nu}^a = \partial_\mu h_\nu^a - \partial_\nu h_\mu^a. \qquad (8.55)$$

(c) Gravitational action.

A real physical difference between Yang–Mills gravity and TG shows up in the gravitational actions.[16,d,e] The gravitational Lagrangian L_ϕ in Yang–Mills gravity is given in equations (8.16) and (8.17). In TG, the gravitational action and Lagrangian L_h are given by Hayashi and Shirafuji,[13,14]

$$S_{TG} = \int L_h \sqrt{-det\, g_{\mu\nu}}\; d^4x, \qquad (8.56)$$

$$L_h = a_1(t_{\lambda\mu\nu}t^{\lambda\mu\nu}) + a_2(v^\mu v_\mu) + a_3(a^\mu a_\mu), \qquad (8.57)$$

$$t_{\lambda\mu\nu} = \frac{1}{2}(T_{\lambda\mu\nu} + T_{\mu\lambda\nu}) + \frac{1}{6}(g_{\nu\lambda}v_\mu + g_{\nu\mu}v_\lambda) - \frac{1}{3}g_{\lambda\mu}v_\nu, \qquad (8.58)$$

$$v_\mu = T^\lambda_{\;\lambda\mu}, \qquad a_\mu = \frac{1}{6}\epsilon_{\mu\nu\rho\sigma}T^{\nu\rho\sigma}, \qquad (8.59)$$

$$T^\lambda_{\;\mu\nu} = h^\lambda_a(\partial_\nu h^a_\mu - \partial_\mu h^a_\nu), \qquad g_{\mu\nu}(x) = \eta_{ab}h^a_\mu(x)h^b_\nu(x), \qquad (8.60)$$

where $T^\lambda_{\;\mu\nu}$ is the torsion tensor. It appears to be impossible to adjust the three parameters a_1, a_2 and a_3 in (8.57) to make S_{TG} in (8.56) equal to the $S = \int L_\phi d^4x$ in Yang–Mills gravity for both inertial and non-inertial frames. In general, if the tetrad in a field theory involves dynamical fields, it leads to complicated vertices in the Feynman rules due to the presence of $\sqrt{-det\, g_{\mu\nu}}$ in the action, if the field can be quantized.

Yang–Mills gravity is a local gauge field theory for the microscopic world in flat space-time and hence, is not related to the equivalence principle. However, in the classical limit (i.e., geometric-optics limit), the fermion wave equation reduces to the relativistic Einstein–Grossmann equation (8.40), which describes the free-fall motion of a classical object in inertial frames. In this sense, Yang–Mills gravity is approximately compatible with the equivalence principle in the geometric-optics limit.

---

[d]A. A. Logunov, M. A. Mestvirishvili and N. Wu discussed a theory of gravity based on flat space-time. However, in their formulation, the graviton still involved an N-vertex of self-coupling, and hence, their formulation is more complicated than the maximum 4-vertex for the self-coupling of gravitons in Yang–Mills gravity (see Refs. 1 and 15). All these formulations of gravity are different from Yang–Mills gravity.
[e]D. Fine has pointed out that, from the viewpoint of the geometric picture, Yang–Mills gravity and teleparallel gravity are very similar.

References

1. A. A. Logunov, *The Theory of Gravity*, Trans. by G. Pontecorvo (Moscow, Nauka, 2001).
2. J.-P. Hsu, *Phys. Lett. B* **119**, 328 (1982).
3. J. P. Hsu and L. Hsu, in *A Broader View of Relativity: General Implications of Lorentz and Poincaré Invariance* (World Scientific, 2006), Appendix D; J. P. Hsu, *Intl. J. Mod Phys. A* **21**, 5119 (2006).
4. R. Utiyama and T. Fukuyama, *Prog. Theor. Phys.* **45**, 612 (1971); Y. M. Cho, *Phys. Rev. D* **14**, 2521 (1976).
5. Y. M. Cho, *Phys. Rev. D* **14**, 3341 (1976).
6. K. Hayashi and T. Shirafuji, *Phys. Rev. D* **19**, 3524 (1979).
7. F. J. Dyson, in *100 Years of Gravity and Accelerated Frames: The Deepest Insights of Einstein and Yang–Mills*, eds. J. P. Hsu and D. Fine (World Scientific, 2005), p. 348.
8. L. Landau and E. Lifshitz, in *The Classical Theory of Fields*, Trans. by M. Hamermesh (Addison-Wesley, 1951), pp. 29–30, 136–137 and 268–270.
9. J. D. Bjorken and S. D. Drell, in *Relativistic Quantum Fields* (McGraw-Hill, 1965), p. 386.
10. M. Fierz, *Helv. Phys. Acta* **12**, 3 (1939).
11. M. Fierz and W. Pauli, *Proc. Roy. Soc.* **173**, 211 (1939).
12. S. Weinberg, *The Quantum Theory of Fields*, Vol. 2 (Cambridge University Press, 1996), pp. 6–7.
13. K. Hayashi and T. Shirafuji, *Phys. Rev. D* **19**, 3524 (1979).
14. V. C. de Andrade and J. G. Pereira, *Phys. Rev. D* **56**, 4689 (1997).
15. A. A. Logunov and M. A. Mestvirishvili, *Prog. Theor. Phys.* **74**, 31, (1985); N. Wu, *Commun. Theor. Phys. (Beijing, China)* **42**, 543 (2004).
16. D. Fine, Private correspondence (2012).

Chapter 9

Experimental Tests of Classical Yang–Mills Gravity

9.1. Motion in a Central Gravitational Field

In physics, the ultimate test of any theory is its ability to explain and predict experimental phenomena. In this chapter, we test the classical approximation of Yang–Mills gravity against the results of a number of well-established experiments, such as the perihelion shift of the planet Mercury, the deflection of light in a gravitational field, and the gravitational quadruple energy radiation from systems such as binary stars. In particular, we find that a more accurate measurement of the deflection of light in a gravitational field could provide a test of Yang–Mills gravity versus Einstein gravity (see Section 9.3).

We begin by considering the motion of a classical object in a central gravitational force field. Before we can use the Einstein–Grossmann equation of motion (8.40) to determine the path of an object, however, we must first calculate the effective metric tensors $G^{\mu\nu} = P_{\alpha\beta}J^{\alpha\mu}J^{\beta\nu}$ in (8.40). Thus, we first solve the field equation (8.26) to obtain the static potential $\phi_{\mu\nu}$ and hence, $J_{\mu\nu}$.

The only non-vanishing component $S^{00} = m\delta^3(\mathbf{r})$ from equations (8.32) and (8.33) leads to the well-known result, $g\phi^{00} = g^2 m/(8\pi r)$, in the Newtonian limit. As usual, $G^{00} = 1 + 2Gm/r$ holds in this limit in the Einstein–Grossman equation (8.40), where G is the gravitational constant. Based on these results, together with $G^{\mu\nu}$ in (8.40) and $I_{\mu\nu}$ in (8.47), we have

$$g = \sqrt{8\pi G}, \quad \text{and} \quad g\phi^{00} = g\phi^{11} = Gm/r, \quad g\phi^{22} = Gm/r^3, \qquad (9.1)$$

to the first-order approximation in an inertial frame. These results can be obtained by solving the linearized field equation (8.33) in spherical coordinate, $x^\mu = (w, r, \theta, \phi)$.

Because the perihelion shift of Mercury is sensitive to the coefficient of the second-order term of G^{00} or I_{00} and the coefficient of the first-order term of

G^{11}, G^{22} and G^{33}, we shall also calculate the second-order terms of the components of the effective metric tensor $G^{\mu\nu}$. We solve the nonlinear gauge field equations by the method of successive approximation and carry out the related post-Newtonian approximation to the second order. For the gauge field equations to be well defined, it is convenient to use the gauge field equation (8.26) with the gauge parameter ξ.

For simplicity, we consider an inertial frame and a static, spherically symmetric system, in which the tensor gauge fields are produced by a spherical object at rest with mass m. Based on symmetry considerations,[1] the non-vanishing components of the exterior solutions $\phi^{\mu\nu}(r)$ are $\phi^{00}(r), \phi^{11}(r), \phi^{22}(r)$ and $\phi^{33}(r) = \phi^{22}/\sin^2\theta$, where $x^\mu = (w, r, \theta, \phi)$. To solve the static gauge field, we write

$$J^{00} = J_0^0 = S(r), \quad -J^{11} = J_1^1 = R(r),$$
$$-r^2 J^{22} = J_2^2 = -r^2 \sin^2\theta J_{33} = J_3^3 = T(r). \tag{9.2}$$

The metric tensor is given by $P_{\mu\nu} = (1, -1, -r^2, -r^2\sin^2\theta)$. In this coordinate system, the non-vanishing components of the Christoffel symbol $\Gamma^\alpha_{\mu\nu}$ are

$$\Gamma^1_{22} = -r, \quad \Gamma^1_{33} = -r\sin^2\theta, \quad \Gamma^2_{12} = 1/r,$$
$$\Gamma^2_{33} = -\sin\theta\cos\theta, \quad \Gamma^3_{13} = 1/r, \quad \Gamma^3_{23} = \cot\theta. \tag{9.3}$$

After some tedious but straightforward calculations (see Appendix C for more details), the gauge field equation (8.26) with $(\mu, \nu) = (0,0), (1,1), (2,2), (3,3)$ can be written as

$$\frac{d}{dr}\left(R^2\frac{dS}{dr}\right) + \frac{2}{r}R^2\frac{dS}{dr} - \left(R\frac{d}{dr} + \frac{dR}{dr} + \frac{2R}{r}\right)$$
$$\times\left(R\left[\frac{dS}{dr} + 2\frac{dT}{dr}\right] + \frac{2T^2}{r} - \frac{2TR}{r}\right) + \xi\left[\frac{1}{4}\frac{d^2}{dr^2}(S - R + 2T)\right]$$
$$+ \frac{\xi}{r}\left[\frac{d}{dr}(S - R + 2T)\right] = 0, \tag{9.4}$$

$$R\left(\frac{dS}{dr}\right)^2 + 2r^3\frac{d(T/r)}{dr}\left[\frac{R}{r}\frac{d(T/r)}{dr} + \frac{T^2}{r^3}\right] + \left(\frac{dS}{dr} + 2\frac{dT}{dr} - \frac{2T}{r}\right)$$
$$\times\left(-R\left[\frac{dS}{dr} + 2\frac{dT}{dr}\right] - \frac{2}{r}T^2 + \frac{2}{r}TR\right) + \xi\left[\frac{1}{4}\frac{d^2}{dr^2}(S - R + 2T)\right] = 0, \tag{9.5}$$

$$\left(R\frac{d}{dr} + \frac{dR}{dr} + \frac{5R}{r} - \frac{2T}{r} \right) \left[\frac{R}{r}\frac{d(T/r)}{dr} + \frac{T^2}{r^3} \right]$$

$$+ \left[\frac{1}{r^2}\left(R\frac{d}{dr} + \frac{dR}{dr} \right) + \frac{3R}{r^3} - \frac{2T}{r^3} \right]\left[-R\left(\frac{dS}{dr} + 2\frac{dT}{dr} \right) - \frac{2}{r}T^2 + \frac{2}{r}TR \right]$$

$$+ \xi\left[\frac{1}{4r^2}\frac{d^2}{dr^2}(S - R + 2T) \right] + \xi\left[\frac{1}{2r^3}\frac{d}{dr}(S - R + 2T) \right] = 0, \qquad (9.6)$$

respectively. The equation for $(\mu, \nu) = (3,3)$ is the same as that in (9.6).

We can solve the gauge field equations (9.4)–(9.6) to the second-order approximation by setting

$$S = 1 + a_1/r + a_2/r^2, \quad R = 1 + b_1/r + b_2/r^2, \quad T = 1 + c_1/r + c_2/r^2. \qquad (9.7)$$

We find

$$a_1 = Gm, \quad a_2 = G^2m^2/2, \quad b_1 = -Gm,$$

$$b_2 = G^2m^2\left(-\frac{5}{6} + \frac{2}{3\xi} \right), \quad c_1 = -Gm, \quad c_2 = -G^2m^2\left(\frac{2}{3} + \frac{2}{3\xi} \right). \qquad (9.8)$$

From (9.2), (9.7), (9.8) and $J^{\mu\nu} = P^{\mu\nu} + g\phi^{\mu\nu}$, we obtain

$$g\phi^{00} = S - 1 = \frac{Gm}{r} + \frac{G^2m^2}{2r^2}, \quad g\phi^{11} = 1 - R = \frac{Gm}{r} + \frac{G^2m^2}{r^2}\left(\frac{5}{6} - \frac{2}{3\xi} \right),$$

$$g\phi^{22} = \frac{1}{r^2}(1 - T) = \frac{1}{r^2}\left[\frac{Gm}{r} + \frac{G^2m^2}{r^2}\left(\frac{2}{3} + \frac{2}{3\xi} \right) \right], \quad g\phi^{33} = \frac{g\phi^{22}}{\sin^2\theta}. \qquad (9.9)$$

Thus, only the second-order terms in ϕ^{11}, ϕ^{22} and ϕ^{33} depend on the gauge parameter ξ. As expected, the first-order terms and the second-order terms in ϕ^{00}, i.e., all those crucial to the observable quantities in measuring the perihelion shift of Mercury, are independent of the gauge parameter ξ.

From the result (9.9) and the relations $G^{\mu\nu} = P_{\alpha\beta}J^{\alpha\mu}J^{\beta\nu}$ and $P_{\mu\nu} = (1, -1, -r^2, -r^2 \sin^2\theta)$, we can now write down the effective metric tensor,

$$G^{00}(r) = 1 + \frac{2Gm}{r} + \frac{2G^2m^2}{r^2},$$

$$G^{11}(r) = -\left[1 - \frac{2Gm}{r} + \frac{G^2m^2}{r^2}\left(\frac{-2}{3} + \frac{4}{3\xi}\right)\right],$$

$$G^{22}(r) = -\frac{1}{r^2}\left(1 - \frac{2Gm}{r} - \frac{G^2m^2}{r^2}\left(\frac{1}{3} + \frac{4}{3\xi}\right)\right), \quad G^{33}(r) = G^{22}(r)/\sin^2\theta.$$

$$(9.10)$$

These results are well defined in the limit $\xi \to \infty$ and this particular choice of gauge may be called the "static gravity gauge." In this gauge, the effective metric tensors are

$$G^{00}(r) = 1 + \frac{2Gm}{r} + \frac{2G^2m^2}{r^2}, \quad G^{11}(r) = -\left[1 - \frac{2Gm}{r} - \frac{2G^2m^2}{3}\right],$$

$$G^{22}(r) = -\frac{1}{r^2}\left(1 - \frac{2Gm}{r} - \frac{G^2m^2}{3r^2}\right), \quad G^{33}(r) = G^{22}(r)/\sin^2\theta.$$

$$(9.11)$$

In the calculation of the perihelion shift, we shall use the metric in its usual form in spherical coordinates $x^\mu = (w, \rho, \theta, \phi)$ with $G^{22}(\rho) = -1/\rho^2$. This can be accomplished by a change of variable

$$\rho^2 = r^2\left(1 - \frac{2Gm}{r} - \frac{G^2m^2}{r^2}\left[\frac{1}{3} + \frac{4}{3\xi}\right]\right)^{-1}.$$

$$(9.12)$$

From (9.12), we solve for r to obtain

$$r = \rho B, \quad B \equiv \left[1 - \frac{Gm}{\rho} - \frac{G^2m^2}{\rho^2}\left(\frac{5}{3} + \frac{2}{3\xi}\right)\right],$$

$$dr = d\rho\left[1 + \frac{G^2m^2}{\rho^2}\left(\frac{5}{3} + \frac{2}{3\xi}\right)\right].$$

$$(9.13)$$

The Einstein–Grossmann equation of motion (8.40) can be expressed in terms of ρ,

$$G^{00}(\rho)\frac{\partial S}{\partial w}\frac{\partial S}{\partial w} + G^{11}(\rho)\frac{\partial S}{\partial \rho}\frac{\partial S}{\partial \rho} + G^{22}(\rho)\frac{\partial S}{\partial \theta}\frac{\partial S}{\partial \theta} + G^{33}(\rho)\frac{\partial S}{\partial \phi}\frac{\partial S}{\partial \phi} - m^2 = 0,$$

$$G^{00}(\rho) = G^{00}(r)_{r=\rho B}, \quad G^{11}(\rho) = G^{11}(r)\left(\frac{d\rho}{dr}\right)^2_{r=\rho B},$$

$$(9.14)$$

$$G^{22}(\rho) = G^{22}(r)_{r=\rho B}, \quad G^{33}(\rho) = G^{33}(r)_{r=\rho B}.$$

In spherical coordinates $x^\mu = (w, \rho, \theta, \phi)$, the effective metric tensor $G^{\mu\nu}(\rho)$ in the Einstein–Grossmann equation (9.14) is

$$G^{00}(\rho) = 1 + 2Gm/\rho + 4G^2m^2/\rho^2,$$
$$G^{11}(\rho) = -\left[1 - 2Gm/\rho - 6G^2m^2/\rho^2\right], \qquad (9.15)$$
$$G^{22}(\rho) = -1/\rho^2, \quad G^{33}(\rho) = -1/(\rho^2 \sin^2\theta),$$

where we have used (9.13) and (9.14). The gauge parameter ξ in $G^{11}(r)$ from (9.10) and in $(d\rho/dr)^2$ from (9.13) cancel so that $G^{11}(\rho)$ in (9.15) is ξ-independent, in agreement with the gauge invariance of Yang–Mills gravity. Therefore, all components of $G^{\mu\nu}(\rho)$ in spherical coordinates are independent of the gauge parameter ξ to the second-order approximation.

This ξ-independent property of $G^{\mu\nu}(\rho)$ is notable because the complete Yang–Mills theory of gravity at the quantum level is T_4 gauge invariant. However, in the geometric-optics limit, results such as the Einstein–Grossmann equation of motion for classical objects may not, in general, be gauge invariant because the equation of motion (9.14) and the result (9.15) are only approximations. Thus, it is not unreasonable that the effective metric tensors may not be ξ-independent to all orders of Gm/ρ.

9.2. The Perihelion Shift of Mercury

To compare the predictions of the classical approximation of Yang–Mills gravity to the observations of the perihelion shift of Mercury, we choose $\theta = \pi/2$ in (9.15) so that the Einstein–Grossmann equation (9.14) for a planet with mass m_p is

$$G^{00}(\rho)\left(\frac{\partial S}{\partial w}\right)^2 + G^{11}(\rho)\left(\frac{\partial S}{\partial \rho}\right)^2 + G^{33}(\rho)\left(\frac{\partial S}{\partial \phi}\right)^2 - m_p^2 = 0. \qquad (9.16)$$

To solve (9.16), we follow the general procedure for solving the Hamilton–Jacobi equation, writing the solution in the form $S = -E_o w + M\phi + f(\rho)$.[2] Solving for $f(\rho)$, we obtain

$$S = -E_o w + M\phi + \int \frac{1}{\sqrt{|G^{11}|(\rho)}} \sqrt{E_o^2 G^{00}(\rho) - m_p^2 - \frac{M^2}{\rho^2}}\, d\rho, \qquad (9.17)$$

where E_o and M are the constant energy and angular momentum of the planet, respectively. The trajectory is determined by $\partial S/\partial M = $ constant, so that

$$\phi = \int \frac{(M/\rho^2)d\rho}{\sqrt{E_o^2 G^{00}|G^{11}| - m_p^2|G^{11}| - M^2|G^{11}|/\rho^2}}. \qquad (9.18)$$

To find the trajectory, it is convenient to write (9.18) as a differential equation of $\sigma = 1/\rho$. By differentiating equation (9.18) with respect to ϕ, we obtain

$$\frac{d^2\sigma}{d\phi^2} = \frac{1}{P} - \sigma(1 + Q) + 3Gm\sigma^2,$$

$$P = \frac{M^2}{m_p^2 Gm}, \quad Q = \frac{6Gm}{P}\left(\frac{E_o^2 - m_p^2}{m_p^2}\right),$$

$$(9.19)$$

where m is the solar mass. Thus, the equation for the trajectory (9.19) in Yang–Mills gravity differs slightly from the corresponding result in general relativity by a correction term Q. This correction term Q is of the order of $(Gm/P)\beta^2$ which is undetectable by present observations because the velocity β of the planet is very small compared to the speed of light, $\beta \ll 1$.

By the usual method of successive approximations,[a] we obtain the solution

$$\sigma = \frac{1}{P(1+Q)}\left[1 + e\cos\left(\phi\left(1 - \frac{3Gm}{P} + \frac{Q}{2}\right)\right)\right]. \quad (9.20)$$

The advance of the perihelion for one revolution of the planet is given by

$$\delta\phi \approx \frac{6\pi Gm}{P}\left(1 - \frac{(E_o^2 - m_p^2)}{4m_p^2} - \frac{2Gm}{3P}\right). \quad (9.21)$$

The second and the third terms on the right-hand side of (9.21) differentiate Yang–Mills gravity from Einstein's general relativity. However, these correction terms are only on the order of 10^{-7} for Mercury and the predictions of both theories fall well within the range of experimental observations (the observational accuracy of the perihelion shift of Mercury is about 1%).[b] Thus, the predicted perihelion shift in Yang–Mills gravity is independent of the gauge parameter ξ that appears in the second-order approximation of the solution of $g\phi^{\mu\nu}$.

9.3. Deflection of Light in a Gravitational Field

The deflection of light can be derived from the equation for the propagation of a light ray in an inertial frame in the geometric-optics limit. In the presence of a gravitational field, the path of a light ray is determined by the eikonal equation

---

[a] For the procedure of calculations, see Ref. 2.
[b] For a recent light deflection experiment, see Ref. 3.

with the form,

$$G^{\mu\nu}\partial_\mu\Psi\partial_\nu\Psi = 0, \tag{9.22}$$

where $G^{\mu\nu}$ will be replaced by $G_L^{\mu\nu}$ if the gauge condition $\partial_\mu A^\mu = 0$ is imposed. If the Yang–Mills gravity were to preserve the U_1 gauge symmetry, one can choose a gauge condition such as $\partial_\mu A^\mu = 0$ for calculations. In this case, $G^{\mu\nu} = J^{\mu\alpha}J^{\nu\beta}\eta_{\alpha\beta}$ in (9.22) is replaced by $G_L^{\mu\nu} = G^{\mu\nu} - (g/4)\phi_\lambda^\mu J^{\lambda\nu}$.

Following the usual convention, we assume that the motion of the light ray is in the plane defined by $\theta = \pi/2$ in spherical coordinates (w, r, θ, ϕ) (the x–y plane in Cartesian coordinates). Using $G^{\mu\nu}$ and $G_L^{\mu\nu}$ from (8.36), the static solution (9.9), and $x^\mu = (w, r, \theta, \phi)$ and going through the steps (9.10)–(9.15), the eikonal equation (9.22) can be written as

$$G^{00}\left(\frac{\partial\Psi}{\partial w}\right)^2 + G^{11}\left(\frac{\partial\Psi}{\partial\rho}\right)^2 - \frac{1}{\rho^2}\left(\frac{\partial\Psi}{\partial\phi}\right)^2 = 0,$$

$$G^{00} = 1 + \frac{2Gm}{\rho}, \quad G^{11} = -1 + \frac{2Gm}{\rho} \quad \text{(without } \partial_\mu A^\mu = 0), \tag{9.23}$$

$$G^{00} = 1 + \frac{7Gm}{4\rho} = G_L^{00}, \quad G^{11} = -1 + \frac{7Gm}{4\rho} = G_L^{11} \quad \text{(with } \partial_\mu A^\mu = 0),$$

where $G^{33} = G_L^{33} = -1/\rho^2$ and m is the mass of the sun. Similar to the general procedure for solving (9.16) in a spherical symmetric tensor field, we look for the eikonal Ψ in the form[2]

$$\Psi = -E_0 w + M\phi + f(\rho). \tag{9.24}$$

One can then determine $f(\rho)$ and solve for the trajectory of the ray, which is similar to (9.17) with $m_p \to 0$ and E_0 replaced by $\omega_o = -\partial\Psi/\partial w$ $(c = 1)$. We have

$$\frac{d^2\sigma}{d\phi^2} = -\sigma(1 + Q_o) + Q_1\sigma^2, \quad \sigma = \frac{1}{\rho} \tag{9.25}$$

$$Q_1 = 3Gm \quad \text{(without } \partial_\mu A^\mu = 0),$$
$$Q_1 = 21Gm/8 \quad \text{(with } \partial_\mu A^\mu = 0), \tag{9.26}$$

where the correction term $Q_o = 49G^2 m^2\omega_o^2/16M^2$ (with $\partial_\mu A^\mu = 0$) is extremely small and negligible. Following the usual procedure for calculating the deflection of light due to the sun,[2] we find the following results:

$$\Delta\phi \approx \frac{4Gm}{R} \approx 1.75'' \quad (\text{without } \partial_\mu A^\mu = 0),$$

$$(9.27)$$

$$\Delta\phi \approx \frac{7Gm}{2R} \approx 1.53'' \quad (\text{with } \partial_\mu A^\mu = 0),$$

for the deflection of a light ray passing through the spherically symmetric tensor field generated by the sun at a distance $R = M/\omega_0$ from the center of the sun to the first-order approximation. Because the accuracy of measurements of the deflection of light (with optical frequencies) by the sun are no better than (10–20)%, both results in (9.27) are consistent with experimental observations of $\Delta\phi_{\text{exp}} \approx 1.75''$.[3]

If the electromagnetic U_1 gauge symmetry is not violated by Yang–Mills gravity, one can impose the gauge condition $\partial_\mu A^\mu = 0$ to predict the angle of deflection as $\Delta\phi \approx 1.53''$ because the physical results are independent of gauge conditions. The fact that we have two different results in (9.27) indicates that electrodynamics is not gauge invariant in the presence of gravity. Furthermore, quantum Yang–Mills gravity predicts that the electric charge is not absolutely conserved in the presence of gravity. (See discussions in Section 14.4 in Chapter 14.)

Comparisons of results in Yang–Mills gravity and in general relativity should be made with caution because the calculations in Yang–Mills gravity are carried out in an inertial frame, while the corresponding results in Einstein gravity are not calculated in inertial frames.[4c]

If the accuracy of the measurements of the deflection of light rays (with optical frequencies) by the sun can be improved to the level of a few percents, then one might be able to rule out $\Delta\phi \approx 1.53''$, which is obtained on the basis of the electromagnetic U_1 gauge invariance with the gauge condition $\partial_\mu A^\mu = 0$. This can test unambiguously whether the electromagnetic U_1 gauge symmetry is violated by gravity (with T_4 gauge invariance). The significance of such an experiment cannot be over emphasized since it would determine whether electric charge is absolutely conserved. According to quantum Yang–Mills gravity, the universal coupling of the gravitational field to all fields in nature implies that all internal gauge symmetries will have very small violations in the presence of gravity. In other words, all internal charges such as the color charges of quarks, the baryon

---

[c] Wigner said: "Evidently, the usual statements about future positions of particles, as specified by their coordinates, are not meaningful statements in general relativity. This is a point which cannot be emphasized strongly enough... Expressing our results in terms of the values of coordinates became a habit with us to such a degree that we adhere to this habit also in general relativity, where values of coordinates are not per se meaningful." (See Ref. 4.)

charge (i.e., baryon number) and the lepton charge (i.e., lepton number) are not absolutely conserved in the presence of gravity. In the future, such experiments could determine whether the Yang–Mills idea of gauge symmetry for all interactions is consistent with experiments.

It is important to note that the result (9.27) is valid only for electromagnetic waves in the (high frequency) geometric-optics limit. The eikonal equation (9.22) has the simple effective metric tensor only for high-frequency waves such as visible light. The results are not valid for experiments involving electromagnetic radiation at radio frequencies. Radar echo experiments to test time dilation due to a gravitational field have used radiation with frequencies in the thousands of megahertz, roughly 10^9 (1/sec),[1] which are too small in comparison with the frequencies of visible light, $\approx 10^{14}$ (1/sec), and may not be suitable as tests of Yang–Mills gravity.

Finally, we note that the gravitational redshift in Yang–Mills gravity is identical to the conventional result. A photon in a gravitational field has kinetic energy $\hbar\omega$ and potential energy $\hbar\omega g\phi^{00}$.[5] In Yang–Mills gravity, the conservation of energy, $\hbar\omega + \hbar\omega g\phi^{00} = $ constant, and the static potentials in (9.9) lead to the usual gravitational redshift,

$$\omega_2/\omega_1 = (1 + g\phi_1^{00})/(1 + g\phi_2^{00}), \qquad (9.28)$$

which has been experimentally confirmed to 1% accuracy.[3]

9.4.   Gravitational Quadrupole Radiation

Although Yang–Mills gravity is formulated for both inertial and non-inertial frames,[6,7] for simplicity, we choose inertial frames, where $P_{\mu\nu} = \eta_{\mu\nu} = (+, -, -, -)$ and $D_\mu = \partial_\mu$, in which to discuss the gravitational quadrupole radiation. We note that the calculations of gravitational quadrupole radiation performed using Einstein gravity are not based in an inertial frame because the framework of Einstein gravity does not allow for inertial frames.

Ordinary matter consists of fermions such as protons and neutrons (or up- and down-quarks). The Lagrangian L involving the gravitational tensor field $\phi_{\mu\nu}$ and a fermion field ψ,

$$L = \frac{1}{2g^2}(C_{\mu\alpha\beta}C^{\mu\beta\alpha} - C_{\mu\alpha}{}^{\alpha}C^{\mu\beta}{}_{\beta})$$

$$+ \frac{i}{2}[\overline{\psi}\gamma^\mu(\Delta_\mu\psi) - (\Delta_\mu\overline{\psi})\gamma^\mu\psi] - m\overline{\psi}\psi, \qquad (9.29)$$

$$C^{\mu\nu\alpha} = J^{\mu\lambda}(\partial_\lambda J^{\nu\alpha}) - J^{\nu\lambda}(\partial_\lambda J^{\mu\alpha}), \quad C_{\mu\alpha\beta}C^{\mu\beta\alpha} = (1/2)C_{\mu\alpha\beta}C^{\mu\alpha\beta},$$

$$\Delta_\mu\psi = J_{\mu\nu}\partial^\nu\psi, \quad J_{\mu\nu} = \eta_{\mu\nu} + g\phi_{\mu\nu} = J_{\nu\mu}, \quad c = \hbar = 1. \tag{9.30}$$

For the gravitational quadrupole radiation, we consider only the tensor field $\phi_{\mu\nu}$. Moreover, it suffices to calculate the gravitation radiation to just the second order in $g\phi_{\mu\nu}$. As usual, we impose the gauge condition

$$\partial^\mu\phi_{\mu\nu} = (1/2)\partial_\nu\phi_\lambda^\lambda, \tag{9.31}$$

which is always possible.[1] The gauge invariant action with the Lagrangian (9.29) leads to the gravitational tensor field equation in inertial frames,[7]

$$H^{\mu\nu} = g^2 S^{\mu\nu}, \tag{9.32}$$

$$H^{\mu\nu} \equiv \partial_\lambda(J_\rho^\lambda C^{\rho\mu\nu} - J_\alpha^\lambda C^{\alpha\beta}{}_\beta\eta^{\mu\nu} + C^{\mu\beta}{}_\beta J^{\nu\lambda})$$
$$- C^{\mu\alpha\beta}\partial^\nu J_{\alpha\beta} + C^{\mu\beta}{}_\beta\partial^\nu J_\alpha^\alpha - C^{\lambda\beta}{}_\beta\partial^\nu J_\lambda^\mu, \tag{9.33}$$

where μ and ν should be made symmetric and we have used the identities in (8.14). It is not necessary to write this symmetry of μ and ν in (9.32) explicitly in the following discussions of gravitational radiation. The source tensor $S^{\mu\nu}$ in the gravitational field equation (9.32) is given by

$$S^{\mu\nu} = (1/2)[\overline{\psi}i\gamma^\mu\partial^\nu\psi - i(\partial^\nu\overline{\psi})\gamma^\mu\psi]. \tag{9.34}$$

For weak fields in inertial frames with the gauge condition (9.31), the field equation can be linearized as follows:

$$\partial_\lambda\partial^\lambda\phi^{\mu\nu} - \partial^\mu\partial_\lambda\phi^{\lambda\nu} + \partial^\mu\partial^\nu\phi_\lambda^\lambda - \partial^\nu\partial_\lambda\phi^{\lambda\mu} = g\left(S^{\mu\nu} - \frac{1}{2}\eta^{\mu\nu}S_\lambda^\lambda\right), \tag{9.35}$$

where we have used $J^{\mu\nu} = \eta^{\mu\nu} + g\phi^{\mu\nu}$. With the help of the gauge condition (9.31), (9.35) can be written as

$$\partial_\lambda\partial^\lambda\phi_{\mu\nu} = g\left(S_{\mu\nu} - \frac{1}{2}\eta_{\mu\nu}S_\lambda^\lambda\right) \equiv gS_{\mu\nu}^o, \quad g = \sqrt{8\pi G}, \tag{9.36}$$

where G is the Newtonian gravitational constant. To the first-order approximation, we have the static solution,[5]

$$g\phi_{00} = g\phi_{11} = \frac{Gm}{r}, \text{ etc.} \tag{9.37}$$

Note that the source tensor $S^{\mu\nu}$ given by (9.34) in Yang–Mills gravity is independent of $\phi^{\mu\nu}$. It satisfies the conservation law,

$$\partial_\mu S^{\mu\nu} = 0, \tag{9.38}$$

in the weak field approximation.

From equation (9.36), one has the usual retarded potential

$$\phi_{\mu\nu}(\mathbf{x}, t) = \frac{g}{4\pi} \int d^3x' \frac{S^o_{\mu\nu}(\mathbf{x}', t - |\mathbf{x} - \mathbf{x}'|)}{|\mathbf{x} - \mathbf{x}'|}, \tag{9.39}$$

$$\mathbf{x} \equiv \mathbf{r}, \quad \mathbf{x}' \equiv \mathbf{r}', \quad x^\mu = (w, x, y, z), \quad w = ct = t,$$

which is generated by the source $S_{\mu\nu}$ in (9.36). This equation is usually used to discuss the gravitational radiation. When discussing the radiation in the wave zone at a distance much larger than the dimension of the source, the solution can be approximated by a plane wave,[1]

$$\phi_{\mu\nu}(x) = e_{\mu\nu} \exp(-ik_\lambda x^\lambda) + e^*_{\mu\nu} \exp(ik_\lambda x^\lambda), \tag{9.40}$$

where $e_{\mu\nu}$ is the polarization tensor. The plane wave property and the usual gauge condition (9.31) lead to

$$k_\mu k^\mu = 0, \quad k_\mu e^\mu_\nu = (1/2) k_\nu e^\mu_\mu, \quad k^\mu = \eta^{\mu\nu} k_\nu. \tag{9.41}$$

For the symmetric polarization tensor, $e_{\mu\nu} = e_{\nu\mu}$, of the massless tensor field in flat spacetime, there are only two physical states with helicity ± 2 that are invariant under the Lorentz transformation.

Let us write $S_{\mu\nu}(\mathbf{x}, t)$ in terms of a Fourier integral,[1]

$$S_{\mu\nu}(\mathbf{x}, t) = \int_0^\infty S_{\mu\nu}(\mathbf{x}, \omega) e^{-i\omega t} d\omega + c.c. \tag{9.42}$$

The retarded field emitted by a single Fourier component $S_{\mu\nu}(\mathbf{x}, t) = [S_{\mu\nu}(\mathbf{x}, \omega) e^{-i\omega t} + c.c.]$ is given by

$$\phi_{\mu\nu}(\mathbf{x}, t) = \frac{g}{4\pi} \int \frac{d^3x'}{\mathbf{x} - \mathbf{x}'} S^o_{\mu\nu}(\mathbf{x}', \omega) \exp(-i\omega t + i\omega|\mathbf{x} - \mathbf{x}'|) + c.c. \tag{9.43}$$

$$S^o_{\mu\nu}(\mathbf{x}, \omega) = S_{\mu\nu}(\mathbf{x}, \omega) - (1/2)\eta_{\mu\nu} S^\lambda{}_\lambda(\mathbf{x}, \omega).$$

The source tensor $s^{\mu\nu}$ of gravitation is defined by the exact field equation (9.32) written in the following form:

$$\partial^\lambda \partial_\lambda \phi^{\mu\nu} = g(S^{\mu\nu} - s^{\mu\nu}). \tag{9.44}$$

Thus we have

$$
\begin{aligned}
s^{\mu\nu} = \frac{1}{g^2}[&C^{\rho\mu\nu}\partial_\lambda J^\lambda_\rho + \partial_\rho(g\phi^{\rho\lambda}\partial_\lambda J^{\mu\nu}) + g\phi^\lambda_\rho \partial_\lambda(J^{\rho\alpha}\partial_\alpha J^{\mu\nu}) \\
&- J^\lambda_{\ \rho}\partial_\lambda(J^{\mu\alpha}\partial_\alpha J^{\rho\nu}) - C^{\mu\beta\alpha}\partial^\nu J_{\alpha\beta} - \eta^{\mu\nu}\partial_\lambda(J^\lambda_{\ \rho}C^{\rho\beta}_{\ \ \beta}) + \partial_\lambda(C^{\mu\beta}_{\ \ \beta}J^{\nu\lambda}) \\
&+ C^{\mu\beta}_{\ \ \beta}\partial^\nu J^\lambda_{\ \lambda} - C^{\lambda\beta}_{\ \ \beta}\partial^\nu J^\mu_{\ \lambda}]. \tag{9.45}
\end{aligned}
$$

To the second order, then, we obtain the "source tensors" of the gravitational field,

$$s^{\mu\nu} = s_1^{\mu\nu} + s_2^{\mu\nu}, \tag{9.46}$$

$$
\begin{aligned}
s_1^{\mu\nu} = &(\partial_\lambda\phi)\partial^\lambda\phi^{\mu\nu} - \frac{1}{2}(\partial_\lambda\phi)\partial^\mu\phi^{\lambda\nu} + 2\phi^{\lambda\sigma}\partial_\lambda\partial_\sigma\phi^{\mu\nu} - (\partial_\lambda\phi^{\mu\sigma})\partial_\sigma\phi^{\lambda\nu} \\
&- \phi^\lambda_\rho\partial_\lambda\partial^\mu\phi^{\rho\nu} - \frac{1}{2}\phi^{\mu\sigma}\partial_\sigma\partial^\nu\phi - (\partial^\mu\phi^{\beta\alpha})\partial^\nu\phi_{\alpha\beta} + (\partial^\beta\phi^{\mu\alpha})\partial^\nu\phi_{\alpha\beta}, \tag{9.47}
\end{aligned}
$$

$$
\begin{aligned}
s_2^{\mu\nu} = &-\frac{3}{4}(\partial_\lambda\phi)\partial^\lambda\phi\eta^{\mu\nu} - \phi^{\lambda\sigma}\partial_\lambda\partial_\sigma\phi\eta^{\mu\nu} + (\partial_\lambda\phi^{\beta\sigma})\partial_\sigma\phi^\lambda_{\ \beta}\eta^{\mu\nu} \\
&+ \frac{1}{2}(\partial^\nu\phi^{\mu\lambda})\partial_\lambda\phi + \frac{1}{2}\phi^{\nu\lambda}\partial^\mu\partial_\lambda\phi + \phi^{\mu\sigma}\partial_\sigma\partial^\nu\phi \\
&- (\partial^\nu\phi^{\beta\sigma})\partial_\sigma\phi^\mu_{\ \beta} - \phi^{\beta\sigma}\partial_\sigma\partial^\nu\phi^\mu_{\ \beta} + \frac{3}{4}(\partial^\nu\phi)\partial^\mu\phi, \tag{9.48}
\end{aligned}
$$

where $\phi \equiv \phi^\lambda_\lambda$, and we have used the gauge condition (9.31). The gravitational source tensors $s_1^{\mu\nu}$ and $s_2^{\mu\nu}$ arise from the first and the second quadratic gauge-curvatures $C_{\mu\alpha\beta}C^{\mu\beta\alpha}$ and $-C_{\mu\alpha}^{\ \ \alpha}C^{\mu\beta}_{\ \ \beta}$, respectively, in the Lagrangian (9.29). One can then use the plane wave solution (9.40) and the gauge condition (9.41) to calculate the gravitational source tensors (9.46) in an inertial frame. This complicated result can be simplified by taking the average of $s^{\mu\nu}$ over a region of space and time much larger than the wavelengths of the radiated waves.[1] After such an averaging process, one obtains the following results:

$$
\begin{aligned}
\langle s_1^{\mu\nu} \rangle &= -2k^\mu k^\nu e^{\lambda\rho}e^*_{\lambda\rho} + (1/2)k^\mu k^\nu e^\lambda_\lambda e^{*\alpha}_\alpha, \\
\langle s_2^{\mu\nu} \rangle &= (1/2)k^\mu k^\nu e^\lambda_\lambda e^{*\alpha}_\alpha. \tag{9.49}
\end{aligned}
$$

In the wave zone approximation, one can write the polarization tensor in terms of the Fourier transform of $S_{\mu\nu}$:

$$e_{\mu\nu}(\mathbf{x}, \omega) = \frac{g}{4\pi r}\left[S_{\mu\nu}(\mathbf{k}, \omega) - \frac{1}{2}\eta_{\mu\nu}S^\lambda_\lambda(\mathbf{k}, \omega)\right], \qquad (9.50)$$

$$S_{\mu\nu}(\mathbf{k}, \omega) \equiv \int d^3x' S_{\mu\nu}(\mathbf{x}', \omega)]\exp(-i\mathbf{k}\cdot\mathbf{x}'), \qquad (9.51)$$

$$\phi_{\mu\nu}(\mathbf{x}, t) \approx e_{\mu\nu}(\mathbf{x}, \omega)\exp(-ik_\lambda x^\lambda) + c.c., \qquad (9.52)$$

where we have used (9.41) and (9.42) with the wave zone approximation $|\mathbf{x}-\mathbf{x}'| \approx r - \mathbf{x}'\cdot\mathbf{x}/|\mathbf{x}|$ and $\mathbf{k} = \omega\mathbf{x}/|\mathbf{x}|$. Thus, the average source tensor of a gravitational plane wave can be written as

$$\langle s^{\mu\nu}\rangle = -\frac{G}{\pi r^2}k^\mu k^\nu\left(S^{\lambda\rho}(\mathbf{k}, \omega)S^*_{\lambda\rho}(\mathbf{k}, \omega) - \frac{1}{2}S^\lambda_\lambda(\mathbf{k}, \omega)S^{*\sigma}_\sigma(\mathbf{k}, \omega)\right). \qquad (9.53)$$

The power P_o emitted per unit solid angle in the direction $\mathbf{x}/|\mathbf{x}|$ is[1]

$$\frac{dP_o}{d\Omega} = r^2\frac{x^i\langle s_{i0}\rangle}{|\mathbf{x}|}, \qquad (9.54)$$

which can be written in terms of $S_{\mu\nu}(\mathbf{k}, \omega)$ in (9.51),

$$\frac{dP_o}{d\Omega} = \frac{G\omega^2}{\pi}\left(S^{\lambda\rho}(\mathbf{k}, \omega)S^*_{\lambda\rho}(\mathbf{k}, \omega) - \frac{1}{2}S^\lambda_\lambda(\mathbf{k}, \omega)S^{*\sigma}_\sigma(\mathbf{k}, \omega)\right). \qquad (9.55)$$

Although the energy–momentum tensor of the gravitational field (9.46) in Yang–Mills gravity is quite different from that in general relativity, the result (9.55) for the power emitted per solid angle turns out to be the same as that obtained in general relativity and consistent with experimental observations.[1,8] Following the usual method and using the standard approximations, one can calculate the power radiated by a body rotating around one of the principal axes of the ellipsoid of inertia.[1] At twice the rotating frequency Ω, i.e., $\omega = 2\Omega$, the total power $P_o(\omega)$ emitted by a rotating body is

$$P_o(2\Omega) = [32G\Omega^6 I^2 e_q^2/5], \qquad c = 1, \qquad (9.56)$$

where I and e_q are the moment of inertia and equatorial ellipticity, respectively. Thus, to the second-order approximation, the gravitational quadrupole radiation (9.56) predicted by Yang–Mills gravity is also the same as that predicted by general relativity.[1]

References

1. S. Weinberg, in *Gravitation and Cosmology* (John Wiley & Sons, 1972), pp. 79–80, pp. 175–210 and pp. 251–272.

2. L. Landau and E. Lifshitz, in *The Classical Theory of Fields*, Trans. by M. Hamermesh (Addison-Wesley, Cambridge, MA, 1951), pp. 276–277 and pp. 312–316; J. P. Hsu, *Eur. Phys. J. Plus* **128**, 31 (2013), DOI 10.1140/epjp/i2013-1303-3.

3. W.-T. Ni, in *100 Years of Gravity and Accelerated Frames: The Deepest Insights of Einstein and Yang–Mills*, eds. J. P. Hsu and D. Fine (World Scientific, 2005), p. 478 and p. 481; E. B. Fomalont and S. M. Kopeikin, *Astrophys. J.* **598**, 704 (2003).

4. E. P. Wigner, in *Symmetries and Reflections, Scientific Essays* (The MIT Press, 1967), pp. 52–53.

5. S. Weinberg, in *Gravitation and Cosmology* (John Wiley & Sons, 1972), p. 85.

6. J. P. Hsu and L. Hsu, in *A Broader View of Relativity: General Implications of Lorentz and Poincaré Invariance* (World Scientific, 2006), Appendix D.

7. J. P. Hsu, *Int. J. Mod. Phys. A* **21**, 5119 (2006).

8. J. H. Taylor, *Rev. Mod. Phys.* **66**, 771 (1994).

Chapter 10

The S-Matrix in Yang–Mills Gravity

10.1. The Gauge-Invariant Action and Gauge Conditions

Beginning with this chapter, we discuss a quantum theory of gravity under the Yang–Mills–Utiyama–Weyl framework. Although Yang–Mills gravity can be formulated in both inertial and non-inertial frames and in the presence of both fermion and scalar fields,[1] it is difficult to discuss quantum field theory and particle physics even in the simplest non-inertial frame.[2-4] For that reason, we consider quantum Yang–Mills gravity in inertial frames with the Minkowski metric tensor $\eta_{\mu\nu} = (1, -1, -1, -1)$ and without involving other fields. In this chapter, we demonstrate some basic physical properties of quantum Yang–Mills gravity, namely, that the S-matrix satisfies both the conditions of unitarity and gauge invariance. The action S_{pg} for pure gravity (i.e., involving only the space-time gauge field $\phi_{\mu\nu}$ with the gauge-fixing terms $L_{\xi\zeta}$) is assumed to be[1]

$$S_{pg} = \int (L_\phi + L_{\xi\zeta})d^4x, \tag{10.1}$$

$$L_\phi = \frac{1}{4g^2}(C_{\mu\nu\alpha}C^{\mu\nu\alpha} - 2C_{\mu\alpha}{}^\alpha C^{\mu\beta}{}_\beta), \tag{10.2}$$

$$L_{\xi\zeta} = \frac{\xi}{2g^2}\left[\partial^\mu J_{\mu\alpha} - \frac{\zeta}{2}\partial_\alpha J\right]\left[\partial^\nu J_{\nu\beta} - \frac{\zeta}{2}\partial_\beta J\right]\eta^{\alpha\beta}, \tag{10.3}$$

$$C^{\mu\nu\alpha} = J^{\mu\sigma}\partial_\sigma J^{\nu\alpha} - J^{\nu\sigma}\partial_\sigma J^{\mu\alpha}, \quad J_{\mu\nu} = \eta_{\mu\nu} + g\phi_{\mu\nu} = J_{\nu\mu},$$

$$J = J^\lambda_\lambda = \delta^\lambda_\lambda - g\phi, \quad \phi = \phi^\lambda_\lambda, \quad c = \hbar = 1, \tag{10.4}$$

where $C^{\mu\alpha\beta}$ is the T_4 gauge curvature for inertial frames. In quantum Yang–Mills gravity, it is necessary to include the gauge fixing terms $L_{\xi\zeta}$ in the action functional (10.1), where $L_{\xi\zeta}$ takes the form (10.3) and may involve two arbitrary gauge parameters ξ and ζ in general. We note that the gauge-fixing term $L_{\xi\zeta}$ in (10.3) is the same (i.e., involves ordinary partial derivatives) in both inertial and non-inertial frames, so that $L_{\xi\zeta}$ is not gauge invariant.[a,b,c] However, in a non-inertial frame, the constant metric tensor $\eta_{\mu\nu}$ and ∂_μ in the Lagrangian L_ϕ given by (10.2) are replaced by space-time-dependent Poincaré metric tensor $P_{\mu\nu}$ and the associated covariant derivative D_μ, respectively. In this way, the Lagrangian $L_\phi\sqrt{-P}$ in a general frame changes only by a divergence under T_4 gauge transformations and hence, the action $S_\phi = \int L_\phi\sqrt{-P}d^4x$ is T_4 gauge invariant.[1]

The Lagrangian $L_{\xi\zeta}$ in (10.3) corresponds to a class of gauge conditions of the following form:

$$\partial^\lambda J_{\rho\lambda} - \frac{\zeta}{2}\partial_\rho J = Y_\rho, \tag{10.5}$$

where $Y_\rho = Y_\rho(x)$ is independent of the fields and the gauge function Λ^μ. For simplicity, we concentrate on two cases, $\zeta = 1$ and $\zeta = 0$ (with ξ arbitrary) in the following discussion. The Lagrangian for pure gravitational field, $L_{pg} = L_\phi + L_{\xi\zeta}$, can be expressed in terms of space-time gauge fields $\phi_{\mu\nu}$

$$L_{pg} = L_2 + L_3 + L_4 + L_{\xi\zeta}, \tag{10.6}$$

where

$$L_2 = \frac{1}{2}(\partial_\lambda\phi_{\alpha\beta}\partial^\lambda\phi^{\alpha\beta} - \partial_\lambda\phi_{\alpha\beta}\partial^\alpha\phi^{\lambda\beta} - \partial_\lambda\phi\partial^\lambda\phi + 2\partial_\lambda\phi\partial^\beta\phi_\beta^\lambda - \partial_\lambda\phi_\mu^\lambda\partial^\beta\phi_{\mu\beta}), \tag{10.7}$$

$$L_{\xi\zeta} = \frac{\xi}{2g^2}\left(\partial_\lambda J^{\lambda\alpha} - \frac{\zeta}{2}\partial^\alpha J\right)\left(\partial^\rho J_{\rho\alpha} - \frac{\zeta}{2}\partial_\alpha J\right)$$

$$= \frac{\xi}{2}\left[(\partial_\lambda\phi^{\lambda\alpha})\partial^\rho\phi_{\rho\alpha} - \zeta(\partial_\lambda\phi^{\lambda\alpha})\partial_\alpha\phi + \frac{\zeta^2}{4}(\partial^\alpha\phi)\partial_\alpha\phi\right]. \tag{10.8}$$

The Lagrangians L_2 and $L_{\xi\zeta}$ involve terms quadratic in $\phi_{\mu\nu}$ and determine the propagator of the graviton in Yang–Mills gravity. The Lagrangians L_3 and L_4

---

[a]The gauge-fixing terms are not covariant because they are introduced to fix the gauge, i.e., to remove the ambiguity in the tensor field equation due to the T_4 gauge symmetry.

[b]Without this gauge fixing term (10.3), one cannot find a well-defined static solution, as one can see from equation (9.9). The solution (9.9) diverges as $\xi \to 0$, i.e., in the absence of the gauge-fixing terms (10.3).

[c]In electrodynamics, where the Maxwell equations have U_1 gauge symmetry, one can remove the ambiguity in their solution by choosing a particular gauge, e.g., the Coulomb gauge.

correspond to the interactions of 3 and 4 gravitons, respectively, and will be discussed in Chapter 11.

10.2. Feynman–DeWitt–Mandelstam Ghost Fields in Yang–Mills Gravity

Because Yang–Mills gravity has nonlinear field equations, the gauge condition (10.5) cannot be imposed for all time and the theory will have gauge non-invariant amplitudes. One must introduce "ghost fields" to cancel these unwanted gauge non-invariant amplitudes, so that the theory, and in particular the S-matrix, is gauge invariant. The ghost fields and their interactions with the graviton in Yang–Mills gravity can be obtained by the Faddeev–Popov method.[5,6] Let us choose the gauge condition (10.5) for the following discussions. It is important to symmetrize the indices of the symmetric tensor field. Thus, we write the gauge condition (10.5) in the following form where the indices μ and v of $J_{\mu v}$ have been explicitly symmetrized:

$$Y_\lambda = \frac{1}{2}(\delta_\lambda^\mu \partial^v + \delta_\lambda^v \partial^\mu - \zeta \eta^{\mu v} \partial_\lambda) J_{\mu v}. \qquad (10.9)$$

The vacuum-to-vacuum amplitude is defined as the probability amplitude that the physical system will be in the vacuum state at time $w = \infty$, when it is known to be in the vacuum state at time $w = -\infty$. This is similar to that in the usual gauge theory.[6,7] It is convenient to use the method of path integrals to calculate the vacuum-to-vacuum amplitude, which completely determines the interactions of a physical system.[7] The vacuum-to-vacuum amplitude (with an external source $j^{\mu v}$) of quantum Yang–Mills gravity is given by

$$W_Y[j] \equiv \int d[J_{\rho\sigma}] \left(\exp\left[i \int d^4x \, (L_\phi + J_{\mu v} j^{\mu v}) \right] \right.$$

$$\left. \times (\det U) \prod_{\alpha, x} \delta(\partial^\lambda J_{\lambda\alpha} - \frac{1}{2}\partial_\alpha J_\lambda^\lambda - Y_\alpha) \right). \qquad (10.10)$$

The functional determinant $\det U$ in (10.10) is determined by

$$\frac{1}{\det U} = \int d[\Lambda^\lambda] \prod_{x,v} \delta\left(\frac{1}{2}(\delta_\lambda^\mu \partial^\mu + \delta_\lambda^v \partial^\mu - \zeta \eta^{\mu v} \partial_\lambda) J_{\mu v}^\$ - Y_\lambda \right),$$

$$J_{\mu v}^\$ = J_{\mu v} - \Lambda^\lambda \partial_\lambda J_{\mu v} - (\partial_\mu \Lambda^\lambda) J_{\lambda v} - (\partial_v \Lambda^\lambda) J_{\mu\lambda}. \qquad (10.11)$$

It follows from (10.11) that

$$U_{\mu\nu} = (\partial^\lambda E_{\mu\nu\lambda}) + E_{\mu\nu\lambda}\partial^\lambda, \tag{10.12}$$

where

$$E_{\mu\nu\lambda} = J_{\mu\nu}\partial_\lambda + J_{\nu\lambda}\partial_\mu + (\partial_\nu J_{\mu\lambda}) - \zeta\eta_{\mu\lambda}J_{\nu\sigma}\partial^\sigma - \frac{\zeta}{2}\eta_{\mu\lambda}(\partial_\nu J).$$

Since the vacuum-to-vacuum amplitude $W_Y[j]$ in (10.10) is invariant under an infinitesimal change of $Y^\alpha(x)$ for all $Y^\alpha(x)$, we may write (10.10) in the form

$$W[j] = \int d[Y^\nu]W_Y[j]\exp\left[i\int d^4x\left(\frac{\xi}{2g^2}Y^\mu Y_\mu\right)\right]$$

$$= \int d[J_{\alpha\beta}](\det U)\exp\left[i\int d^4x(L_\phi + L_{\xi\zeta} + J_{\mu\nu}j^{\mu\nu})\right], \tag{10.13}$$

to within an unimportant multiplicative factor.

As usual, the functional determinant $(\det U)$ in (10.11) can be expressed in terms of a Lagrangian L_{gho} for ghost fields V^μ and $\overline{V}^\mu$,

$$\det(U_{\mu\nu}) = \int \exp\left(i\int L_{\text{gho}}\, d^4x\right) d[V(x)^\lambda, \overline{V}(x)^\sigma], \tag{10.14}$$

$$\int L_{\text{gho}}\, d^4x = \int \overline{V}^\mu U_{\mu\nu} V^\nu\, d^4x = -\int (\partial_\lambda \overline{V}^\mu) E_{\mu\nu\lambda} V^\nu\, d^4x$$

$$= \int \left[\overline{V}^\mu\left[\partial^2\eta_{\mu\nu} + (1-\zeta)\partial_\mu\partial_\nu\right]V^\nu - g(\partial^\lambda \overline{V}^\mu)\phi_{\mu\nu}\partial_\lambda V^\nu\right.$$

$$- g(\partial^\lambda\overline{V}^\mu)\phi_{\lambda\nu}\partial_\mu V^\nu - g(\partial^\lambda\overline{V}^\mu)(\partial_\nu\phi_{\mu\lambda})V^\nu + \zeta g(\partial_\mu\overline{V}^\mu)\phi_{\sigma\nu}\partial^\sigma V^\nu$$

$$\left. + \frac{\zeta}{2} g(\partial_\mu\overline{V}^\mu)(\partial_\nu\phi_\lambda^\lambda)V^\nu\right] d^4x, \tag{10.15}$$

where $U_{\mu\nu}$ is given by (10.12) and $\overline{V}^\nu$ is considered to be an independent field.

The quanta of the ghost fields V^μ and $\overline{V}^\nu$ in the Lagrangian L_{gho} are the ghost (or fictitious) particles in Yang–Mills gravity. These ghost fields are vector fields and are quantized in the same way as fermion fields. Such "ghost vector-fermions" obey Fermi–Dirac statics, and were discussed in quantum Einstein gravity by Feynman, DeWitt and Mandelstam.[8] By definition, they do not exist in initial and final states of a physical process. They interact with themselves and other physical particles and can only appear in the intermediate states of a process.

Their interactions contribute extra amplitudes to remove the unwanted amplitudes due to the interaction of unphysical components of gauge fields.[d] Therefore, they are crucial for unitarity and gauge invariance of the S-matrix in any quantum theory of gravity and may be called Feynman–DeWitt–Mandelstam (FDM) ghosts in quantum Yang–Mills gravity.

Thus, the vacuum-to-vacuum amplitude (10.13) with (det U) given by (10.14) indicates that the Yang–Mills gravity, including the gravitational and the ghost fields, is completely determined by the Lagrangian L_{tot},

$$L_{tot} = L_\phi + L_{\xi\zeta} + \overline{V}^\mu U_{\mu\nu} V^\nu. \tag{10.16}$$

The Feynman–Dyson rules in Yang–Mills gravity for Feynman diagrams can be derived from this total Lagrangian. The last term $\overline{V}^\mu U_{\mu\nu} V^\nu$ is the FDM ghost Lagrangian L_{gho}, which is obtained by considering the T_4 gauge transformation of the gauge condition as shown in (11.12).

10.3. Unitarity and Gauge Invariance of the S-Matrix and FDM Ghost Particles

We now give some arguments and a proof for the unitarity of the S-matrix in Yang–Mills gravity based on the total Lagrangian (10.16), similar to those of Fradkin and Tyutin for the S-matrix in Einstein gravity.[9] With the gauge condition in (10.9), one can write the FDM ghost field $V^\mu(x)$ in the following form[9]:

$$V^\mu(x) = \int d^4 y \mathrm{D}_\nu^\mu(x, y, \phi_{\alpha\beta}) \hat{V}^\nu(y). \tag{10.17}$$

The total Lagrangian (10.16) implies that the FDM ghost field V^μ satisfies the equation

$$U_{\mu\nu} V^\nu = 0, \tag{10.18}$$

where $U_{\mu\nu}$ is given in (10.12). Clearly, in the limit of zero coupling strength, $g \rightarrow 0$, one has $J_{\mu\nu} \rightarrow \eta_{\mu\nu}$. Thus, the operator $U_{\mu\nu}$ reduces to a non-singular differential

---

[d]In general, the unphysical components of a gauge field depend on gauge conditions or gauge parameters. Their interactions generate amplitudes that depend on gauge parameters and also upset the unitarity of the S-matrix.

operator in this limit,

$$U_{\mu\nu} \to \eta_{\mu\nu}\partial_\lambda \partial^\lambda \equiv U^0_{\mu\lambda}, \quad \zeta = 1. \tag{10.19}$$

This limiting property can be seen from (10.12). One can choose the function $D^\mu_\nu(x, y, \phi_{\alpha\beta})$ in equation (10.17) to have the specific form

$$\mathbf{D}^\mu_\nu = [U^{-1}]^{\mu\lambda}\, \overleftarrow{U}^0_{\lambda\nu}, \tag{10.20}$$

so that $\hat{V}^\mu$ satisfies the free field equation,

$$U^0_{\lambda\mu} \hat{V}^\mu = [\partial^\sigma \partial_\sigma \eta_{\lambda\mu}]\hat{V}^\mu = 0. \tag{10.21}$$

In gauge invariant gravity, the generating functional for Green's functions can be defined after the gauge condition is specified. Similarly, in Yang–Mills gravity, the generating functional for the Green's functions (or the vacuum-to-vacuum amplitude) (10.13) can be written as

$$W[j] = \int d[J_{\alpha\beta}] \exp\left[i \int d^4x \left(L_\phi + \frac{\xi}{2g^2}F_\mu F_\nu \eta^{\mu\nu} + J_{\mu\nu}j^{\mu\nu}\right) + \mathrm{Tr}\ln U\right], \tag{10.22}$$

$$F_\alpha \equiv \eta^{\rho\lambda}\partial_\rho J_{\lambda\alpha} - \frac{1}{2}\eta^{\lambda\rho}\partial_\alpha J_{\lambda\rho}, \quad \det U = \exp(\mathrm{Tr}\ln U),$$

where the external sources $j^{\mu\nu}$ are arbitrary functions. It follows from (10.21) and (10.22) that the S-matrix corresponding to the generating functional (10.22) is unitary and gauge invariant.[9,10,e] There is no physical difference between $\mathrm{Tr}\ln U$ and $\mathrm{Tr}\ln U(\overleftarrow{U}^0)^{-1}$ because U^0 is a free operator. The presence of U^0 in, for example, (10.22) contributes only an unimportant multiplicative factor to the generating functional $W[j]$.

The last term in (10.22) can be written in terms of the FDM ghost fields $V^\alpha(x)$ and $\overline{V}^\beta(x)$,[10,11]

$$\exp(\mathrm{Tr}\ln U) = \int d[V^\alpha, \overline{V}^\beta] \exp\left(i \int L_{\mathrm{gho}} d^4x\right), \tag{10.23}$$

where the ghost Lagrangian L_{gho} (given in (10.15)) describes the FDM ghosts associated with the gauge specified in (10.9). Note that $\overline{V}^\mu$ is considered to be an independent field. The quanta of the fields V^μ and $\overline{V}^\mu$ in the Lagrangian L_{gho} are the FDM ghost particles, which are vector-fermions. By definition of the

---

[e]Since the vacuum-to-vacuum amplitude (10.22) is unitary, it can only be a phase factor, which is the generating functional for connected Green's functions.

physical state subspace for the S-matrix, these FDM ghost particles do not exist in the external states. They can only appear in the intermediate steps of a physical process.[10,11] The T_4 gauge field equations, $H^{\mu\nu} = 0$, for gravitational fields also hold in the physical subspace.

10.4. Discussion

For a more intuitive and physical understanding of the process by which ghost fields can be used to restore the unitarity and gauge invariance of the action, let us consider quantum electrodynamics (QED) with a more complicated nonlinear gauge condition involving an artificial "new" photon coupling,[11]

$$\partial_\mu A^\mu - \beta' A_\mu A^\mu = a(x), \quad a(x) : \text{a suitable function.} \tag{10.24}$$

Roughly speaking, the unitarity of the S-matrix in QED hinges on whether the unphysical component, $\chi = \partial_\mu A^\mu$, of the gauge field A^μ satisfies the free equation

$$\partial_\mu \partial^\mu \chi = 0. \tag{10.25}$$

This is indeed the case if one imposes the usual linear gauge condition in QED for quantization. However, suppose one instead imposes the nonlinear gauge condition (10.24), which one is free to do. Now, $\chi = \partial_\mu A^\mu$ no longer satisfies the free equation. Instead,

$$(\partial_\mu - 2\beta' A_\mu)\partial^\mu a(x) \equiv Ma(x) = 0. \tag{10.26}$$

This equation implies that the unphysical components of A^μ have a new interaction with A_μ. In other words, the nonlinear gauge condition (10.24) cannot be imposed for all times.[11] Such an unwanted new interaction produces extra unphysical amplitudes that violate gauge invariance and hence, the unitarity of the S-matrix. These unphysical amplitudes must be removed from the theory to restore the gauge invariance and unitarity of the S-matrix in QED. The removal of these unphysical amplitudes is accomplished by the presence of

$$\det M = \exp[\text{Tr} \ln M] \tag{10.27}$$

in the generating functional of Green's function. In essence, the non-free operator M is "subtracted" from QED with the nonlinear gauge condition (10.24). As a result, there are no more amplitudes that violate unitarity and gauge invariance in QED. Effectively, the unphysical components of the photon satisfy the free equation, so that the S-matrix in QED with a nonlinear gauge condition becomes unitary and gauge invariant again.

In quantum Yang–Mills gravity, the delta function in (10.10) may be understood as a way to impose the gauge condition (10.9) at all times. This is equivalent to the presence of det U, given by (10.11), in the theory to remove gauge non-invariant amplitudes and to restore the unitarity of the S-matrix.

For the quantization of gauge fields with internal gauge groups and the unitarity of the S-matrix, it is convenient to use the method of Lagrange multipliers, which may be understood as the unphysical components of gauge fields.[9,11] If a Lagrange multiplier satisfies a free equation, then the unphysical components of gauge fields do not interact and the S-matrix is unitary. Equivalently, the chosen gauge condition can be imposed for all times. In this case, the ghost particles in the theory satisfy the free equations and can be ignored. QED with a linear gauge condition is a well-known gauge theory in which ghost particles can be ignored.

If, on the other hand, the Lagrange multipliers do not obey free equations, then the unphysical components produce extra unphysical amplitudes to violate unitarity. In this case, gauge invariance will also be violated because the unphysical components of the gauge fields are gauge dependent. Consequently, one needs to use FDM ghosts to restore unitarity and gauge invariance.[f]

References

1. J. P. Hsu and L. Hsu, in *A Broader View of Relativity: General Implications of Lorentz and Poincaré Invariance* (World Scientific, 2006), Appendix D. Available online: "google books, a broader view of relativity, Hsu"; J. P. Hsu, *Intl. J. Mod Phys. A* **21** 5119 (2006).

2. J. P. Hsu, *Chin. J. Phys.* **40**, 265 (2002).

3. D. Schmidt and J. P. Hsu, *Int. J. Mod. Phys. A* **20**, 5989 (2005).

4. J. P. Hsu and D. Fine, *Int. J. Mod. Phys. A* **20**, 7485 (2005).

5. V. N. Popov, in *Functional Integrals in Quantum Field Theory and Statistical Physics*, Trans. from Russian by J. Niederle and L. Hlavaty (D. Reidel Publishing Co., Boston, 1983), Chapters 2 and 3.

6. L. D. Faddeev and V. N. Popov, in *100 Years of Gravity and Accelerated Frames: The Deepest Insights of Einstein and Yang–Mills*, eds. J. P. Hsu and D. Fine (World Scientific, 2005), p. 325.

---

[f] In Yang–Mills gravity, the direct calculations of the equation of motion for the Lagrange multiplier are much more complicated, making it more difficult to obtain the corresponding FDM ghost Lagrangian. The Faddeev and Popov method used in Sections 10.2 and 10.3 is easier.

7. K. Huang, in *Quarks, Leptons & Gauge Fields* (World Scientific, 1982), pp. 127–147.

8. R. P. Feynman, in *100 Years of Gravity and Accelerated Frames: The Deepest Insights of Einstein and Yang–Mills*, eds. J. P. Hsu and D. Fine (World Scientific, 2005), pp. 272–297; B. S. DeWitt, *ibid.*, pp. 298–324; S. Mandelstam, *Phys. Rev.* **175**, 1604 (1968).

9. F. S. Fradkin and I. V. Tyutin, *Phys. Rev. D* **2**, 2841 (1970).

10. J. P. Hsu and J. A. Underwood, *Phys. Rev. D* **12**, 620 (1975).

11. J. P. Hsu, *Phys. Rev. D* **8**, 2609 (1973); J. P. Hsu and E. C. G. Sudarshan, *Nucl. Phys. B* **91**, 477 (1975).

Chapter 11

Quantization of Yang–Mills Gravity
and Feynman–Dyson Rules

11.1. A Gauge Invariant Action for Gravitons and Fermions and Their Field Equations

In this chapter, we derive rules for Feynman diagrams for quantum Yang–Mills gravity. In the canonical quantization of fields, one uses the Schrödinger field operators and assumes equal-time commutation relations for them. Suppose the eigenstates of a field operator $\phi_{op}(\mathbf{r}) \equiv \phi_{op}^{\mu\nu}(\mathbf{r})$ are denoted by $|\phi\rangle$. One can use the eigenstates to define the transition amplitude, $\langle \phi_2, w_2 | \phi_1, w_1 \rangle$, which determines completely the dynamical interactions of the physical system and hence, one has a quantum field theory. Alternatively, one could also use the method of path integrals to express the transition amplitude $\langle \phi_2, w_2 | \phi_1, w_1 \rangle$ in terms of the classical Lagrangian density without using the Schrödinger field operators and eigenstates of field operators.[1] For the S-matrix of gravitational fields, it suffices to find the vacuum-to-vacuum amplitude, which we have done in the Chapter 10. We now generalize this amplitude to include gravitons, scalar bosons and fermions. As with previous discussions, although Yang–Mills gravity can be formulated in both inertial and non-inertial frames due to the generality of the taiji symmetry framework, we limit our discussion to inertial frames because of the complexity of its formulation in even the simplest accelerated frame.

The gauge invariant action $S_{\phi\psi}$ for spin 2 field $\phi_{\mu\nu} = \phi_{\nu\mu}$, fermion field ψ and scalar field Φ is given by[2]

$$S_{\phi\psi} = \int (L_\phi + L_\psi + L_\Phi + L_{\xi\zeta})d^4x, \quad c = \hbar = 1,$$

$$L_\phi = \frac{1}{2g^2}\left(\frac{1}{2}C_{\mu\nu\alpha}C^{\mu\nu\alpha} - C_{\mu\alpha}{}^\alpha C^{\mu\beta}{}_\beta\right), \tag{11.1}$$

157

$$L_\psi = \frac{i}{2}[\overline{\psi}\gamma_\mu(\Delta^\mu\psi) - (\Delta^\mu\overline{\psi})\gamma_\mu\psi] - m\overline{\psi}\psi, \tag{11.2}$$

$$L_\Phi = \frac{1}{2}[\eta^{\mu\nu}(\Delta_\mu\Phi)(\Delta_\nu\Phi) - m_\Phi^2\Phi^2], \tag{11.3}$$

$$L_{\xi\zeta} = \frac{\xi}{2g^2}\left[\partial^\mu J_{\mu\alpha} - \frac{\zeta}{2}\partial_\alpha J_\lambda^\lambda\right]\left[\partial^\nu J_{\nu\beta} - \frac{\zeta}{2}\partial_\beta J_\lambda^\lambda\right]\eta^{\alpha\beta}, \tag{11.4}$$

$$\Delta_\mu\psi = J_{\mu\nu}\partial^\nu\psi, \quad J_{\mu\nu} = \eta_{\mu\nu} + g\phi_{\mu\nu} = J_{\nu\mu},$$

where the T_4 gauge curvature $C^{\mu\alpha\beta}$ is given by

$$C^{\mu\nu\alpha} = J^{\mu\sigma}\partial_\sigma J^{\nu\alpha} - J^{\nu\sigma}\partial_\sigma J^{\mu\alpha}. \tag{11.5}$$

The gauge-fixing term $L_{\xi\zeta}$, involving up to two arbitrary gauge parameters ξ and ζ, is necessary for the quantization of fields with gauge symmetry. We choose the usual gauge condition of the form

$$\partial^\mu J_{\mu\nu} - \frac{\zeta}{2}\partial_\nu J = Y_\nu, \quad J = J_\lambda^\lambda = \delta_\mu^\lambda - g\phi, \quad \phi = \phi_\lambda^\lambda, \tag{11.6}$$

where Y_ν is a suitable function of space-time.

The Lagrangian for just a gravitational field (or "pure gravity") $L_{pg} = L_\phi + L_{\xi\zeta}$ in terms of the tensor field $\phi_{\mu\nu}$ can be written

$$L_{pg} = L_2 + L_3 + L_4 + L_{\xi\zeta}, \tag{11.7}$$

where

$$L_2 = \frac{1}{2}\left(\partial_\lambda\phi_{\alpha\beta}\partial^\lambda\phi^{\alpha\beta} - \partial_\lambda\phi_{\alpha\beta}\partial^\alpha\phi^{\lambda\beta} - \partial_\lambda\phi\partial^\lambda\phi\right.$$
$$\left. + 2\partial_\lambda\phi\partial^\beta\phi_\beta^\lambda - \partial_\lambda\phi_\mu^\lambda\partial_\beta\phi^{\mu\beta}\right), \tag{11.8}$$

$$L_3 = g[(\partial_\lambda\phi_{\alpha\beta}\partial_\rho\phi^{\alpha\beta})\phi^{\lambda\rho} - (\partial_\lambda\phi_{\alpha\beta}\partial_\rho\phi^{\lambda\beta})\phi^{\alpha\rho} - (\partial_\lambda\phi\partial_\rho\phi)\phi^{\lambda\rho}$$
$$+ (\partial^\lambda\phi\partial_\rho\phi_{\lambda\beta})\phi^{\beta\rho} + (\partial_\lambda\phi\partial^\lambda\phi_\beta^\mu)\phi_\mu^\lambda - (\partial^\lambda\phi_{\lambda\mu}\partial_\rho\phi_\beta^\mu)\phi^{\beta\rho}], \tag{11.9}$$

$$L_4 = \frac{g^2}{2}[(\partial^\lambda\phi_{\alpha\beta}\partial_\rho\phi^{\alpha\beta})\phi_{\lambda\mu}\phi^{\mu\rho} - (\partial_\lambda\phi_{\alpha\beta}\partial_\rho\phi^{\mu\beta})\phi_\mu^\lambda\phi^{\alpha\rho}$$
$$+ (\partial_\lambda\phi\partial_\rho\phi_\beta^\mu)\phi_\mu^\lambda\phi^{\beta\rho} + (\partial_\lambda\phi\partial_\rho\phi_\beta^\mu)\phi_\mu^\rho\phi^{\beta\lambda}$$
$$- (\partial_\lambda\phi\partial_\rho\phi)\phi_\mu^\lambda\phi^{\mu\rho} - (\partial_\lambda\phi_{\alpha\mu}\partial_\rho\phi_\beta^\mu)\phi^{\lambda\alpha}\phi^{\beta\rho}], \tag{11.10}$$

$$L_{\xi\zeta} = \frac{\xi}{2}\left[(\partial_\lambda\phi^{\lambda\alpha})\partial^\rho\phi_{\rho\alpha} - \zeta(\partial_\lambda\phi^{\lambda\alpha})\partial_\alpha\phi + \frac{\zeta^2}{4}(\partial^\alpha\phi)\partial_\alpha\phi\right]. \tag{11.11}$$

The action (11.1) with the Lagrangians given in (11.2), (11.3) and (11.4) leads to the T_4 field equation in inertial frames,

$$H^{\mu\nu} + \xi A^{\mu\nu} = g^2 S^{\mu\nu},$$

$$H^{\mu\nu} \equiv \text{Sym}[\partial_\lambda(J_\rho^\lambda C^{\rho\mu\nu} - J_\alpha^\lambda C^{\alpha\beta}{}_\beta \eta^{\mu\nu} + C^{\mu\beta}{}_\beta J^{\nu\lambda})$$

$$- C^{\mu\alpha\beta}\partial^\nu J_{\alpha\beta} + C^{\mu\beta}{}_\beta \partial^\nu J_\alpha^\alpha - C^{\lambda\beta}{}_\beta \partial^\nu J_\lambda^\mu],$$

$$A^{\mu\nu} = \text{Sym}[\partial^\mu \partial_\sigma J^{\sigma\nu} - \frac{\zeta}{2}\partial^\mu \partial^\nu J - \frac{\zeta}{2}\eta^{\mu\nu}\partial^\lambda \partial^\sigma J_{\lambda\sigma} + \frac{\zeta^2}{4}\eta^{\mu\nu}\partial^2 J],$$

(11.12)

where "Sym" denotes that μ and ν should be made symmetric, e.g., $L_{\mu\nu} = (1/2)[L_{\mu\nu} + L_{\nu\mu}]$.

The identities for the T_4 gauge curvature are

$$C_{\mu\nu\alpha} = -C_{\nu\mu\alpha}, \quad C_{\mu\nu\alpha} + C_{\nu\alpha\mu} + C_{\alpha\mu\nu} = 0.$$

(11.13)

For simplicity, we ignore the scalar field Φ in the following discussion of the energy–momentum tensor. The symmetric "source tensor" $S^{\mu\nu}$ in equation (11.12) is given by

$$S^{\mu\nu} = \frac{1}{4}\left[\overline{\psi}i\gamma^\mu \partial^\nu \psi - i(\partial^\nu \overline{\psi})\gamma^\mu \psi + \overline{\psi}i\gamma^\nu \partial^\mu \psi - i(\partial^\mu \overline{\psi})\gamma^\nu \psi\right],$$

(11.14)

which acts as the fermion source for generating the gravitational field $\phi_{\mu\nu}$. For weak fields in inertial frames, the field equation (11.12) can be linearized as follows:

$$\partial_\lambda \partial^\lambda \phi^{\mu\nu} - \partial^\mu \partial_\lambda \phi^{\lambda\nu} - \eta^{\mu\nu}\partial^2 \phi + \eta^{\mu\nu}\partial_\alpha \partial_\beta \phi^{\alpha\beta}$$

$$+ \partial^\mu \partial^\nu \phi - \partial^\nu \partial_\lambda \phi^{\lambda\mu} + \xi\partial^\mu \partial_\lambda \phi^{\lambda\nu} - \frac{\xi\zeta}{2}\partial^\mu \partial^\nu \phi$$

$$- \frac{\xi\zeta}{2}\eta^{\mu\nu}\partial_\alpha \partial_\beta \phi^{\alpha\beta} + \frac{\xi\zeta^2}{4}\eta^{\mu\nu}\partial^2 \phi = gS^{\mu\nu},$$

(11.15)

where g is related to Newtonian constant G by $g = \sqrt{8\pi G}$. In this weak field limit, the linearized equation (11.15) is the same as that in Einstein gravity. In momentum space, (11.15) can be written as

$$A^{\mu\nu\rho\sigma}(k)\phi_{\rho\sigma}(k) = 0,$$

$$A^{\mu\nu\rho\sigma}(k) = \frac{k^2}{2}(\eta^{\mu\rho}\eta^{\nu\sigma} + \eta^{\mu\sigma}\eta^{\nu\rho}) - \left(1 - \frac{\xi}{2}\right)k^\nu k^\sigma \eta^{\mu\rho} - \left(1 - \frac{\xi}{2}\right)k^\mu k^\sigma \eta^{\nu\rho}$$

$$- \left(1 - \frac{\xi\zeta^2}{4}\right)k^2 \eta^{\mu\nu}\eta^{\rho\sigma} + \left(1 - \frac{\xi\zeta}{2}\right)k^\rho k^\sigma \eta^{\mu\nu}$$

$$+ \left(1 - \frac{\xi\zeta}{2}\right)k^\mu k^\nu \eta^{\rho\sigma}, \tag{11.16}$$

when $g = 0$. Equation (11.16) will be useful for deriving the Feynman propagator for gravitons in Section 11.2.

In the presence of a space-time gauge field, the Dirac equation of a fermion can be derived from the fermion Lagrangian (11.2),

$$i(\gamma^\nu + g\gamma_\mu \phi^{\mu\nu})\partial_\nu \psi + \frac{ig}{2}(\partial_\nu \phi^{\mu\nu})\gamma_\mu \psi - m\psi = 0,$$

$$\tag{11.17}$$

$$i(\partial^\mu + g\phi^{\mu\nu}\partial_\nu)\overline{\psi}\gamma_\mu + \frac{ig}{2}\overline{\psi}\gamma_\mu(\partial_\nu \phi^{\mu\nu}) + m\overline{\psi} = 0.$$

It follows from these two equations that the modified fermion current j^ν in the presence of the gravitational field is conserved,

$$\partial_\nu j^\nu = 0, \tag{11.18}$$

where

$$j^\nu = \overline{\psi}\gamma^\nu \psi + g\phi^{\mu\nu}\overline{\psi}\gamma_\mu \psi = J^{\mu\nu}\overline{\psi}\gamma_\mu \psi.$$

The conserved total energy–momentum tensor $T^{\mu\nu}(\phi\psi)$ of the graviton–fermion system is given by the Lagrangian $L_{\phi\psi} = L_\phi + L_\psi$,

$$T^{\mu\nu}(\phi\psi) = T(\phi)^{\mu\nu} + T(\psi)^{\mu\nu},$$

$$T(\phi)^{\mu\nu} = \frac{\partial L_\phi}{\partial(\partial_\mu \phi_{\alpha\beta})}\partial^\nu \phi_{\alpha\beta} - \eta^{\mu\nu}L_\phi, \tag{11.19}$$

$$T(\psi)^{\mu\nu} = \frac{\partial L_\psi}{\partial(\partial_\mu \psi)}\partial^\nu \psi + \partial^\nu \overline{\psi}\frac{\partial L_\psi}{\partial(\partial_\mu \overline{\psi})} - \eta^{\mu\nu}L_\psi,$$

according to Noether's theorem. Note that L_ψ does not contain $\partial_\mu \phi_{\alpha\beta}$ and L_ϕ does not involve $\partial_\mu \psi$ and $\partial_\mu \overline{\psi}$. From equation (11.19), we obtain

$$T(\phi)^{\mu\nu} = \frac{1}{g}(J_\lambda^\mu C^{\lambda\alpha\beta} - \eta^{\alpha\beta} J_\lambda^\mu C^{\lambda\sigma}{}_\sigma + J^{\beta\mu} C^{\alpha\sigma}{}_\beta) \partial^\nu \phi_{\alpha\beta}$$

$$- \eta^{\mu\nu} \frac{1}{4g^2} (C_{\mu\alpha\beta} C^{\mu\alpha\beta} - 2C_{\mu\alpha}{}^\alpha C^{\mu\beta}{}_\beta), \tag{11.20}$$

$$T(\psi)^{\mu\nu} = \frac{i}{2}[\overline{\psi}\gamma_\alpha J^{\alpha\mu} \partial^\nu \psi - (\partial^\nu \overline{\psi}) J^{\alpha\mu} \gamma_\alpha \psi] - \eta^{\mu\nu} L_\psi. \tag{11.21}$$

These conserved energy–momentum tensors for the fermion and gauge fields are not inherently symmetric, but they can be symmetrized.[3,4]

In Yang–Mills gravity, it is the symmetric source tensor $S^{\mu\nu}$ in (11.12) and (11.14) that produces the gravitational field $\phi_{\alpha\beta}$, rather than the non-symmetric energy–momentum tensor $T(\psi)^{\mu\nu}$ for fermions. Thus, the source tensor $S^{\mu\nu}$ differs from the energy–momentum tensor in Einstein gravity.

From the linearized equation (11.15) for weak fields, one can derive the conservation law

$$\partial_\mu S^{\mu\nu} = 0.$$

It is gratifying that the source tensor $S^{\mu\nu}$ in the field equation (11.12) does not involve the tensor gauge field, as shown in (11.14), in contrast to the Noether tensor $T^{\mu\nu}(\psi)$ in (11.21). This makes the coupling of fields in Yang–Mills gravity simpler than in Einstein gravity.

11.2. The Feynman–Dyson Rules for Quantum Yang–Mills Gravity

We now use the results from the previous section to write down the Feynman–Dyson rules for Yang–Mills gravity. Although these are conventionally referred to as the Feynman rules, we choose to refer to them as the Feynman–Dyson rules in this case because the intuitive Feynman rules for calculating the invariant amplitude (a Lorentz scalar) are incomplete. It was Dyson's rigorous derivation and elaboration of the rules that allowed their application to higher order processes, such as Kinoshita's higher order corrections to the magnetic moment of the electron.[5,a]

Let us consider the general propagator for the graviton corresponding to the general gauge-fixing terms in (11.4) in Yang–Mills gravity. This general Feynman

---

[a]See also Ref. 12.

propagator involves two arbitrary parameters, ξ and ζ, and can be derived from the free equation (11.16) in momentum space. The graviton propagator $G_{\alpha\beta\rho\sigma}(k)$ is defined as the inverse of $A^{\mu\nu\rho\sigma}(k)$ in (11.16) as follows:

$$G_{\alpha\beta\rho\sigma}(k)A^{\mu\nu\rho\sigma}(k) = \frac{i}{2}(\delta_\alpha^\mu \delta_\beta^\nu + d_\alpha^\nu \delta_\beta^\mu), \tag{11.22}$$

with a factor i. Its general structure can be inferred from Lorentz and Poincaré covariance. We can write the graviton propagator $G_{\alpha\beta\rho\sigma}(k)$ in the following form:

$$G_{\alpha\beta\rho\sigma} = A(\eta_{\alpha\beta}\eta_{\rho\sigma} - \eta_{\rho\alpha}\eta_{\sigma\beta} - \eta_{\rho\beta}\eta_{\sigma\alpha}) - B(k_\alpha k_\beta \eta_{\rho\sigma} + k_\rho k_\sigma \eta_{\alpha\beta})$$

$$+ C(k_\sigma k_\beta \eta_{\rho\alpha} + k_\alpha k_\sigma \eta_{\rho\beta} + k_\rho k_\beta \eta_{\sigma\alpha} + k_\rho k_\alpha \eta_{\sigma\beta}) - Dk_\alpha k_\beta k_\rho k_\sigma,$$

where the unknown functions $A(k^2)$, $B(k^2)$, $C(k^2)$, $D(k^2)$ are to be determined by equation (11.22). It follows from (11.16), (11.22) and the above expression for $G_{\alpha\beta\rho\sigma}$ that

$$G_{\alpha\beta\rho\sigma} = \frac{-i}{2k^2}\left(\left[\eta_{\alpha\beta}\eta_{\rho\sigma} - \eta_{\rho\alpha}\eta_{\sigma\beta} - \eta_{\rho\beta}\eta_{\sigma\alpha}\right] - \frac{1}{k^2}\frac{(2\zeta - 2)}{(\zeta - 2)}\right.$$

$$\times (k_\alpha k_\beta \eta_{\rho\sigma} + k_\rho k_\sigma \eta_{\alpha\beta}) + \frac{1}{k^2}\frac{(\xi - 2)}{\xi}(k_\sigma k_\beta \eta_{\rho\alpha} + k_\alpha k_\sigma \eta_{\rho\beta} + k_\rho k_\beta \eta_{\sigma\alpha}$$

$$\left. + k_\rho k_\alpha \eta_{\sigma\beta}) - \frac{1}{k^4}\frac{(2\zeta - 2)}{\xi(\zeta - 2)}\left[\frac{4 + 2\xi - 4\xi\zeta}{\zeta - 2} - 4 + 2\xi\right]k_\alpha k_\beta k_\rho k_\sigma\right),$$

$$\tag{11.23}$$

where the usual prescription of $i\epsilon$ for the Feynman propagator (11.23) is understood. For example, $k^2 = k_0^2 - \mathbf{k}^2$ in (11.23) is understood as $k^2 + i\epsilon$. This is related to the property that the integral $\int d^4x(L_\phi + J_{\mu\nu}j^{\mu\nu})$ in the vacuum-to-vacuum amplitude (10.10) is defined by a passage to Euclidean space-time and then a continuation back to Minkowski space-time. This procedure gives the correct $i\epsilon$ prescription in the Feynman propagator (11.23). In this book, it should always be understood as being done, even when the $i\epsilon$ is not explicitly included in the Feynman propagators.

In the calculation of higher order Feynman diagrams such as the self-energy of the graviton or gravitational vertex corrections, the general propagator (11.23) with two arbitrary parameters leads to a complicated result because the divergent terms are gauge dependent. It is very time consuming to calculate, even using symbolic computing. Thus, it is useful to limit oneself to a special case such as $\zeta = 1$ and arbitrary ξ for an explicit calculation. In this case, the propagator (11.23) reduces to the simpler form

$$
G_{\alpha\beta\rho\sigma} = \frac{-i}{2k^2} \Big[(\eta_{\alpha\beta}\eta_{\rho\sigma} - \eta_{\rho\alpha}\eta_{\sigma\beta} - \eta_{\rho\beta}\eta_{\sigma\alpha})
$$
$$
+ \frac{1}{k^2}\frac{(\xi-2)}{\xi}(k_\sigma k_\beta \eta_{\rho\alpha} + k_\alpha k_\sigma \eta_{\rho\beta} + k_\rho k_\beta \eta_{\sigma\alpha} + k_\rho k_\alpha \eta_{\sigma\beta}) \Big]. \quad (11.24)
$$

In the special case of $\zeta = 0$ and arbitrary ξ, the graviton propagator (11.23) reduces to the form

$$
G_{\alpha\beta\rho\sigma} = \frac{-i}{2k^2} \Big([\eta_{\alpha\beta}\eta_{\rho\sigma} - \eta_{\rho\alpha}\eta_{\sigma\beta} - \eta_{\rho\beta}\eta_{\sigma\alpha}]
$$
$$
- \frac{1}{k^2}[k_\alpha k_\beta \eta_{\rho\sigma} + k_\rho k_\sigma \eta_{\alpha\beta}] + \frac{1}{k^2}\Big[\frac{\xi-2}{\xi}\Big](k_\sigma k_\beta \eta_{\rho\alpha} + k_\alpha k_\sigma \eta_{\rho\beta}
$$
$$
+ k_\rho k_\beta \eta_{\sigma\alpha} + k_\rho k_\alpha \eta_{\sigma\beta}) - \frac{1}{k^4}\Big[\frac{\xi-6}{\xi}\Big]k_\alpha k_\beta k_\rho k_\sigma \Big). \quad (11.25)
$$

In Yang–Mills gravity, the simple gauge specified by $\xi = 2$ and $\zeta = 1$ has been discussed by DeWitt and may be called the DeWitt gauge.[5] In this gauge, the covariant graviton propagator takes the simplest form, $G_{\alpha\beta\rho\sigma} = (-i/2k^2)(\eta_{\alpha\beta}\eta_{\rho\sigma} - \eta_{\rho\alpha}\eta_{\sigma\beta} - \eta_{\rho\beta}\eta_{\sigma\alpha})$ and is consistent with that obtained in previous works by Leibbrandt, Capper and others in general relativity.[6–8] As an aside, the overall factor of $(1/2)$ in the graviton propagators (11.23) is necessary for Yang–Mills gravity to be consistent with its T_4 gauge identity — a generalized Ward–Takahashi identity for the Abelian group T_4 with ghosts. (Cf. Section 13.2 in Chapter 13.) The free Lagrangian (11.8) in Yang–Mills gravity is mathematically the same as that in the Hilbert–Einstein Lagrangian. In quantum Einstein gravity as formulated and discussed by Feynman and DeWitt,[5] this factor of $(1/2)$ for the covariant graviton propagator is necessary in order for it to be consistent with the Slavnov–Taylor identity, as shown in Refs. 5–8.

We now consider the Feynman–Dyson rules for graviton interactions corresponding to iL_3. The momenta in the Feynman–Dyson rules in this chapter are in-coming to the vertices. Suppose the tensor-indices and momenta of the 3-vertex are denoted by $[\phi^{\mu\nu}(p)\phi^{\sigma\tau}(q)\phi^{\lambda\rho}(k)]$. The graviton 3-vertex is

$$
ig\,\text{Sym}\,P_6(-p^\lambda q^\rho \eta^{\sigma\mu}\eta^{\tau\nu} + p^\sigma q^\rho \eta^{\lambda\mu}\eta^{\tau\nu} + p^\lambda q^\rho \eta^{\sigma\tau}\eta^{\mu\nu}
$$
$$
- p^\sigma q^\rho \eta^{\mu\nu}\eta^{\tau\lambda} - p^\rho q^\sigma \eta^{\mu\nu}\eta^{\tau\lambda} + p^\mu q^\rho \eta^{\nu\sigma}\eta^{\lambda\tau}). \quad (11.26)
$$

As before, we use the symbol Sym to denote that a symmetrization is to be performed on each index pair $(\mu\nu)$, $(\sigma\tau)$ and $(\lambda\rho)$ in $[\phi^{\mu\nu}(p)\phi^{\sigma\tau}(q)\phi^{\lambda\rho}(k)]$. The symbol P_n denotes a summation to be carried out over permutations of the momentum-index triplets, and the subscript gives the number of permutations in each case. For example,

$$P_6(p^\lambda q^\rho \eta^{\sigma\mu}\eta^{\tau\nu}) = (p^\lambda q^\rho \eta^{\sigma\mu}\eta^{\tau\nu} + p^\sigma k^\tau \eta^{\lambda\mu}\eta^{\rho\nu} + p^\rho q^\lambda \eta^{\mu\sigma}\eta^{\nu\tau}$$
$$+ p^\tau k^\sigma \eta^{\mu\lambda}\eta^{\nu\rho} + q^\mu k^\nu \eta^{\lambda\sigma}\eta^{\rho\tau} + k^\mu q^\nu \eta^{\sigma\lambda}\eta^{\tau\rho}),$$

which is the result that one obtains directly from the first term in the Lagrangian L_3 in (11.9) by following the usual method for obtaining the vertex using the Feynman–Dyson method.[9–11]

The 4-vertex (i.e., iL_4 with $\phi^{\mu\nu}(p)\phi^{\sigma\tau}(q)\phi^{\lambda\rho}(k)\phi^{\alpha\beta}(l)$) is given by

$$\frac{ig^2}{2}\mathrm{Sym}\ P_{24}(-p^\rho q^\beta \eta^{\mu\sigma}\eta^{\nu\tau}\eta^{\lambda\alpha} + p^\rho q^\beta \eta^{\lambda\sigma}\eta^{\nu\tau}\eta^{\mu\alpha}$$
$$-p^\rho q^\beta \eta^{\lambda\sigma}\eta^{\tau\alpha}\eta^{\mu\nu} - p^\beta q^\rho \eta^{\lambda\sigma}\eta^{\tau\alpha}\eta^{\mu\nu}$$
$$+p^\rho q^\beta \eta^{\mu\nu}\eta^{\sigma\tau}\eta^{\lambda\alpha} + p^\rho q^\beta \eta^{\lambda\nu}\eta^{\tau\alpha}\eta^{\mu\sigma}). \tag{11.27}$$

The fermion propagator has the usual form

$$\frac{i}{\gamma^\mu p_\mu - m}. \tag{11.28}$$

The fermion–graviton 3-vertex (i.e., iL_ψ^{int} with $\overline{\psi}(q)\psi(p)\phi_{\mu\nu}(k)$) takes the form

$$\mathrm{Sym}\ \frac{i}{2}g\gamma_\mu(p_\nu + q_\nu), \tag{11.29}$$

where $\psi(p)$ is treated as an annihilation operator of a fermion with momentum p_ν, and $\overline{\psi}(q)$ is a creation operator of a fermion with the momentum q_ν.

To obtain the Feynman–Dyson rules for the general propagator (involving an arbitrary gauge parameter ζ) for the ghost particle and the interaction vertices of the ghost particle, it is convenient to write the Lagrangian L_{eff} in equation (10.15) in the following form:

$$L_{\text{eff}} = \overline{V}^\mu[\eta_{\mu\nu}\partial_\lambda\partial^\lambda + (1-\zeta)\partial_\mu\partial_\nu]V^\nu$$
$$- g[(\partial^\alpha\overline{V}^\beta)V^\nu\partial_\nu\phi_{\alpha\beta} + (\partial^\lambda\overline{V}^\mu)(\partial_\lambda V^\nu)\phi_{\mu\nu} + (\partial^\lambda\overline{V}^\mu)(\partial_\mu V^\nu)\phi_{\lambda\nu}]$$
$$+ g\zeta(\partial^\lambda\overline{V}_\lambda)(\partial^\mu V^\nu)\phi_{\mu\nu} + g(\zeta/2)(\partial_\mu\overline{V}^\mu)V^\nu(\partial_\nu\phi_{\alpha\beta})\eta^{\alpha\beta}. \tag{11.30}$$

The propagator of the FDM ghosts can be derived from the Lagrangian L_{eff} in (11.30) for ghost fields,

$$G^{\mu\nu} = \frac{-i}{k^2} \left(\eta^{\mu\nu} - \frac{k^\mu k^\nu}{k^2} \frac{(1-\zeta)}{(2-\zeta)} \right). \tag{11.31}$$

Similarly, the ghost–ghost–graviton vertex (denoted by $\overline{V}^\mu(p) V^\nu(q) \phi^{\alpha\beta}(k)$) is found to be

$$\frac{ig}{2} [p^\alpha k^\nu \eta^{\mu\beta} + p^\beta k^\nu \eta^{\mu\alpha} + p \cdot q (\eta^{\mu\alpha} \eta^{\nu\beta} + \eta^{\mu\beta} \eta^{\nu\alpha}) + p^\alpha q^\mu \eta^{\nu\beta}$$
$$+ p^\beta q^\mu \eta^{\nu\alpha} - \zeta p^\mu k^\nu \eta^{\alpha\beta} - \zeta p^\mu q^\beta \eta^{\nu\alpha} - \zeta p^\mu q^\alpha \eta^{\nu\beta}]. \tag{11.32}$$

We have used the convention that all momenta are incoming to the vertices. All ghost-particle vertices are bilinear in the FDM ghost,[12] as shown in the effective Lagrangian (11.30). Thus, the FDM ghosts appear, by definition of the physical subspace, only in closed loops in the intermediate states of a physical process, and there is a factor of -1 for each vector-fermion loop in Yang–Mills gravity.

The Feynman–Dyson rules for the coupling between the graviton and a scalar particle can be obtained from the Lagrangian L_Φ in (11.3). As usual, the propagator $G(k)$ for the scalar particle is given by

$$G(k) = \frac{i}{k^2 - m_\Phi^2}, \tag{11.33}$$

where the $i\epsilon$ prescription is understood. The 3-vertex coupling between graviton and the scalar particle, $\phi_{\mu\nu}(k)\Phi(p)\Phi(q)$ is

$$-ig(p_\mu q_\nu + q_\mu p_\nu). \tag{11.34}$$

The 4-vertex, $\phi_{\alpha\beta}(l)\phi_{\mu\nu}(k)\Phi(p)\Phi(q)$, is given by

$$\frac{-ig^2}{4} [\eta_{\alpha\mu}(p_\beta q_\nu + p_\nu q_\beta) + \eta_{\beta\mu}(p_\alpha q_\nu + p_\nu q_\alpha)$$
$$+ \eta_{\alpha\nu}(p_\beta q_\mu + p_\mu q_\beta) + \eta_{\beta\nu}(p_\alpha q_\mu + p_\mu q_\alpha)], \tag{11.35}$$

where the symmetrization for each index pair $(\mu\nu)$, $(\alpha\beta)$ has been carried out explicitly. Again, all momenta are, by definition, incoming into the vertices. There is a factor $(1/2)$ for each closed loop containing only two graviton lines. For example, this factor $(1/2)$ applies to the process $\Phi\Phi \to \phi_{\alpha\beta}\phi_{\mu\nu} \to \Phi\Phi$.

References

1. K. Huang, in *Quarks, Leptons & Gauge Fields* (World Scientific, 1982), pp. 122–135.
2. J. P. Hsu, *Int. J. Mod. Phys. A* **21**, 5119 (2006); J. P. Hsu and L. Hsu, in *A Broader View of Relativity: General Implications of Lorentz and Poincaré Invariance* (World Scientific, 2006), Appendix D. Available online: "google books, a broader view of relativity, Hsu."
3. G. Wentzel, in *Quantum Theory of Fields*, Trans. from German by C. Houtermans and J. M. Jauch (Interscience, 1943). Appendix I, pp. 203–211.
4. V. A. Fock, in *The Theory of Space Time and Gravitation*, Trans. by N. Kemmer (Pergamon Press, 1959), pp. 158–166; L. Landau and E. Lifshitz, in *The Classical Theory of Fields*, Trans. by M. Hamermesh (Addison-Wesley, 1951), pp. 293–296 and pp. 316–318.
5. R. P. Feynman, in *Theory of Fundamental Processes* (W. A. Benjamin, 1962), p. 168; F. J. Dyson, in *Lorentz and Poincaré Invariance*, eds. J. P. Hsu and Y. Z. Zhang (World Scientific, 2001), p. 316; T. Kinoshita, Talk at Nishina Hall, RIKEN (2010). (Google search: Kinoshita, Magnetic moment.); B. S. DeWitt, *Phys. Rev.* **162**, 1239 (1967).
6. D. M. Capper and M. Ramon Medrano, *Phys. Rev. D* **9**, 1641 (1974); G. Leibbrandt, *Rev. Mod. Phys.* **47**, 849 (1975).
7. D. M. Capper and M. A. Namazie, *Nucl. Phys. B* **142**, 535 (1978).
8. N. Grillo, *Class. Quantum Grav.* **18**, 141 (2001).
9. J. J. Sakurai, in *Advanced Quantum Mechanics* (Addison-Wesley, 1967), p. 313.
10. N. N. Bogoliubov and D. V. Shirkov, in *Introduction to the Theory of Quantized Fields*, Trans. by G. M. Volkov (Interscience, 1959), pp. 206–226.
11. J. P. Hsu and J. A. Underwood, *Phys. Rev. D* **12**, 620 (1975).
12. R. P. Feynman and B. S. DeWitt in *100 Years of Gravity and Accelerated Frames: The Deepest Insights of Einstein and Yang–Mills*, eds. J. P. Hsu and D. Fine (World Scientific, 2005), pp. 272–324; S. Mandelstam, *Phys. Rev.* **175**, 1604 (1968).

Chapter 12

Gravitational Self-Energy of the Graviton

12.1. Graviton Self-Energy in the DeWitt Gauge

Yang–Mills gravity with the space-time translational gauge symmetry has some interesting properties. On the one hand, the graviton self-coupling has a maximum of four gravitons at a vertex, which resembles the situation in renormalizable gauge theories such as quantum chromodynamics (QCD) and electroweak theory. On the other hand, the total Lagrangian (10.16) appears not to be renormalizable by the criterion of power counting because the coupling strength g has the dimension of length, in contrast to usual gauge theories. In order to shed some light on how Yang–Mills gravity might be made renormalizable despite this difference from the usual gauge theories, we now consider the graviton self-energy at the one-loop level and compare the result in Yang–Mills gravity with other known renormalizable theories. In the process, we will employ the D-dimensional regularization to preserve the space-time translational gauge symmetry of physical amplitudes and to define the divergent quantities.

The divergent self-energy of the graviton has contributions from several Feynman diagrams. The most important of these corresponds to the following process:

$$\phi_{\mu\nu}(p) \rightarrow \phi_{\rho\sigma}(q)\phi_{\gamma\eta}(p-q) \rightarrow \phi_{\alpha\beta}(p). \tag{12.1}$$

Its amplitude, denoted by $S_1^{\mu\nu\alpha\beta}(p)$, can be calculated using the Feynman rules with $\xi = 2$ and $\zeta = 1$ in the gauge-fixing Lagrangian (11.4), for simplicity. Process (12.1) leads to the following amplitude:

$$S_1^{\mu\nu\alpha\beta} = \int \frac{d^D q}{(2\pi)^D} V^{\mu\nu\rho\sigma\gamma\lambda}(p, q, k) G_{\rho\sigma\rho'\sigma'}(q)$$

$$\times G_{\tau\eta\gamma\lambda}(k_1) V^{\alpha\beta\rho'\sigma'\tau\eta}(p_1, q_1, k_1), \tag{12.2}$$

where $D \to 4$ at the end of calculations. Since all momenta are, by definition, incoming to the vertex, we have $p + q + k = 0, q_1 = -q, k_1 = -k = p + q$, and $p_1 = -p$. The graviton 3-vertex is given by (11.26), while the graviton propagator in the DeWitt gauge is given by (11.23) with $\xi = 2$ and $\zeta = 1$, i.e.,

$$G_{\rho\sigma,\rho'\sigma'}(q) = \frac{-i}{2q^2}(\eta_{\rho\sigma}\eta_{\rho'\sigma'} - \eta_{\rho\rho'}\eta_{\sigma\sigma'} - \eta_{\rho\sigma'}\eta_{\sigma\rho'}). \tag{12.3}$$

Using FeynCalc[a] to evaluate the integral (12.2) in the limit, $D \to 4$, we obtain

$$\begin{aligned}
S_1^{\mu\nu\alpha\beta} = {}& \frac{ig^2}{24(4\pi)^2}\Gamma(0)[p^4(8\eta^{\alpha\nu}\eta^{\beta\mu} + 8\eta^{\alpha\mu}\eta^{\beta\nu} + 31\eta^{\alpha\beta}\eta^{\mu\nu}) \\
& + 56p^\alpha p^\beta p^\mu p^\nu - 30p^2(p^\alpha p^\beta \eta^{\mu\nu} + p^\mu p^\nu \eta^{\alpha\beta}) \\
& - 7p^2(p^\alpha p^\mu \eta^{\beta\nu} + p^\beta p^\mu \eta^{\alpha\nu} + p^\alpha p^\nu \eta^{\beta\mu} + p^\beta p^\nu \eta^{\alpha\mu}),
\end{aligned} \tag{12.4}$$

where $p^4 = (p^2)^2 = (p_\mu p^\mu)^2$, and $\Gamma(0)$ is the Euler function $\Gamma(n)$ with $n = 0$. The factor of $1/2$ for the Feynman diagram for the process (12.1) is included in (12.4), which is consistent with that of the symbolic computation using xAct (see footnote a).

A second contribution to the graviton self-energy comes from a diagram involving a ghost-loop, which corresponds to the process

$$\phi_{\mu\nu}(p) \to V_\rho(q)V'_\lambda(p - q) \to \phi_{\alpha\beta}(p), \tag{12.5}$$

where the vector ghost-loop is associated with a factor of (-1) due to the fermion property of the Feynman–DeWitt–Mandelstam (FDM) ghost. The virtual process (12.5) leads to the self-energy amplitude,

$$\begin{aligned}
S_2^{\mu\nu\alpha\beta}(p) = {}& \int \frac{d^D q}{(2\pi)^D} V^{\mu\nu\rho\lambda}(p, q, k)G_{\rho\rho'}(q) \\
& \times G_{\lambda\lambda'}(k)V^{\alpha\beta\rho'\lambda'}(p_1, q_1, k_1).
\end{aligned} \tag{12.6}$$

The FDM ghost propagator $G^{\mu\nu}(q)$ and the graviton–ghost–ghost vertex $\phi^{\mu\nu}V'^\rho(k)V^\lambda(q)$ are given by (11.31) and (11.32), respectively, with $\zeta = 1$.

---

[a]We have employed R. Mertig's "FeynCalc" and J. M. Martin-Garcia's package "xAct" to carry symbolic computing and to check consistency. The symbolic computing was carried out in collaboration with S. H. Kim.

With the help of FeynCalc, (12.6) gives the result

$$
\begin{aligned}
S_2^{\mu\nu\alpha\beta} = \frac{-i}{48} \frac{g^2}{(4\pi)^2} \Gamma(0) p^2 [& 7p^4 \eta^{\alpha\beta} \eta^{\mu\nu} + 2p^4 (\eta^{\alpha\mu} \eta^{\beta\nu} + \eta^{\alpha\nu} \eta^{\beta\mu}) \\
& - p^2 (p^\alpha p^\mu \eta^{\beta\nu} + p^\beta p^\mu \eta^{\alpha\nu} + p^\alpha p^\nu \eta^{\beta\mu} + p^\beta p^\nu \eta^{\alpha\mu}) \\
& - 6p^2 (p^\alpha p^\beta \eta^{\mu\nu} + p^\mu p^\nu \eta^{\alpha\beta}) + 8p^\mu p^\nu p^\alpha p^\beta].
\end{aligned}
\tag{12.7}
$$

At the one-loop level, there are three more tadpole diagrams involving (i) a graviton-loop and a graviton 4-vertex, (ii) a graviton-loop and two graviton 3-vertices, and (iii) a ghost-loop with one graviton 3-vertex and a graviton–ghost–ghost vertex. When we use dimensional regularization[1,b] to calculate the diagram corresponding to (i) with a factor of 1/2 for this Feynman diagram, we obtain

$$
S_3^{\mu\nu\alpha\beta} = \frac{1}{2} \int \frac{d^D q}{(2\pi)^D} V^{\mu\nu\sigma\tau\rho\lambda\alpha\beta}(p, q, k, l) G_{\sigma\tau\lambda\rho}(q),
\tag{12.8}
$$

where the 4-vertex $V^{\mu\nu\sigma\tau\rho\lambda\alpha\beta}(p, q, k, l)$ involves terms of the form

$$
p^\rho q^\beta \eta^{\mu\sigma} \eta^{\nu\tau} \eta^{\lambda\alpha}, \text{ etc.}
\tag{12.9}
$$

Each term has one momentum q^β and a D-dimensional integration, so that it vanishes,

$$
\int \frac{d^D q}{(2\pi)^D} q^\beta f(q^2) = 0,
\tag{12.10}
$$

because it is an odd function in q^α.

The other two tadpole diagrams (ii) and (iii) also have vanishing loop-integrals under dimensional regularizations due to (12.10) and

$$
\int \frac{d^D q}{(2\pi)^D} (q^2)^{\beta-1} = 0, \quad \beta = 0, 1, 2, 3, \dots.
\tag{12.11}
$$

These two tadpole diagrams have an additional complication because the loop and the external graviton line $\phi_{\mu\nu}(p) \to \phi_{\alpha\beta}(p')$ are connected by a graviton propagator involving a factor $1/(p - p')^2$, which diverges in the limit $p \to p'$. Thus, these two tadpole amplitudes have the form $0/0$, which is undefined. To overcome this complication, we adopt the dimensional regularization method with the prescription that one first calculates the D-dimensional integrals before

---

[b] For a brief review of different regularizations, see pp. 103–109 in Muta's book *Foundations of Quantum Chromodynamics*.

carrying out the other limiting procedures such as $p \to p'$. In this way, the gravitational tadpoles do not contribute. In gauge theories with internal gauge groups, one might include counterterms to remove the contributions from such tadpole diagrams.[2]

The total divergent graviton self-energy amplitude $S^{\mu\nu\alpha\beta}(p)$ is thus the sum of (12.4) and (12.7),

$$
\begin{aligned}
S^{\mu\nu\alpha\beta} = S_1^{\mu\nu\alpha\beta} + S_2^{\mu\nu\alpha\beta} &= \frac{ig^2}{48(4\pi)^2}\Gamma(0) \\
&\times [p^4(14\eta^{\alpha\nu}\eta^{\beta\mu} + 14\eta^{\alpha\mu}\eta^{\beta\nu} + 55\eta^{\alpha\beta}\eta^{\mu\nu}) \\
&+ 104p^\alpha p^\beta p^\mu p^\nu - 54p^2(p^\alpha p^\beta \eta^{\mu\nu} + p^\mu p^\nu \eta^{\alpha\beta}) \\
&- 13p^2(p^\alpha p^\mu \eta^{\beta\nu} + p^\beta p^\mu \eta^{\alpha\nu} + p^\alpha p^\nu \eta^{\beta\mu} + p^\beta p^\nu \eta^{\alpha\mu})].
\end{aligned}
\tag{12.12}
$$

This self-energy amplitude in the DeWitt gauge is particularly simple. The calculation of this amplitude is done first for a general consistency check of symbolic computing. This calculation also paves the way to much more complicated calculations of self-energy in general gauge involving two gauge parameters ξ and ζ. The physical implications of this result will be discussed in Section 12.3.

12.2.   Graviton Self-Energy in a General Gauge

In general, the self-energy of a particle in gauge theories depends on gauge conditions or the gauge parameters. To wit, let us consider a general gauge-fixing Lagrangian involving two parameters, ξ and ζ:

$$
L_{\xi\zeta} = \frac{\xi}{2g^2}\left[\partial^\mu J_{\mu\alpha} - \frac{\zeta}{2}\partial_\alpha J_\lambda^\lambda\right]\left[\partial_\nu J^{\nu\alpha} - \frac{\zeta}{2}\partial^\alpha J_\lambda^\lambda\right].
\tag{12.13}
$$

In this case, the corresponding graviton propagator also involves two parameters,

$$
\begin{aligned}
G_{\alpha\beta\rho\sigma}(k) = \frac{-i}{2}\bigg[&\frac{1}{k^2}(\eta_{\alpha\beta}\eta_{\rho\sigma} - \eta_{\rho\alpha}\eta_{\sigma\beta} - \eta_{\rho\beta}\eta_{\sigma\alpha}) \\
&+ \frac{1}{k^4}\frac{\xi - 2}{\xi}(k_\sigma k_\beta \eta_{\rho\alpha} + k_\alpha k_\sigma \eta_{\rho\beta} + k_\rho k_\beta \eta_{\sigma\alpha} + k_\rho k_\alpha \eta_{\sigma\beta}) \\
&- \frac{2(\zeta - 1)}{k^4(\zeta - 2)}(k_\alpha k_\beta \eta_{\rho\sigma} + k_\rho k_\sigma \eta_{\alpha\beta}) \\
&- \frac{4(\zeta - 1)}{k^6\xi(\zeta - 2)^2}(6 - \xi - 2\xi\zeta - 2\zeta)k_\alpha k_\beta k_\rho k_\sigma\bigg],
\end{aligned}
\tag{12.14}
$$

where again, the $i\epsilon$ prescription for the Feynman propagators is understood. In the limit, $D \to 4$, FeynCalc (see footnote a) gives the self-energy $S_1^{\mu\nu\alpha\beta}$ due to the graviton loop as

$$S_1^{\mu\nu\alpha\beta} = \frac{ig^2}{(4\pi)^2} \frac{\Gamma(0)}{240\xi^2(\zeta-2)^4} [S_a p^4 (\eta^{\alpha\nu}\eta^{\beta\mu} + \eta^{\alpha\mu}\eta^{\beta\nu})$$

$$+ S_b p^4 \eta^{\alpha\beta}\eta^{\mu\nu} + S_c p^\alpha p^\beta p^\mu p^\nu + S_d p^2 (p^\alpha p^\beta \eta^{\mu\nu} + p^\mu p^\nu \eta^{\alpha\beta})$$

$$+ S_e p^2 (p^\alpha p^\mu \eta^{\beta\nu} + p^\beta p^\mu \eta^{\alpha\nu} + p^\alpha p^\nu \eta^{\beta\mu} + p^\beta p^\nu \eta^{\alpha\mu})], \qquad (12.15)$$

where

$$S_a = [(69\xi^2 + 80\xi + 92)\zeta^4 - 8(54\xi^2 + 47\xi + 92)\zeta^3$$

$$+ 8(165\xi^2 + 52\xi + 252)\zeta^2 - 32(60\xi^2 - \xi + 68)\zeta$$

$$+ 4(273\xi^2 - 60\xi + 196)], \qquad (12.16)$$

$$S_b = 8[(3\xi^2 + 20\xi + 19)\zeta^4 + (101\xi^2 - 62\xi - 152)\zeta^3$$

$$+ (-425\xi^2 + 112\xi + 312)\zeta^2 + 8(85\xi^2 - 67\xi - 4)\zeta$$

$$- 381\xi^2 + 620\xi - 212], \qquad (12.17)$$

$$S_c = 16[(17\xi^2 + 10\xi + 36)\zeta^4 - 3(17\xi^2 + 6\xi + 96)\zeta^3$$

$$+ 2(75\xi^2 - 56\xi + 404)\zeta^2 - 8(35\xi^2 - 22\xi + 116)\zeta$$

$$+ 186\xi^2 - 40\xi + 392], \qquad (12.18)$$

$$S_d = -2[(17\xi^2 + 60\xi + 76)\zeta^4 + 8(58\xi^2 - 11\xi - 76)\zeta^3$$

$$- 8(225\xi^2 + 34\xi - 156)\zeta^2 + 16(155\xi^2 - 44\xi - 8)\zeta$$

$$- 4(301\xi^2 - 380\xi + 212)], \qquad (12.19)$$

$$S_e = [(21\xi^2 - 120\xi - 92)\zeta^4 + (-68\xi^2 + 776\xi + 736)\zeta^3$$

$$- 32(10\xi^2 + 53\xi + 68)\zeta^2 + 16(65\xi^2 + 98\xi + 176)\zeta$$

$$- 4(193\xi^2 + 100\xi + 356)], \qquad (12.20)$$

where we have used (12.2), (12.3) and (12.14).

For the general gauge-fixing Lagrangian (12.13), the graviton–ghost–ghost vertex $\overline{V}^\mu(p)V^\nu(q)\phi^{\alpha\beta}(k) \equiv U^{\mu\nu\alpha\beta}(p,q,k)$ and the ghost propagator $G^{\mu\nu}(q)$ are both dependent on the gauge parameter ζ and are given by

$$U^{\mu\nu\alpha\beta}(p,q,k) = \frac{ig}{2}[p^\alpha k^\nu \eta^{\mu\beta} + p^\beta k^\nu \eta^{\alpha\mu} + p \cdot q(\eta^{\mu\alpha}\eta^{\nu\beta} + \eta^{\mu\beta}\eta^{\nu\alpha})$$
$$+ p^\alpha q^\mu \eta^{\beta\nu} + p^\beta q^\mu \eta^{\alpha\nu} - \zeta p^\mu k^\nu \eta^{\alpha\beta} - \zeta p^\mu q^\beta \eta^{\alpha\nu} - \zeta p^\mu q^\alpha \eta^{\beta\nu}],$$

$$\text{(12.21)}$$

$$G^{\mu\nu}(q) = \frac{-i}{q^2}\left(\eta^{\mu\nu} - \frac{q^\mu q^\nu}{q^2}\frac{(1-\zeta)}{(2-\zeta)}\right), \quad \zeta \neq 2, \quad \text{(12.22)}$$

respectively. FeynCalc gives the following contribution of the ghost-loop to the graviton self-energy:

$$S_2^{\mu\nu\alpha\beta} = -\frac{ig^2}{(4\pi)^2}\frac{\Gamma(0)}{480(\zeta-2)^2}[(15\zeta^2 - 46\zeta + 51)\,p^4(\eta^{\alpha\nu}\eta^{\beta\mu} + \eta^{\alpha\mu}\eta^{\beta\nu})$$
$$+ (5\zeta^2 + 14\zeta + 51)\,p^4\eta^{\alpha\beta}\eta^{\mu\nu} + 8(15\zeta^2 - 46\zeta + 41)\,p^\alpha p^\beta p^\mu p^\nu$$
$$+ 2(18\zeta - 23)\,p^2(p^\alpha p^\mu \eta^{\beta\nu} + p^\beta p^\mu \eta^{\alpha\nu} + p^\alpha p^\nu \eta^{\beta\mu} + p^\beta p^\nu \eta^{\alpha\mu})$$
$$- 2(5\zeta^2 + 22\zeta + 3)\,p^2(p^\alpha p^\beta \eta^{\mu\nu} + p^\mu p^\nu \eta^{\alpha\beta})], \quad \text{(12.23)}$$

where the ghost-loop is associated with a factor (-1) due to the fermion nature of FDM ghost. Finally, we can now obtain the total self-energy amplitude $T^{\mu\nu\alpha\beta}$ for the graviton[3]:

$$T^{\mu\nu\alpha\beta} = S_1^{\mu\nu\alpha\beta} + S_2^{\mu\nu\alpha\beta}$$
$$= \frac{ig^2}{(4\pi)^2}\frac{\Gamma(0)}{480\xi^2(\zeta-2)^4}[T_1\,p^4(\eta^{\alpha\nu}\eta^{\beta\mu} + \eta^{\alpha\mu}\eta^{\beta\nu}) + T_2\,p^4\eta^{\alpha\beta}\eta^{\mu\nu}$$
$$+ T_3\,p^\alpha p^\beta p^\mu p^\nu + T_4\,p^2(p^\alpha p^\beta \eta^{\mu\nu} + p^\mu p^\nu \eta^{\alpha\beta})$$
$$+ T_5\,p^2(p^\alpha p^\mu \eta^{\beta\nu} + p^\beta p^\mu \eta^{\alpha\nu} + p^\alpha p^\nu \eta^{\beta\mu} + p^\beta p^\nu \eta^{\alpha\mu})], \quad \text{(12.24)}$$

where the coefficients $T_1, T_2, \ldots, T_5$ are given by

$$T_1 = (123\xi^2 + 160\xi + 184)\zeta^4 - 2(379\xi^2 + 376\xi + 736)\zeta^3$$
$$+ (2345\xi^2 + 832\xi + 4032)\zeta^2 - 4(863\xi^2 - 16\xi + 1088)\zeta$$
$$+ 4(495\xi^2 - 120\xi + 392), \quad \text{(12.25)}$$

$$T_2 = (43\xi^2 + 320\xi + 304)\zeta^4 + 2(811\xi^2 - 496\xi - 1216)\zeta^3$$
$$- (6815\xi^2 - 1792\xi - 4992)\zeta^2 + 4(2757\xi^2 - 2144\xi - 128)\zeta$$
$$- 6300\xi^2 + 9920\xi - 3392, \tag{12.26}$$

$$T_3 = 8[(53\xi^2 + 40\xi + 144)\zeta^4 - 2(49\xi^2 + 36\xi + 576)\zeta^3$$
$$+ (315\xi^2 - 448\xi + 3232)\zeta^2 - 4(193\xi^2 - 176\xi + 928)\zeta$$
$$+ 4(145\xi^2 - 40\xi + 392)], \tag{12.27}$$

$$T_4 = -2[(29\xi^2 + 120\xi + 152)\zeta^4 + (926\xi^2 - 176\xi - 1216)\zeta^3$$
$$- (3535\xi^2 + 544\xi - 2496)\zeta^2 + (4884\xi^2 - 1408\xi - 256)\zeta$$
$$- 4(605\xi^2 - 760\xi + 424)], \tag{12.28}$$

$$T_5 = 2[(21\xi^2 - 120\xi - 92)\zeta^4 - (86\xi^2 - 776\xi - 736)\zeta^3$$
$$- (225\xi^2 + 1696\xi + 2176)\zeta^2 + 4(219\xi^2 + 392\xi + 704)\zeta$$
$$- 8(85\xi^2 + 50\xi + 178)]. \tag{12.29}$$

The self-energy has a very complicated dependence on the gauge parameters ξ and ζ. In the special case $\xi = 2$ and $\zeta = 1$, the general result (12.24) reduces to the graviton self-energy in the DeWitt gauge given by (12.12), as expected.

12.3. Discussion

The physical graviton propagator is different from the free propagator (12.3) in a general gauge because a graviton can disintegrate virtually into two gravitons and an FDM ghost pair for a small interval of time, as indicated in (12.1) and (12.5), respectively. A similar situation also occurs in quantum Einstein gravity.[4] After dimensional regularization, the ultraviolet divergence of the graviton self-energy contains only the simple pole terms in (12.24). In this sense, the ultraviolet divergence in Yang–Mills gravity is no worse than that in quantum electrodynamics (QED) at the one-loop level. Yang–Mills gravity and Einstein gravity also have another type of divergence at low energies, i.e., the infrared divergence, which is found in other massless gauge theories, which will not be discussed here.

In QED, it is easy to see how such a self-interaction affects the photon self-energy (or mass) by summing up the free photon propagator and its one-loop correction.[5] The situation is much more complicated for the graviton, however. For simplicity, let us consider the graviton propagator in the DeWitt gauge, i.e., (12.3). We have

$$G_{\alpha\beta\rho\sigma}(p) = -i\left[\frac{1}{2p^2}(\eta_{\alpha\beta}\eta_{\rho\sigma} - \eta_{\rho\alpha}\eta_{\sigma\beta} - \eta_{\rho\beta}\eta_{\sigma\alpha})\right], \tag{12.30}$$

which is independent of the gauge parameters and corresponds to the Feynman gauge in QED. The one-loop correction to $G_{\mu\nu\alpha\beta}$ is

$$G^{(1)}_{\mu\nu\alpha\beta} = G_{\mu\nu\mu'\nu'}S^{\mu'\nu'\alpha'\beta'}G_{\alpha'\beta'\alpha\beta}, \tag{12.31}$$

using (12.12) and (12.30), so that

$$
\begin{aligned}
G^{(1)}_{\mu\nu\alpha\beta} = \frac{-ig^2\Gamma(0)}{48(4\pi)^2 p^4}[&14p^4(\eta^{\alpha\nu}\eta^{\beta\mu} + \eta^{\alpha\mu}\eta^{\beta\nu} + \eta^{\alpha\beta}\eta^{\mu\nu})\\
&+ 104p^\alpha p^\beta p^\mu p^\nu + 28p^2(p^\alpha p^\beta \eta^{\mu\nu} + p^\mu p^\nu \eta^{\alpha\beta})\\
&- 13p^2(p^\alpha p^\mu \eta^{\beta\nu} + p^\beta p^\mu \eta^{\alpha\nu} + p^\alpha p^\nu \eta^{\beta\mu} + p^\beta p^\nu \eta^{\alpha\mu})].
\end{aligned}
\tag{12.32}
$$

Unlike the equivalent situation for photons in QED, the form of (12.32) differs from that of a free graviton propagator (12.30). This difference appears to suggest that the quantum correction of the graviton propagator does not affect the graviton mass.

We have thus seen that the divergences in the graviton self-energy at the one-loop level resemble those in QED, i.e., they all involve simple poles $\Gamma(0)$, corresponding to a logarithmic divergence. Thus, it is indeed possible to remove those divergences using gauge-dependent terms, even if the structure of the counterterms necessary to remove the divergences in the graviton self-energy is more complicated than the corresponding terms for the photon self-energy in QED. Presumably, these differences in the structure of the counterterms reflect the difference in the gauge symmetry groups of Yang–Mills gravity and QED: the external space-time translational (non-compact) group T_4 and the internal (compact) group U_1. To see more dramatic differences in the divergences, it is likely necessary to calculate the graviton self-energy at the 2-loop level.

The motivation for investigating the high-energy behavior of quantum gravity is to have a complete understanding of the whole of field theory, rather than just its experimental verifications. Because of the weakness of the coupling strength of the gravitational interaction ($\approx 10^{-39}$), the experimental effects of quantum gravity may never be detected in Earth-bound laboratories. However, we would like to make sure that observable results from QCD, the electroweak theory, and all other calculations at relatively low energies[6,7] will not be upset by higher order corrections due to the gravitational interaction. This is important theoretically because all physical particles in nature participate in gravitational interactions.[c]

References

1. C. G. Bollini and J. J. Giambiagi, *Phys. Lett. B* **40**, 566 (1972); *Nuovo Cimento B* **12**, 20 (1972); G. 't Hooft and M. T. Veltman, *Nucl. Phys. B* **44**, 189 (1972); J. F. Ashmore, *Lett. Nuovo Cimento* **4**, 289 (1972); G. Cicuta and E. Montaldi, *Lett. Nuovo Cimento* **4**, 329 (1972); T. Muta, in *Foundations of Quantum Chromodynamics* (World Scientific, 1987), pp. 103–109.

2. J. P. Hsu and J. A. Underwood, *Phys. Rev. D* **15**, 1668 (1977).

3. J. P. Hsu and S. H. Kim, *Eur. Phys. J. Plus* **127**, 146 (2012), DOI: 10.1140/epjp/i2012-12146-3; arXiv:1210.4503v1.

4. D. M. Capper and M. Ramon Medrano, *Phys. Rev. D* **9**, 1641 (1974); G. Leibbrandt, *Rev. Mod. Phys.* **47**, 849 (1975); D. M. Capper and M. A. Namazie, *Nucl. Phys. B* **142**, 535 (1978).

5. J. J. Sakurai, in *Advanced Quantum Mechanics* (Addison-Wesley, 1967), pp. 273–275.

6. Q. Wang and Y. P. Kuang, in *Symposium on the Frontiers of Physics at Millennium*, eds. Y. L. Wu and J. P. Hsu (World Scientific, 2001), pp. 51–59 and pp. 344–354; Y. P. Kuang and B. Fu, *Phys. Rev. D* **42**, 2300 (1990).

7. K. C. Chou and Y. L. Wu, *Phys. Rev. D* **53**, R3492 (1996) (Rapid Communication).

---

[c]Thus, strictly speaking, unless one has a finite quantum theory of gravity, one cannot say that the observed physical mass of say, an electron, is finite and independent of the gauge condition.

Chapter 13

Space-Time Gauge Identities
and Finite-Loop Renormalization

13.1. Space-Time (T_4) Gauge Identities

We continue our discussion of renormalizability in Yang–Mills gravity. In quantum electrodynamics (QED), the Abelian U_1 gauge symmetry implies that a particular linear gauge condition can be imposed for all times.[a] This local gauge symmetry leads to the conservation of the electric charge and the Ward–Takahashi–Fradkin (WTF) identities,[2] which imply relations between different renormalization constants. In non-Abelian gauge theories with internal gauge groups, such as the Yang–Mills theory however, a particular gauge condition cannot be imposed for all times. If one imposes a gauge condition at a certain time, it does not hold at a later time. Faddeev and Popov introduced a δ-function involving the gauge condition to the vacuum-to-vacuum amplitude in gauge theory, as shown in equation (10.10). As a result, there emerges a new Lagrangian involving additional couplings of gauge fields with anticommuting scalar fields (or ghost scalar-fermions). These ghosts are called Faddeev–Popov ghosts, and appear only in the intermediate steps of a physical process. The non-Abelian gauge symmetry leads to Slavnov–Taylor identities, which are a generalization of the WTF identities.[3]

In Yang–Mills gravity, the situation is different from either of the cases described. It has both the Abelian T_4 gauge symmetry and a nonlinear gauge field equation, so that a particular gauge condition for the Abelian T_4 gauge field cannot be imposed for all times. In order to restore gauge invariance and unitarity of the

---

[a]However, if one chooses a non-linear gauge condition, the gauge condition cannot be imposed at all times, even in quantum electrodynamics. In this case, one needs to introduce scalar ghosts to restore gauge invariance and unitarity in electrodynamics (see, e.g., Ref. 1).

S-matrix, there must be additional couplings between the T_4 gauge fields $\phi_{\mu\nu}$ and the anticommuting vector fields $V_\mu(x)$ and $\overline{V}_\nu(x)$, which are Feynman–DeWitt–Mandelstam (FDM) ghosts in quantum Yang–Mills gravity.[4–6]

In this first part of this chapter, we derive generalized WTF identities in Yang–Mills gravity for a class of gauge conditions involving two gauge parameters ξ and ζ, as shown in the gauge-fixing Lagrangian (11.4). By explicit calculation in the lowest order (i.e., at the lowest tree-level), we show that the T_4 gauge identities are satisfied by a general propagator of the graviton involving ξ and ζ, and by FDM ghosts involving the parameter ζ.

Let us derive the general T_4 gauge identities for the class of gauge conditions

$$F_\lambda(J) = Y_\lambda(x), \quad J \equiv J_{\mu\nu} = \eta_{\mu\nu} + g\phi_{\mu\nu}, \tag{13.1}$$

$$F_\lambda(J) = \frac{1}{2}(\delta_\lambda^\mu \partial^\nu + \delta_\lambda^\nu \partial^\mu - \zeta\eta^{\mu\nu}\partial_\lambda)J_{\mu\nu} \equiv h_\lambda^{\mu\nu}J_{\mu\nu}. \tag{13.2}$$

In Yang–Mills gravity, we shall consistently use the covariant tensor $J_{\mu\nu}$ in (13.1) to express the T_4 gauge curvature $C_{\mu\nu\alpha} = J_{\mu\sigma}\partial^\sigma J_{\nu\alpha} - J_{\nu\sigma}\partial^\sigma J_{\mu\alpha}$, the gravitational action $S_\phi = \int d^4x L_\phi$, the gauge conditions, etc., and to perform variations and calculations.

First, we consider the vacuum-to-vacuum amplitude with the gauge conditions (13.1). It is given by

$$W = \int d[J](\det U)\exp\left[i\int d^4x(L_\phi + L_{\xi\zeta})\right] \tag{13.3}$$

$$= \int d[J,\overline{V},V]\exp\left[i\int d^4x(L_\phi + L_{\xi\zeta} + \overline{V}^\mu U_{\mu\nu}V^\nu)\right], \tag{13.4}$$

$$L_{\xi\zeta} = \frac{\xi}{2g^2}F_\alpha F_\beta \eta^{\alpha\beta}, \quad F_\alpha = h_\alpha^{\mu\nu}J_{\mu\nu}, \tag{13.5}$$

where L_ϕ and $U_{\mu\nu}$ are given by (10.2) and (10.12), respectively. To generate Green's functions,[7,8] we introduce an external source term $j_{\mu\nu}$ for the T_4 gauge fields in the vacuum-to-vacuum amplitudes (13.3),

$$W[j] \equiv \int d[J](\det U)\exp\left[i\int d^4x(L_\phi + L_{\xi\zeta} + J_{\mu\nu}j^{\mu\nu})\right], \tag{13.6}$$

where $W[j]$, $\{d[J] \det U\}^b$ and $\int d^4x L_\phi$ are invariant under the infinitesimal T_4 gauge transformations (8.5)–(8.6),

$$T_{\mu\nu} \to (T_{\mu\nu})^\$ = T_{\mu\nu} - \Lambda^\lambda \partial_\lambda T_{\mu\nu} - T_{\mu\alpha} \partial_\nu \Lambda^\alpha - T_{\alpha\nu} \partial_\mu \Lambda^\alpha, \tag{13.7}$$

$$Q^{\mu\nu} \to (Q^{\mu\nu})^\$ = Q^{\mu\nu} - \Lambda^\lambda \partial_\lambda Q^{\mu\nu} + Q^{\lambda\nu} \partial_\lambda \Lambda^\mu + Q^{\mu\lambda} \partial_\lambda \Lambda^\nu. \tag{13.8}$$

In the derivation of the T_4 gauge identities, we impose the following constraint for the gauge function $\Lambda_\lambda(x)$:

$$\Lambda_\lambda = [U^{-1}(J)]_{\lambda\sigma} \rho^\sigma \equiv \int [U^{-1}(x,y)]_{\lambda\sigma} \rho^\sigma(y) d^4y, \tag{13.9}$$

where ρ^σ is an arbitrarily infinitesimal function independent of $\phi_{\mu\nu}(x)$.

Under the infinitesimal T_4 gauge transformation (13.7)–(13.8), we have

$$J_{\mu\nu}^\$ = J_{\mu\nu} + \delta J_{\mu\nu}, \quad F_\nu^\$ = F_\nu + \delta F_\nu, \quad W[j]^\$ = W[j], \tag{13.10}$$

where

$$\delta J_{\mu\nu} = f_{\mu\nu\lambda} \Lambda^\lambda = f_{\mu\nu\lambda} [U^{-1}(J)]^{\lambda\alpha} \rho_\alpha, \tag{13.11}$$

$$\delta F_\lambda(J) = h_{\mu\nu\lambda} f^{\mu\nu\alpha} \Lambda_\alpha = h_{\mu\nu\lambda} f^{\mu\nu\alpha} [U^{-1}(J)]_{\alpha\beta} \rho^\beta = \rho_\lambda, \tag{13.12}$$

where we have used (13.9) and

$$f_{\mu\nu\lambda} = -[(\partial_\lambda J_{\mu\nu}) + J_{\lambda\nu} \partial_\mu + J_{\mu\lambda} \partial_\nu],$$

$$h_\lambda^{\mu\nu} = \frac{1}{2}(\delta_\lambda^\mu \partial^\nu + \delta_\lambda^\nu \partial^\mu - \zeta \eta^{\mu\nu} \partial_\lambda). \tag{13.13}$$

The operator $U_{\rho\lambda}$ is given by

$$U_{\rho\lambda}(J) = -h_\rho^{\mu\nu} f_{\mu\nu\lambda}, \tag{13.14}$$

which is the same as that in equation (10.12) and consistent with (10.15) in Chapter 10.

The invariance of $d[J] \det U$ has been well-established for all compact Lie groups. However, for noncompact and infinite continuous groups, the formal proof of its invariance in the literature is not completely satisfactory. Cf. equations (13.56)–(13.57) and discussions below.

From (13.6) and (13.10), we have

$$\frac{\delta W[j]}{\delta \rho_\alpha} = \int d[\phi] (\det U) \exp\left[i \int d^4x (L_\phi + L_{\xi\zeta} + J_{\mu\nu} j^{\mu\nu})\right] Z^\alpha = 0, \quad (13.15)$$

where

$$Z^\alpha = \frac{\delta}{\delta \rho_\alpha} \int d^4x \left(\frac{\xi}{g^2} F_\alpha \delta F_\beta \eta^{\alpha\beta} + j^{\mu\nu} \delta J_{\mu\nu}\right). \quad (13.16)$$

Equation (13.16) implies

$$Z^\alpha = \frac{\xi}{g^2} F^\alpha(J) + j^{\mu\nu} f_{\mu\nu\lambda} [U^{-1}(J)]^{\lambda\alpha}. \quad (13.17)$$

Thus, the T_4 identities (13.15) can also be written in the form

$$\left[\left(\frac{\xi}{g^2} F^\alpha(\Delta)\right)_y + \int d^4z \, j^{\mu\nu}(z)(f_{\mu\nu\lambda})_z [U^{-1}(\Delta)]^{\lambda\alpha}_{zy}\right] W[j] = 0,$$
$$\Delta \equiv \frac{1}{i} \frac{\delta}{\delta j}. \quad (13.18)$$

For those identities to be useful, we must know $[U^{-1}]W[j]$. Using (13.4) with an external source, let us consider[7]

$$W^{\mu\nu}[j] = \int d[\phi, \overline{V}, V] V^\mu \overline{V}^\nu \exp\left[i \int d^4x (L_{\text{tot}} + j^{\mu\nu} J_{\mu\nu})\right], \quad (13.19)$$

$$L_{\text{tot}} = L_\phi + L_{\xi\zeta} + \overline{V}^\mu U_{\mu\nu} V^\nu, \quad (13.20)$$

where $\det U$ in (13.6) and (13.15) can be described as loops generated by FDM ghost fields $\overline{V}^\mu$ and V^ν, as shown in (10.14) and (10.15).

This Green's function for FDM ghosts V^μ and $\overline{V}^\nu$ in the presence of external source satisfies

$$[U(\Delta)]_{\mu\nu} W^{\nu\lambda}[j] = \delta^\lambda_\mu W[j]. \quad (13.21)$$

For the operator $U(\Delta)$, we have

$$[U^{-1}(\Delta)]^{\alpha\mu} [U(\Delta)]_{\mu\nu} = \delta^\alpha_\nu. \quad (13.22)$$

It follows from (13.18), (13.21) and (13.22) that

$$\int \exp\left[i\int d^4x(L_{\text{tot}} + J_{\mu\nu}(x)j^{\mu\nu}(x))\right]$$

$$\times \left[\frac{\xi}{g^2}(h_{\mu\nu\lambda}J^{\mu\nu})_y + \int d^4z\, \overline{V}^\nu(y)j^{\mu\nu}(z)(f_{\mu\nu\lambda}V^\lambda)_z\right] d[\phi, V, \overline{V}] = 0,$$

$$(13.23)$$

where

$$\left(f_{\mu\nu\lambda}V^\lambda\right)_z = \left[-\left(\frac{\partial}{\partial z^\lambda}J_{\mu\nu}(z)\right) - J_{\mu\lambda}(z)\frac{\partial}{\partial z^\nu} - J_{\lambda\nu}(z)\frac{\partial}{\partial z^\mu}\right]V^\lambda(z), \quad \text{etc.}$$

These are the gravitational space-time gauge identities for the generating functional $W[j]$ for the wide class of gauge conditions specified by (13.1) and (13.2).

13.2.   Space-Time Gauge Identities and a General Graviton Propagator

In general, the graviton and ghost propagators in the Feynman–Dyson rules depend on the specific form of the gauge condition and gauge parameters. For example, if one chooses a gauge condition specified by the gauge-fixing Lagrangian (13.5) and (13.2) involving two arbitrary parameters ξ and ζ, the graviton propagator will depend on the two gauge parameters,[9]

$$G_{\alpha\beta\rho\sigma} = \frac{-i}{2k^2}\left(\left[\eta_{\alpha\beta}\eta_{\rho\sigma} - \eta_{\rho\alpha}\eta_{\sigma\beta} - \eta_{\rho\beta}\eta_{\sigma\alpha}\right] - \frac{1}{k^2}\frac{(2\zeta - 2)}{(\zeta - 2)}\right.$$

$$\times (k_\alpha k_\beta\eta_{\rho\sigma} + k_\rho k_\sigma\eta_{\alpha\beta}) + \frac{1}{k^2}\frac{(\xi - 2)}{\xi}(k_\sigma k_\beta\eta_{\rho\alpha} + k_\alpha k_\sigma\eta_{\rho\beta} + k_\rho k_\beta\eta_{\sigma\alpha}$$

$$+ k_\rho k_\alpha\eta_{\sigma\beta}) - \frac{1}{k^4}\frac{(2\zeta - 2)}{\xi(\zeta - 2)}\left[\frac{4 + 2\xi - 4\xi\zeta}{\zeta - 2} - 4 + 2\xi\right]k_\alpha k_\beta k_\rho k_\sigma\right),$$

$$(13.24)$$

where the $i\epsilon$ prescription for the Feynman propagator is understood.

The propagator $G^{\mu\nu}$ of the FDM ghost particle can be derived from the effective Lagrangian (13.20) and the relation (13.14) for $U_{\mu\nu}$. We obtain

$$G^{\mu\nu} = \frac{-i}{k^2}\left(\eta^{\mu\nu} - \frac{k^\mu k^\nu}{k^2}\frac{(1 - \zeta)}{(2 - \zeta)}\right), \qquad (13.25)$$

where ζ is not equal to 2.

Differentiating the gauge identities (13.23) with respect to $j_{\mu\nu}$ and letting all external sources vanish, we obtain the simplest identity

$$\xi\langle T\,(h^{\lambda\mu\nu}\phi_{\mu\nu})_x\phi^{\alpha\beta}(y)\rangle = -\langle T\,\overline{V}^\lambda(x)(f^{\alpha\beta\sigma}V_\sigma)_y\rangle, \qquad (13.26)$$

where T denotes chronological ordering of the fields. This is a generalized WTF identity for an Abelian T_4 gauge group because (13.26) contains FDM ghosts.

In the lowest order, the identity (13.26), together with (13.2) and (13.13), leads to the identity,

$$\xi k^\beta G_{\alpha\beta\rho\sigma}(k) - \frac{\xi\zeta}{2}k_\alpha\eta^{\mu\nu}G_{\mu\nu\rho\sigma}(k) = -k_\sigma G_{\alpha\rho}(k) - k_\rho G_{\alpha\sigma}, \qquad (13.27)$$

where $G_{\alpha\beta\rho\sigma}(k)$ and $G_{\alpha\rho}(k)$ are general propagators of the graviton (13.24) and the FDM ghost (13.25), respectively.

Direct calculations give the following results:

$$A_{\alpha\rho\sigma} \equiv \xi k^\beta G_{\alpha\beta\rho\sigma}(k) = \frac{-i}{2k^2}\left[-\xi k_\sigma\eta_{\rho\alpha}\frac{\zeta}{\zeta-2} - 2(k_\sigma\eta_{\rho\alpha} + k_\rho\eta_{\sigma\alpha})\right.$$
$$\left. + \frac{k_\alpha k_\rho k_\sigma}{k^2}\left(2\xi - 4 + \frac{2\zeta - 2}{\zeta-2}\left[-3\xi + 4 + \frac{4\xi\zeta - 4 - 2\xi}{\zeta-2}\right]\right)\right], \qquad (13.28)$$

$$B_{\alpha\rho\sigma} \equiv -\frac{\xi\zeta}{2}k^\alpha\eta^{\mu\nu}G_{\mu\nu\rho\sigma}(k) = \frac{-i}{2k^2}\left[-\xi k_\alpha\eta_{\rho\sigma}\frac{\zeta}{\zeta-2} + \frac{k_\alpha k_\rho k_\sigma}{k^2}\right.$$
$$\left. \times\left(-2\xi\zeta + 4\zeta + \frac{\zeta(2\zeta-2)}{\zeta-2}\left[-3\xi - 2 - \frac{2\xi\zeta - 2 - \xi}{\zeta-2}\right]\right)\right]. \qquad (13.29)$$

From (13.28) and (13.29), we obtain

$$A_{\alpha\rho\sigma} + B_{\alpha\rho\sigma} = \frac{-i}{k^2}\left[-(k_\sigma\eta_{\rho\alpha} + k_\rho\eta_{\sigma\alpha}) + \frac{2\zeta-2}{\zeta-2}\left(\frac{k_\alpha k_\rho k_\sigma}{k^2}\right)\right]. \qquad (13.30)$$

Using (13.25), the right-hand side of (13.26) gives the result,

$$C_{\alpha\rho\sigma} \equiv -k_\sigma G_{\alpha\rho}(k) - k_\rho G_{\alpha\sigma}$$
$$= \frac{-i}{k^2}\left[-k_\sigma\eta_{\alpha\rho} - k_\rho\eta_{\alpha\sigma} + \frac{2k_\sigma k_\alpha k_\rho(1-\zeta)}{k^2(2-\zeta)}\right]. \qquad (13.31)$$

The left-hand side of (13.27) given by (13.30) is the same as the right-hand side of (13.27) given by (13.31). Thus, we have verified the space-time gauge identity (13.27) for general gauge conditions by explicit calculation using a very general graviton propagator involving two arbitrary gauge parameters, ξ and ζ.

The calculation of identities for the next order with a one-loop correction is formidable and requires the aid of a computer.

13.3.   Gauge Identities in QED with a Nonlinear Gauge Condition

From the explicit calculations of the space-time gauge identities (13.27)–(13.31), one can see cancellations of a large number of terms due to the space-time gauge symmetry. The situation is similar to that in non-Abelian gauge field theory: When one calculates the imaginary part of the amplitudes coming from the interactions of say, the unphysical component of a vector boson and scalar ghost to verify unitarity of physical amplitudes, one sees many cancellations of unphysical amplitudes involving gauge parameters, so that unitarity of the S-matrix is preserved. This is the power of gauge symmetry of internal groups.[8]

One might think that quantum chromodynamics has ghost particles solely because its gauge group $[SU_3]_{color}$ is non-Abelian and that QED, with the Abelian group U_1, does not have ghosts. However, this is true if and only if one imposes a linear gauge condition in QED. As a quick illustration of the power of gauge symmetry as instantiated by the generalized WTF identities, we consider the case of QED with a class of nonlinear gauge condition (see footnote a)

$$F(A) \equiv \partial_\mu A^\mu - \beta' A_\mu A^\mu = a(x), \quad \beta' \neq 0, \quad (13.32)$$

where $a(x)$ is a real function independent of A_μ and $\Lambda(x)$.

In the absence of electrons, the gauge invariant Lagrangian L_A for photons implies that photons do not couple with themselves and are free particles. However, when one chooses a nonlinear gauge condition (13.32), there will be new 3-vertex and 4-vertex interactions of photons. These new photon–photon interactions upset the gauge invariance and unitarity of the theory. However, the U_1 gauge symmetry assures that the extra unwanted amplitudes produced by these new vertices will be canceled by the interactions of new electromagnetic ghosts, so that gauge invariance and unitarity of the physical amplitudes are preserved.

The generating functional of the Green's function in the nonlinear gauge can be written as[1]

$$W_\alpha[\eta, \overline{\eta}, \eta_\mu] = \int d[\Omega] (\det M) \exp\left[i \int (L + L_s - \frac{1}{2\alpha} F^2) d^4x \right], \quad (13.33)$$

$$L = -\frac{1}{4} F_{\mu\nu} F^{\mu\nu} + L_\psi, \quad L_s = \eta_\mu A^\mu + \overline{\eta}\psi + \overline{\psi}\eta, \quad [\Omega] \equiv [A_\mu, \overline{\psi}, \psi], \quad (13.34)$$

$$M = (\partial_\mu - 2\beta' A_\mu)\partial^\mu, \quad \det M = \int d[\bar{c}, c] \exp\left(i \int d^4x \, \bar{c}Mc\right), \quad (13.35)$$

where $F_{\mu\nu} = \partial_\mu A_\nu - \partial_\nu A_\mu$, and L_ψ is the usual fermion Lagrangian in QED. The generalized WTF identities are obtained in the form

$$\left[-\frac{1}{\alpha}\left(\partial_\mu \frac{\delta}{i\delta\eta_\mu} + \beta'\frac{\delta^2}{\delta\eta^\mu\delta\eta_\mu} \right) + \eta_\mu\partial^\mu M^{-1} \right.$$
$$\left. - e\frac{\delta}{\delta\eta}\eta M^{-1} - e\bar{\eta}\frac{\delta}{\delta\bar{\eta}}M^{-1} \right] W_\alpha[\eta, \bar{\eta}, \eta_\mu] = 0. \quad (13.36)$$

Here, we have used the gauge transformation $A'_\mu(x) = A_\mu(x) + \partial_\mu\Lambda(x)$ with the constraint $\Lambda(x) = M^{-1}\rho(x)$, where $\rho(x)$ is an arbitrary infinitesimal real function, independent of A_μ. Following similar steps from (13.18) to (13.23), we can express the generalized WTF identities (13.36) in the form

$$\int \exp\left[i \int d^4x(L_{\text{eff}} + A_\mu\eta^\mu + \bar{\eta}\psi + \bar{\psi}\eta) \right]$$
$$\times \left[\frac{1}{\alpha}[(\partial_\mu - \beta' A_\mu)A^\mu]_y + \int d^4z \, \bar{c}(y)\eta^\mu(z)[\partial_\mu c]_z \right] d[\Omega, c, \bar{c}] = 0,$$
$$(13.37)$$

where

$$L_{\text{eff}} = L + \frac{1}{2\alpha}(\partial_\mu A^\mu - \beta' A_\mu A^\mu)^2 + \bar{c}Mc, \quad [\partial_\mu c]_z = \frac{\partial}{\partial z^\mu}c(z), \quad \text{etc.}$$

Thus, we have derived the generalized WTF identities involving electromagnetic ghosts $\bar{c}(x)$ and $c(x)$ for the generating functional $W_\alpha[\eta, \bar{\eta}, \eta_\mu]$ in the nonlinear gauge specified by (13.32).

To the lowest order, the simplest identity obtained from (13.37) is given by

$$\frac{1}{\alpha}k^\mu G_{\mu\nu}(k) + k_\nu G(k) = 0, \quad (13.38)$$

which corresponds to (13.37). The propagators for the photon and the electromagnetic ghost are given by

$$G_{\mu\nu}(k) = \frac{-i}{k^2}\left(\eta_{\mu\nu} - (1-\alpha)\frac{k_\mu k_\nu}{k^2} \right) \quad \text{and} \quad G(k) = \frac{i}{k^2}, \quad (13.39)$$

respectively, where the $i\epsilon$ prescription for the propagator is understood. One can verify that the identity (13.38) is indeed satisfied.

The arbitrary gauge parameter β' does not appear in the lowest-order identity (13.38). It does appear in the higher order identities and also in the additional ghost–photon coupling in the effective Lagrangian L_{eff} in (13.37). However, when one calculates physical amplitudes such as e^+e^- scattering, the terms involving the gauge parameter β' are all canceled,[1] so that the observable results are, as expected, unitary and independent of the parameter β'.

13.4. The Infinite Continuous Group of General Coordinate Transformations in Flat Space-Time

As discussed in Section 7.5, both local translations and the general coordinate transformations in flat space-time can be expressed by

$$x^\mu \to x'^\mu = x^\mu + \Lambda^\mu(x), \tag{13.40}$$

where $\Lambda^\mu(x)$ is an arbitrary infinitesimal vector function. In Chapter 8, we used the generators of space-time translations to formulate Yang–Mills gravity in flat space-time. Now for a detailed discussion of space-time gauge identities, we must also consider the generators and group properties of the general coordinate transformations in flat space-time.

According to the general transformation formula (8.3) for tensors such as scalars, vectors, etc., we have

$$Q(x) \to Q^\$(x) = Q(x) - \Lambda^\lambda(x)\partial_\lambda Q(x), \quad Q = \psi, \overline{\psi}, \Phi, \tag{13.41}$$

$$V_\mu(x) \to V_\mu^\$(x) = V_\mu(x) - \Lambda^\lambda(x)\partial_\lambda V_\mu(x) - V_\lambda(x)\partial_\mu \Lambda^\lambda(x), \tag{13.42}$$

$$J_{\mu\nu}(x) \to J_{\mu\nu}^\$(x) = J_{\mu\nu}(x) - \Lambda^\lambda(x)\partial_\lambda J_{\mu\nu}(x) - J_{\mu\alpha}(x)\partial_\nu \Lambda^\alpha(x)$$
$$- J_{\alpha\nu}(x)\partial_\mu \Lambda^\alpha(x). \tag{13.43}$$

The second term involving $-\Lambda^\lambda(x)\partial_\lambda$ on the right-hand side of (13.41)–(13.43) is the displacement term, which is related to the local space-time displacement. To see the structure of generators associated with the infinite continuous group of general coordinate transformations, we write (13.41)–(13.43) in the following form:

$$\delta Q(x) = Q^\$(x) - Q(x) = -[Q(x)\overleftarrow{\partial}_\lambda]\Lambda^\lambda(x)$$

$$= -\int \partial_\lambda\delta(x-x')Q(x')\Lambda^\lambda(x)dx'$$

$$= -\int \partial_\lambda\delta(x-x')\delta(x-x'')Q(x')\Lambda^\lambda(x'')dx'dx'', \qquad (13.44)$$

$$\delta V_\mu(x) = V_\mu^\$(x) - V_\mu(x) = -V_\alpha(x)\left[\overleftarrow{\partial}_\lambda\delta_\mu^\alpha + \delta_\lambda^\alpha\partial_\mu\right]\Lambda^\lambda(x)$$

$$= -\int [\partial_\lambda\delta(x-x')\delta(x-x'')\delta_\mu^\alpha + \delta_\lambda^\alpha\delta(x-x')\partial_\mu\delta(x-x'')]$$

$$\times V_\alpha(x')\Lambda^\lambda(x'')dx'dx''$$

$$\equiv H_\mu{}^{\alpha'}{}_{\lambda''}V_{\alpha'}\Lambda^{\lambda''}, \qquad (13.45)$$

$$\delta J_{\mu\nu}(x) = -J_{\alpha\beta}(x)\left[\delta_{\mu\nu}^{\alpha\beta}\overleftarrow{\partial}_\lambda + \delta_{\mu\lambda}^{\alpha\beta}\partial_\nu + \delta_{\lambda\nu}^{\alpha\beta}\partial_\mu\right]\Lambda^\lambda(x)$$

$$= -\int \left[\delta_{\mu\nu}^{\alpha\beta}\partial_\lambda\delta(x-x')\delta(x-x'') + \delta_{\mu\lambda}^{\alpha\beta}\delta(x-x')\partial_\nu\delta(x-x'')\right.$$

$$\left.+ \delta_{\lambda\nu}^{\alpha\beta}\delta(x-x')\partial_\mu\delta(x-x'')\right]J_{\alpha\beta}(x')\Lambda^\lambda(x'')dx'dx''$$

$$\equiv H_{\mu\nu}{}^{\alpha'\beta'}{}_{\lambda''}J_{\alpha'\beta'}\Lambda^{\lambda''}, \qquad (13.46)$$

$$\delta(x) \equiv \delta^4(x), \quad dx' \equiv d^4x', \text{ etc.,}$$

where summation and integration over repeated indices (and space-time variables) are understood. Let us compare (13.45) with the SU_2 gauge transformation of the Yang–Mills field b_μ^a,

$$\delta b_\mu^a(x) = \epsilon^{abc}\omega^b(x)b_\mu^c(x) + \frac{1}{f}\partial_\mu\omega_a(x). \qquad (13.47)$$

We see that $H_\mu{}^{\alpha'}{}_{\lambda''}$ in (13.45) corresponds to the structure constant ϵ^{abc} in (13.47). For SU_2, we have $-i\epsilon_{abc} = (\tau_a/2)_{bc}$, where $\tau_a/2$ are the SU_2 generators. Similarly, one may consider $H_\mu{}^{\alpha'}{}_{\lambda''}$ as a matrix element of the generators $H_{\lambda''}$ of the group, which satisfy the commutation relation,[10,c]

$$[H_\mu, H_\nu] = H_{\lambda''}C^{\lambda''}{}_{\mu\nu}, \qquad (13.48)$$

---

[c]DeWitt's discussion of the group of general coordinate transformations is in curved space-time, while our discussion of infinitesimal local translations and general coordinate transformations (13.40) is in flat space-time.

where the structure constants of the group are given by

$$C^{\lambda''}{}_{\mu\nu'} = \delta^\sigma_\mu \partial_\nu \delta(x'' - x)\delta(x'' - x') - \delta^\sigma_\nu \partial_\mu \delta(x'' - x')\delta(x'' - x). \qquad (13.49)$$

The generators of the infinite continuous group are quite different from the internal groups in non-Abelian gauge theories, in which the generators have constant matrix representations.

There is a subtlety related to gauge transformations of the metric $d[J_{\mu\nu}] \det U$ and their relation to the space-time gauge identities. To see it, let us examine the gauge transformation of $W[j^{\mu\nu}]$. When we calculate the Jacobian J_{co} of the gauge transformations (13.46) to the first order in $\Lambda^\lambda(x)$, we have

$$W[j]^\$ = \int d[J^\$_{\alpha\beta}](\mathrm{Det}\ U^\$) \exp\left[i\int d^4x(L_\phi + \frac{\xi}{2g^2}F^\$_\mu F^\$_\nu \eta^{\mu\nu} + J^\$_{\mu\nu}j^{\mu\nu})\right]$$

$$= \int d[J_{\alpha\beta}]J_{co}\mathrm{Det}(U + \delta U) \exp\left[i\int d^4x(L_\phi + \frac{\xi}{2g^2}F_\mu F_\nu \eta^{\mu\nu} + J_{\mu\nu}j^{\mu\nu}\right.$$

$$\left. + \frac{\xi}{g^2}F_\mu \delta F_\nu \eta^{\mu\nu} + j^{\mu\nu}\delta J_{\mu\nu})\right], \qquad (13.50)$$

$$J_{co} = \mathrm{Det}\left(\frac{\delta J^\$_{\mu\nu}}{\delta J_{\alpha\beta}}\right) = \mathrm{Det}\frac{\delta}{\delta J_{\alpha\beta}}\left(J^{\mu\nu} + \delta J_{\mu\nu}\right)$$

$$= \exp\left[\mathrm{Tr}\frac{\delta}{\delta J_{\alpha\beta}}\delta J_{\mu\nu}\right] = \exp\left[\int d^4x \frac{\delta}{\delta J_{\alpha\beta}}\delta J_{\mu\nu}\right]_{\alpha\beta=\mu\nu}$$

$$= 1 + \left[\int d^4x(Q_1 + Q_2)\right], \qquad (13.51)$$

$$Q_1 = \left(\frac{\delta f_{\mu\nu\lambda}}{\delta J_{\mu\nu}}\right)\Lambda^\lambda, \qquad (13.52)$$

$$Q_2 = f_{\mu\nu\lambda}\left(\frac{\delta\Lambda^\lambda}{\delta J_{\mu\nu}}\right), \qquad (13.53)$$

where $J^\$_{\mu\nu} = J_{\mu\nu} + \delta J_{\mu\nu}$ and $\delta J_{\mu\nu}$ are given by (13.43) and (13.46), respectively, and Λ^λ satisfies, by definition, the condition (13.9). The trace "Tr" in (13.51) includes both group indices and space-time labels (summation and integration over repeated indices and labels are always understood). We have also used the relations

$$\mathrm{Det}(1 + Q) = \exp[\mathrm{Tr}\ln(1 + Q)] = \exp[\mathrm{Tr}\ Q], \quad \text{for small } Q. \qquad (13.54)$$

The factor $\mathrm{Det}(U + \delta U)$ in (13.50) can be evaluated as follows:

$$\mathrm{Det}(U + \delta U) = \mathrm{Det}\, U\, \mathrm{Det}(1 + U^{-1}\delta U) = \mathrm{Det}\, U \exp[\mathrm{Tr} \ln (1 + U^{-1}\delta U)]$$

$$= (\mathrm{Det}\, U)\, [1 + Q_3], \quad Q_3 = \mathrm{Tr}\, U^{-1}\delta U. \tag{13.55}$$

Similar to (13.15), we have the T_4 gauge identity

$$\int d[\phi](\det U) \exp\left[i \int d^4x(L_\phi + L_{\xi\zeta} + J_{\mu\nu}j^{\mu\nu}) \right]$$

$$\times \left[\frac{\xi}{g^2} F_\mu \delta F_\nu \eta^{\mu\nu} + j^{\mu\nu}\delta J_{\mu\nu} + Q_1 + Q_2 + Q_3 \right] = 0, \tag{13.56}$$

which is obtained from (13.50) with the help of (13.51)–(13.55) and T_4 gauge invariance, $W[j]^\$ = W[j]$. Comparing (13.56) with (13.16), one can see that the last three terms in (13.56) are new terms arising from the gauge transformations of $d[J](\det U)$. For any compact Lie group, one can show that the new terms corresponding to $Q_1 + Q_2 + Q_3$ in (13.56) do not contribute to the gauge identities.[7] The T_4 gauge group in Yang–Mills gravity is not a compact Lie group. It appears that one cannot evaluate $Q_1 + Q_2 + Q_3$ unambiguously because they involve delta functions and their space-time derivatives. If one uses (13.46) for $\delta J_{\mu\nu}$ to calculate Q_1, one obtains

$$\mathrm{Tr}\, H_{\mu\nu}{}^{\alpha'\beta'}{}_{\lambda''} \frac{\delta J_{\alpha'\beta'}}{\delta J_{\mu\nu}} \Lambda^{\lambda''} = H_{\mu\nu}{}^{\mu\nu}{}_{\lambda''}\Lambda^{\lambda''}, \tag{13.57}$$

which is not well-defined.[d]

In Einstein gravity, the gauge transformation for tensor fields in curved space-time is similar to (13.46). The derivation of gauge identities also leads to result similar to (13.56) in curved space-time. It has been argued that the trace $G_{\mu\nu}{}^{\mu\nu}{}_{\lambda''}$ in Einstein gravity is formally zero,[12] however, such a conclusion based on a formal proof[12] is not completely satisfactory because of the presence of terms such as $0 \times \infty$.

---

[d] It appears that gauge identities (13.56) also contain some sort of "coupling" terms involving gravitons and FDM ghosts, which differ from the usual coupling vertex in the Feynman-Dyson rules. As a result, Yang-Mills gravity may not have gauge identities to all orders, which seems to be supported by explicit one-loop calculations with symbolic computing. The reason may be related to the gauge transformations (14.43) or (14.46), which involve generators of infinite continuous (non-compact) group. The situation resembles to that in Noether's Theorem II, which also involves generators of infinite continuous group (see Ref. 11).

Based on calculations of Feynman diagrams, it appears that one has well-defined T_4 gauge identities (13.23) (or (13.26)) at the tree-level, but the situation at the 1-loop level is not clear and needs further study.

13.5. Remarks on Ultraviolet Divergences and Finite-Loop Renormalization for Gravity

Based on the discussions of Feynman–Dyson rules for quantum Yang–Mills gravity and calculations of higher order Feynman diagrams in Chapters 11 and 12, we attempt to explore a new view of ultraviolet divergences and possible "renormalization" of gravity in this section. Nowadays, renormalization appears to be a necessary procedure for comparing the results of local quantum field theories, which involve ultraviolet divergences related to the idealized point particles, to experimental results.

Historically, the charged electron is assumed to have a finite radius in classical electrodynamics and the electron self-energy is identified with the electron mass. The advance of QED, however, has made this identification improbable.[e] In QED, the electron self-energy is a logarithmically divergent quantity. For QED to make sense and to agree with experiments, one must renormalize the theory. In this process, the electron self-energy is canceled by a counterterm so that the self-energy makes no contribution at all to the electron mass![f] Furthermore, if one looks at the photon self-energy in QED, it diverges quadratically by an "honest" calculation of the Feynman diagrams.[13] This result is incompatible with the gauge invariance of QED and the observed photon mass. Finally, the graviton self-energy in general relativity and Yang–Mills gravity diverges quartically (at the one-loop level). What can one do with all these ultraviolet catastrophes? Thus far, the most effective way for dealing with them is to renormalize the theory. Thus, a crucial property of a gauge theory is whether its ultraviolet divergences can be removed completely by a finite number of counterterms in the Lagrangian to make the S-matrix finite. In all gauge theories with internal gauge groups, this crucial property holds. In these renormalizable theories, all ultraviolet divergent quantities in higher order corrections can be canceled by a finite number of counterterms. Hence, the wonderful success of the electroweak theory and quantum chromodynamics.

---

[e] For a lucid discussion of the renormalization idea, see Ref. 13.

[f] Such a flip-flop in the interpretation of the electron mass from classical to quantum field theories suggests that such a renormalization is not the final story.

In contrast, a theory of gravity with an external gauge symmetry involves a dimensional coupling constant and hence, cannot be renormalized by power counting. The reason is that such a theory has extra momentum in say, the graviton 3-vertex in comparison with the SU_2 gauge theory, as one can see from (11.26). This implies that each successive order has a worse divergence than the previous order. (Ordinarily, this would imply the need for an infinite number of counter terms in the Lagrangian to remove the infinities from a theory, if there were no special symmetry properties to alter the situation.) Thus, it is gratifying that the powerful dimensional regularization technique preserves the gauge invariance of a theory by putting all the different degrees of divergences in higher orders into the form of simple poles, which correspond to a logarithmical divergence.[8] Nevertheless, a new undesirable situation occurs in Einstein gravity with external space-time gauge groups related to the principle of general coordinate invariance. For example, in general relativity, when one calculates the one-loop self-energy of a graviton using Feynman diagrams, one obtains only simple pole terms. Nevertheless, the necessary counterterms to remove them are not part of a generally covariant term such as $\sqrt{-g}\, R^2$, $\sqrt{-g} R_{\mu\nu} R^{\mu\nu}$, etc.[14] Thus, in such a theory of gravity in curved space-time, the ultraviolet divergences cannot be removed completely by generally covariant counterterms in the Hilbert–Einstein Lagrangian to make the S-matrix finite.

Although Yang–Mills gravity is not renormalizable by power counting, the structure of its interactions has a close similarity to the usual renormalizable gauge theories. For example, the maximum number of graviton interactions in a vertex is 4, as shown in the Lagrangian L_ϕ in (10.2). In contrast, there exist vertices with an arbitrary large number of gravitons in Einstein's theory of gravity and in other theories based on curved space-time.[15] It appears unlikely that the graviton coupling of Yang–Mills gravity can be further simplified because its gauge group is already the very simple Abelian group of four-dimensional translation.[h] The problems of renormalizing Yang–Mills gravity is particularly interesting, because (i) Yang–Mills gravity is formulated on the basis of flat space-time and has the

---

[8] In a sense, dimensional regularization effectively makes all couplings "renormalizable," just like those in theories that are renormalizable by power counting. The reason is that all ultraviolet divergences show up as simple poles $\Gamma(0)$ in dimensional regularization due to the relation $\Gamma(-n) = (-1)^n \Gamma(0)/n!$ for positive integers n. This property is quite different from that of Pauli-Villars regularization (cf. Ref. 13).

[h] Because of the translational gauge symmetry and the maximum 4-vertex coupling in Yang–Mills gravity, there is a possibility that the number of gauge invariant counterterms does not increase when the number of loops in Feynman diagrams keeps increasing.

maximum of 4-vertex graviton coupling, and (ii) the gravitational interaction is associated with a dimensional coupling constant. The combination of these two properties suggests a new and practical view of what it means to have a "renormalizable" theory of quantum gravity, which may be simply stated as follows:

In any theory of gravity involving a dimensional coupling constant, suppose that one can write down an action consisting solely of terms allowed by the gauge symmetry. Suppose the divergent terms in the finite-loop diagrams, can be completely removed by counterterms in the action. Then, for all practical purposes, such a theory of gravity with finite-loop renormalization is essentially equivalent to a renormalized gauge theory.[i]

One may wonder: what is the physical meaning of (finite-loop and usual) renormalization? In particular, because of the experimental agreements of renormalization in QED and electroweak theory, the success cannot be pure accidental and probably suggests some truth in the renormalization.[17,j] It is fair to say that the renormalization procedure to make a divergent field theory finite is, strictly speaking, not logical within the framework of local field theories.[18,k] It may appear that physicists tried all kinds of tricks to avoid the divergent difficulties. Nevertheless, it seems reasonable to give the following explanation to the renormalization procedure in a local field theory: A truly satisfactory field theory must not have infinities and must agree with experiments. In such a finite field theory, the truly physical particle is probably not a point particle and, hence, a renormalized local field theory is some sort of "non-local field theory." For example, a physical particle may be more like a fuzzy point particle or associated with a fundamental length L_f or has a non-zero minimum position uncertainty, $(\Delta x_{min} \approx L_f)$.[19,l] However, so far we do not have a suitable mathematics to handle such a non-local quantum field theory and to preserve desirable symmetries, as one can see from the works of Yukawa, Katayama, and others.[20]

Under such a difficult situation, physicists avoid the insurmountable obstacles by constructing a local field theory and using renormalization (with dimensional regularization[15]) to remove unphysical infinities from the theory,

---

[i] Similar ideas have been used and discussed by some physicists (see Ref. 16 and footnote m).

[j] For example, the magnetic moment of the electron has been measured with a fantastic precision: 1.001 159 652 180 76 ± 0.000 000 000 000 27. The difference, i.e., experimental value minus theoretical value, is given by $-2.06(7.72) \times 10^{-12}$.

[k] Local fields are continuous functions of space-time point x^μ and satisfy differential wave equations.

[l] It appears that a departure from the concept of point particles in field theories will lead to a departure from exact Lorentz and Poincaré invariance (or relativity theory) at very high energies.

so that *a field theory can employ simple mathematics with necessary symmetries to approximate the true physical theory.* This feat is indeed ingenious because the renormalized QED has been a wonderful success experimentally. From this viewpoint, the un-renormalized QED with ultraviolet divergences is, logically, a truly local field theory, while the renormalized QED is some sort of effectively non-local field theory, which represents a better approximation to the physical world. Furthermore, whether a local field theory is renormalizable or finite-loop-renormalizable[m] does not seem to make much difference, as long as the theory provides a good description and can make predictions about nature. It is gratifying that gauge-field theorists circumscribe the difficulties encountered by Yukawa and others in a direct assault of the "divergence dragons" in local field theory, and obtained splendid successes in QED and the unified electroweak theory. We hope that Yang–Mills gravity and its unification with other forces (to be discussed later) will have similar results.

References

1. J. P. Hsu, *Phys. Rev. D* **8**, 2609 (1973).

2. J. C. Ward, *Phys. Rev.* **77**, 2931 (1950); F. S. Fradkin, *JETP* **29**, 288 (1955); Y. Takahashi, *Nuovo Cimento* **6**, 370 (1957).

3. A. A. Slavnov, *Theor. Math. Phys.* **10**, 99 (1972); J. C. Taylor, *Nucl. Phys. B* **33**, 436 (1971).

4. R. P. Feynman, *Acta Phys. Polon.* **24**, 697 (1963).

5. B. S. DeWitt, *Phys. Rev.* **162**, 1195, 1239 (1967).

6. S. Mandelstam, *Phys. Rev.* **175**, 1604 (1968).

7. B. W. Lee and J. Zinn-Justin, *Phys. Rev. D* **7**, 1049 (1973).

8. J. P. Hsu and J. A. Underwood, *Phys. Rev. D* **15**, 1668 (1977).

9. J. P. Hsu, *Eur. Phys. J. Plus* **127**, 35 (2012).

10. B. S. DeWitt, in *Dynamical Theory of Groups and Fields* (Gordon and Breach, New York, 1965), pp. 102–105.

---

[m]The bias against needing an infinite number of counterterms to renormalize a Lagrangian seems analogous to earlier controversial views on infinite sets. Poincaré was dismayed by Cantor's infinite sets and said: "Actual infinity does not exist. What we call infinite is only the endless possibility of creating new objects no matter how many exist already." Gauss also had the view that the only access one had to the infinite is through the concept of limits and that one must not treat infinite sets as if they have an existence comparable to the existence of finite sets. However, one could interpret infinite numbers of counterterms as the endless "steps" of approximation to the true physical theory. As long as the approximation gets closer and closer to the experimental value, why should one be dismayed by it? There seems to be no particular reason to believe that an experimentally consistent physical theory must be one that can be approximated by a local field theory with a finite number of "steps."

11. E. Noether, *Goett. Nachr.* **235** (1918). English translation of Noether's paper by M.A. Tavel is online. Google search: M.A. Tavel, Noether's paper.

12. D. M. Capper and M. Ramón Medrano, *Phys. Rev. D* **9**, 1641 (1974).

13. J. J. Sakurai, in *Advanced Quantum Mechanics* (Addison-Wesley, 1984), p. 271, pp. 267–296.

14. P. van Nieuwenhuizen, in *Proc. Marcel Grossmann Meeting on General Relativity*, eds. R. Ruffini, 1978 CERN Preprint TH2473; D. M. Capper and M. A. Namazie, *Nucl. Phys. B* **14**, 535 (1978); D. M. Capper, *J. Phys. A: Math. Gen.* **13**, 199 (1980).

15. R. Utiyama and T. Fukuyama, *Prog. Theor. Phys.* **45**, 612 (1971); Y. M. Cho, *Phys. Rev. D* **14**, 2521 (1976); K. Hayashi and T. Shirafuji, *Phys. Rev. D* **19**, 3524 (1979); A. A. Logunov and M. A. Mestvirishvili, *Progr. Theor. Phys.* **74**, 31, (1985); N. Wu, *Commun. Theor. Phys. (Beijing, China)* **42**, 543 (2004).

16. B. L. Voronov and I. V. Tyutin, *Theore. Math. Phys.* **50**, 218 (1982); **52**, 628 (1982); M. Harada, T. Kugo and K. Yamawaki, *Prog. Theor. Phys.* **91**, 801 (1994); J. Gomis and S. Weinberg, in *The Quantum Theory of Fields* (Cambridge University Press, 1995), Vol. 1, pp. 516–525; Vol. 2, pp. 91–95.

17. T. Kinoshita, Talk at Nishina Hall, RIKEN (2010). (Google search: Kinoshita, Magnetic moment.); Particle data group, *Particle Physics Booklet* (2012).

18. P. A. M. Dirac, *Lectures on Quantum Field Theory* (Academic Press, 1967), p. 2.

19. J. P. Hsu, *Nuovo Cimento B* **88**, 156 (1985); S. Y. Pei and J. P. Hsu, *Nuovo Cimento B* **102**, 347, (1988).

20. Y. Katayama, in *Proc. 1967 Int. Conf. Particles and Fields*, eds. C. R. Hagen, G. Guralnik and V. A. Mathur (Interscience, 1967), p. 157.

Chapter 14

A Unified Gravity-Electroweak Model

14.1. The Gauge Covariant Derivative and Gauge Curvatures of a Unified Gravity-Electroweak Model

The electromagnetic and weak interactions have been "satisfactorily" unified based on a gauge invariant action with the internal gauge group $SU_2 \times U_1$. The strong interaction can be described by quantum chromodynamics with color SU_3 groups. In previous chapters, we have shown that the gravitational interaction can be described by a gauge invariant action with an external space-time translational group. Because all the known interactions can be understood on the basis of gauge symmetries, they can, in principle, be unified. After all these years, the vision of Yang, Mills, Utiyama, Weyl and others may finally be realized.

Before considering a unification of all the fundamental interactions, in this chapter, let us first consider the gravity-electroweak sector of a unified model for simplicity. We would like to show that the gravitational interaction can be unified with the electroweak interaction of the Weinberg–Salam theory,[1,2,a] just as that theory unifies the electromagnetic and weak interactions. Furthermore, such a unification leads to new understandings and predictions, most notably that such a unified model predicts the violation of the electromagnetic U_1 and the weak SU_2 gauge symmetries by the gravitational interaction. This unification is based on the Yang–Mills–Utiyama–Weyl (YMUW) framework, which accommodates both external space-time gauge symmetry and internal gauge symmetries. Although this new unified model can be formulated in both inertial and non-inertial frames, the complications of a non-inertial frame may obscure the points we want to make

---

[a]See also Ref. 5.

and thus, our discussions of this new gravity-electroweak model, its quantization and physical implications will be based on inertial frames with the metric tensor $\eta_{\mu\nu} = (+1, -1, -1, -1)$.

The basic natures of forces due to gauge fields are dictated by the mathematical properties of the generators associated with their gauge groups. Within our gravity-electroweak model, we use the gauge group $T_4 \times SU_2 \times U_1$, which allows us to introduce naturally a new symmetric tensor gauge field $\phi_{\mu\nu} = \phi_{\nu\mu}$ through replacement in the usual Yang–Mills framework,[3]

$$\partial_\mu \rightarrow \partial_\mu - ig\phi_\mu^\nu p_\nu + if W_\mu^a t^a + if' U_\mu t_o$$
$$= \partial_\mu + g\phi_\mu^\nu \partial_\nu + if W_\mu^a t^a + if' U_\mu t_o \equiv d_\mu,$$

where $p_\nu = i\partial/\partial x^\nu$ are the T_4 generators, t^a ($a = 1, 2, 3$) are the generators of SU_2 in general, and t_o (weak hypercharge) is the generator of U_1.[4] The operator d_μ may be called the unified gauge covariant derivative (UGCD). There are three independent gauge coupling constants: f and f' are dimensionless, and g has the dimension of length because the T_4 generator $p_\mu = i\partial/\partial x^\mu$ has the dimension of (1/length) in natural units.

The unified gravity-electroweak model leads to the following conceptual understanding of well-known forces in nature:

(A) Based on the T_4 gauge symmetry, the gravitational forces due to the exchange of virtual gravitons between two electrons, two positrons and between an electron and a positron, must be the same kind of force. The electron and the positron have the same attractive gravitational force characterized by one coupling constant g. The antiparticles are, roughly speaking, related to the complex conjugate of d_μ.

In contrast, the electromagnetic interaction can have both attractive and repulsive forces for matter–matter, matter–antimatter and antimatter–antimatter. These forces are characterized by two dimensionless coupling constants, $-e$ for electrons and $+e$ for positrons. This difference between the gravitational and the electromagnetic forces stems from the absence of an "i" in the T_4 term involving $-ig\phi_\mu^\nu p_\nu = g\phi_\mu^\nu \partial_\nu$ in the UGCD d_μ. (See also discussions in Sec. 0.3.)

(B) The UGCD d_μ can be expressed in terms of the gravitational potential $\phi_{\mu\nu}$ and the electromagnetic potential A_μ,

$$d_\mu = \partial_\mu + g\phi_\mu^\nu \partial_\nu + ieQA_\mu + \cdots,$$

where $Q = [ft_3 \sin\theta_W + f't_o \cos\theta_W]/e = (t_3 + t_o)$.[4] The repulsive electric force between two electrons can be calculated from the exchange of a virtual photon using the formula,[5]

$$V = \frac{i}{(2\pi)^3} \int (-iM_{fi}) exp[i\mathbf{k} \cdot \mathbf{r}] d^3 k, \qquad (14.1)$$

for the effective potential. The result is the well-known Coulomb potential in quantum electrodynamics (QED). Similarly, the gravitational force between two electrons can be calculated from the exchange of a virtual graviton. The invariant amplitude $-iM_{fi}$ can be written using Feynman–Dyson rules (11.23) and the electron–graviton vertex (11.29). We have

$$-iM_{fi} = \left(\frac{ig}{2}\right)^2 \overline{u}(q)\gamma^\alpha u(p)(p^\beta + q^\beta)G_{\alpha\beta\rho\sigma}\overline{u}'(q')\gamma^\rho u'(p')(p'^\sigma + q'^\sigma),$$

where the graviton propagator $G_{\alpha\beta\rho\sigma}$ in the DeWitt gauge is given by (11.23) with $\xi = 2, \zeta = 1$. The invariant amplitude $-iM_{fi}$ is for the scattering process $e^-(p)e^-(p') \to e^-(q)e^-(q')$ with the exchange of one graviton $\phi_{\mu\nu}(k)$, where we have $p + k = q$ and $p' = q' + k$ for two vertices of the Feynman diagram. In the classical and static limit, in which the electrons are at rest, we obtain the attractive potential $V = (-g^2 m^2)/(8\pi r) = -Gm^2/r$ for $r > 0$, where we have used the relation in (9.1).[b] In quantum Yang–Mills gravity, the attractive gravitational force is interlocked with the property that gravitons are quanta of the symmetric tensor field $\phi_{\mu\nu}$.

To see the gauge curvatures associated with each group in the UGCD d_μ, let us calculate the commutators $[d_\mu, d_\nu]$. We obtain

$$[d_\mu, d_\nu] = -iC_{\mu\nu\alpha}p^\alpha + if W^a_{\mu\nu}t^a + if' U_{\mu\nu}t_o, \quad [t^a, t^b] = i\epsilon^{abc}t^c,$$
$$d_\mu = J^\nu_\mu\partial_\nu + if W^a_\mu t^a + if' U_\mu t_o, \quad J^\nu_\mu = \delta^\nu_\mu + g\phi^\nu_\mu, \qquad (14.2)$$

where $J^\nu_\mu\partial_\nu \equiv \Delta_\mu$ is called the T_4 gauge covariant derivative. The T_4 gauge curvature $C_{\mu\nu\alpha}$ in flat space-time is given by (11.5), which is completely different from the Riemann–Christoffel curvature in general relativity. However, there exists

---

[b]The effective potential also involves divergent terms. For example, at $r = 0$, the effective potential due to the exchange of a graviton has a term involving $\delta^3(\mathbf{r})$. Moreover, the minus sign in the effective potential (14.1) occurs when the virtual quantum exchanged is a scalar particle (in Yukawa's attractive nuclear potential) or a tensor particle (in Yang–Mills gravity).

a non-trivial gauge invariant action, which is quadratic in the new T_4 gauge curvatures $C_{\mu\nu\alpha}$, and is consistent with experiments, as discussed in Chapters 8 and 9.

In the unified model, all gauge fields are assumed to appear in the UGCD d_μ on an equal footing just as in the Weinberg–Salam theory, as shown in (14.2). Consequently, the new SU_2 and U_1 gauge curvatures in the presence of the gravitational gauge potential are

$$W_{\mu\nu}^a = J_\mu^\sigma \partial_\sigma W_\nu^a - J_\nu^\sigma \partial_\sigma W_\mu^a - f\epsilon^{abc} W_\mu^b W_\nu^c, \quad \text{and}$$

$$U_{\mu\nu} = J_\mu^\sigma \partial_\sigma U_\nu - J_\nu^\sigma \partial_\sigma U_\mu = \Delta_\mu U_\nu - \Delta_\nu U_\mu, \tag{14.3}$$

respectively. In the absence of gravity, i.e., $g = 0$, the gauge curvatures in (14.3) reduce to those in the Weinberg–Salam theory. We observe that the T_4 gauge potential appears in the SU_2 and U_1 gauge curvatures in (14.3), but that the SU_2 and U_1 gauge potentials do not affect the T_4 gauge curvature in (14.2). This shows the universal nature of the gravitational force, namely that it acts on all particles in the physical world.

14.2.　　The Lagrangian in the Gravity-Electroweak Model

Based on the gauge curvatures in (14.2)–(14.3), the new Lagrangian L_{gew} of the unified gravity-electroweak model (for leptons) can be assumed to take the form,

$$L_{\text{gew}} = L_\phi + L_{\text{WS}},$$

$$L_\phi = \frac{1}{4g^2}(C_{\mu\nu\alpha}C^{\mu\nu\alpha} - 2C_{\mu\alpha}{}^\alpha C^{\mu\beta}{}_\beta),$$

$$L_{\text{WS}} = -\frac{1}{4}W^{a\mu\nu}W_{\mu\nu}^a - \frac{1}{4}U_{\mu\nu}U^{\mu\nu} + \overline{L}i\gamma^\mu d_\mu L + \overline{R}i\gamma^\mu d_\mu R \tag{14.4}$$

$$+ (d_\mu\phi)^\dagger(d^\mu\phi) - V(\phi^\dagger\phi) - \frac{m}{\rho_o}(\overline{L}\phi R + \overline{R}\phi^\dagger L),$$

using the usual notations,[4] i.e., R denotes the right-handed electron iso-singlet with only two independent spinor components, and L is the left-handed lepton iso-doublet, etc. It is possible to include all the quarks and leptons in the three families.[4] We observe that all terms in the original $SU_2 \times U_1$ Lagrangian are affected by the presence of the gravitational field $\phi_{\mu\nu}$ except the last two terms of L_{WS} in (14.4). This property suggests that the spontaneous symmetry breaking and the masses of particles in the model are independent of the gravitational interaction in the gauge invariant Lagrangian (14.4). Moreover, from the gravity-electroweak

Lagrangian (14.4), we see that, similar to renormalizable Yang–Mills fields, the couplings of the gravitational gauge field and electroweak gauge fields are no more complex than a 4-vertex in the Feynman diagrams.

14.3. The Equations of Motion for Quantum and Classical Particles

To explore the macroscopic implications of the gravity-electroweak model, it is more logically coherent to derive the equation of motion for a classical (or macroscopic) object from the wave equations for quantum particles in the model, rather than making a separate assumption about the motion of classical objects. Thus, we investigate the classical limit (i.e., the geometric-optics limit) of the wave equations for gauge bosons, the Higgs boson, the electron, etc.

(A) *The classical equation of motion for gauge bosons*
For the bending of a light ray by the sun, we derived the eikonal equation for a light ray in the presence of gravity from the photon wave equation in Section 8.5, where the classical limit of the electron wave equation was also obtained. Now let us consider the classical limit of the wave equations for the gauge bosons. Similar to L_{em} in (8.34), we consider the following Lagrangian involving the massive gauge boson fields $W_\mu^a, a = 1, 2$, in the presence of gravity in the gravity-electroweak model:

$$L_W = -\frac{1}{4} W_{\mu\nu}^a W^{a\mu\nu} + \frac{1}{2} m_W^2 W_\mu^a W^{a\mu}, \quad a = 1, 2,$$

where the gauge curvature $W_{\mu\nu}^a$ is given by (14.3). This Lagrangian leads to a modified wave equation for the gauge boson with mass m_W in the presence of gravity,

$$\Delta_\mu W^{a\mu\lambda} + (\partial_\alpha J_\mu^\alpha) W^{a\mu\lambda} + m_W^2 W^{a\lambda} = 0, \quad \Delta_\mu = J_\mu^\nu \partial_\nu. \tag{14.5}$$

Using the gauge condition, $\partial_\mu W^{a\mu} = 0$, for vector gauge fields and the limiting expression[6] for the field

$$W^{a\lambda} = I^{a\lambda} \exp(i\Psi),$$

we can derive the "eikonal equation" for massive vector bosons,

$$\left[\delta_\sigma^\lambda \left(G^{\mu\nu} \partial_\mu \Psi \partial_\nu \Psi - m_W^2 \right) - g \phi_\sigma^\mu J^{\lambda\nu} \partial_\mu \Psi \partial_\nu \Psi \right] I^{a\sigma} = 0,$$

$$G^{\mu\nu} = \eta_{\alpha\beta} J^{\alpha\mu} J^{\beta\nu},$$

in the geometric-optics limit (i.e., both the mass m_W and the wave 4-vector (or momentum) $\partial_\mu \Psi$ are very large).[6] Following the steps from (8.35) to (8.36), we obtain

$$\left[G_L^{\mu\nu}(\partial_\mu \Psi)\partial_\nu \Psi - m_W^2 \right] = 0, \tag{14.6}$$

where the effective Riemannian metric tensor for massive vector gauge bosons is

$$G_L^{\mu\nu} = G^{\mu\nu} - \frac{g}{4}\phi_\lambda^\mu J^{\lambda\nu} = \eta_{\alpha\beta}(\eta^{\mu\alpha} + g\phi^{\mu\alpha})\left(\eta^{\nu\beta} + \frac{3g}{4}\phi^{\nu\beta}\right). \tag{14.7}$$

We can show that other gauge bosons such as Z^0 and the photon have the same eikonal equations (14.6), except that m_W^2 is replaced by m_Z^2 and 0 respectively. (Cf. equation (8.36).)

(B) *The classical equations of motion for gravitons*
It is interesting to ask whether there is an "effective metric tensor" in the eikonal equation for gravitons in the geometric-optics limit. To answer this question, we use the free gravitational wave equation, i.e., (8.32) with $S^{\mu\nu} = 0$, and the following limiting expression in inertial frames:

$$\phi^{\mu\nu} = a^{\mu\nu} \exp(i\Phi_G), \quad a^{\mu\nu} = \epsilon^{\mu\nu}(\lambda)b(x),$$

for the tensor field, where the wave 4-vectors $\partial_\mu \Psi_G$ are very large. The polarization tensor $\epsilon^{\mu\nu}(\lambda)$ for gravitons satisfies $\sum_\lambda \epsilon_{\mu\nu}(\lambda)\epsilon_{\alpha\beta}(\lambda) \to -(\eta^{\mu\nu}\eta^{\alpha\beta} - \eta^{\mu\alpha}\eta^{\nu\beta} - \eta^{\mu\beta}\eta^{\nu\alpha})$ by taking the polarization sum, where the factor $-(\eta^{\mu\nu}\eta^{\alpha\beta} - \eta^{\mu\alpha}\eta^{\nu\beta} - \eta^{\mu\beta}\eta^{\nu\alpha})$ is the same as that in the graviton propagator (11.24), similar to the polarization vector for photons.[c] Following the steps (8.35)–(8.36), one can obtain the classical equations of motion for gravitons in flat space-time,

$$\eta^{\mu\nu}\partial_\mu \Psi_G \partial_\nu \Psi_G = 0, \tag{14.8}$$

which does not involve any effective Riemann metric tensor. This property is consistent with the fact that the gravitational field itself $\phi_{\mu\nu}$ is the physical origin of the effective Riemannian metric tensors in the classical equations of motion of all other particles.

---

[c]See discussions in (8.35)–(8.36).

(C) *The classical equations of motion for leptons, quarks and Higgs particles*
The wave equations for fermions (leptons and quarks) lead to the classical equation
(i.e., the Einstein–Grossmann equation),

$$G^{\mu\nu}(\partial_\mu S)(\partial_\nu S) - m^2 = 0, \quad G^{\mu\nu} = \eta_{\alpha\beta} J^{\alpha\mu} J^{\beta\nu}, \tag{14.9}$$

as discussed in Section 8.5. We stress that only in the geometric-optics limit (i.e., the classical limit) of the modified Dirac wave equation (8.37) with the space-time translational gauge symmetry does an "effective Riemann metric tensor" $G^{\mu\nu}$ emerge, as shown in (14.9) or (8.40).

The Lagrangian for the Higgs field $\eta(x)$ in the presence of the gravitational gauge potential is given by

$$L_\eta = \frac{1}{2}\Delta_\mu \eta \Delta^\mu \eta - \frac{1}{2}m_H^2 \eta^2. \tag{14.10}$$

The modified wave equation for the Higgs boson due to the presence of gravity can be derived from (14.10). We can show that it also reduces to the Einstein–Grossmann equation in the classical limit,

$$G^{\mu\nu}(\partial_\mu \Psi)(\partial_\nu \Psi) - m_H^2 = 0, \tag{14.11}$$

where we have used the following relations in the geometric-optics limit,

$$\eta = \eta_0 \exp(i\Psi), \quad \partial_\lambda \partial_\alpha \Psi \ll (\partial_\alpha \Psi)\partial_\lambda \Psi. \tag{14.12}$$

The momentum $\partial_\mu \Psi$ and mass m_H in (14.12) are large quantities in the geometric-optics limit.

Thus, in the unified gravity-electroweak model, there are three different types of classical equations of motion for particles in the geometric-optics limit, in contrast to Einstein gravity:

(a) The eikonal (or Hamilton–Jacobi type) equations (14.6) involving the metric tensor $G^{\mu\nu} = \eta_{\alpha\beta} J^{\alpha\mu} J^{\beta\nu} - (g/4)\phi_\lambda^\mu J^{\lambda\nu}$ for the massive gauge bosons $W^\pm$ and Z^0, and massless photons.
(b) The eikonal equation (14.8) for gravitational rays involving the Minkowski metric tensors.
(c) The Einstein–Grossmann equation (14.9) involving the effective metric tensor $G^{\mu\nu} = \eta_{\alpha\beta} J^{\alpha\mu} J^{\beta\nu}$ for quarks, leptons and scalar particles with masses. All observable matter in the universe is composed of quarks and leptons and obeys the Einstein–Grossmann equation.

Equation (14.9) can describe both the motion of a classical object in free fall and the perihelion shift of the planet Mercury in an inertial frame. Thus, in the geometric-optics limit, Yang–Mills gravity is compatible with the equivalence principle in Einstein gravity.

14.4. Violations of U_1 and SU_2 Gauge Symmetries by Gravity

For gauge transformations in the gravity-electroweak model, we assume that the tensor gauge field $\phi_\mu^\nu(x)$ is an electrically neutral field and an iso-scalar, so that the T_4 gauge potential ϕ_μ^ν does not transform under SU_2 and U_1 gauge transformations. For other fields, the SU_2 and U_1 gauge transformations are the same as those in the Weinberg–Salam theory.[4] For example, in the absence of gravity (i.e., $g = 0$), local gauge transformations for $W_a^\mu \equiv W_a^\mu(x)$, $W_a^{\mu\nu} \equiv W_a^{\mu\nu}(x)$, $U_\mu \equiv U_\mu(x)$ and $U_{\mu\nu} \equiv U_{\mu\nu}(x)$ are given by

$$\delta W_a^\mu = \frac{1}{f}\partial^\mu \omega_a(x) + \epsilon_{abc}\omega_b(x)W_c^\mu, \quad [\delta W_a^{\mu\nu}]_{g=0} = \epsilon_{abc}\omega_b(x)W_c^{\mu\nu}, \quad (14.13)$$

$$\delta U_\mu = \partial_\mu \omega(x), \quad \delta U_{\mu\nu} = 0, \quad (14.14)$$

for SU_2 and U_1, where $\omega_b(x)$ and $\omega(x)$ are infinitesimal functions.[4]

In the presence of gravity (i.e., $g \neq 0$), δU_μ is given by (14.14). However, the U_1 gauge transformations of the gauge curvature $U_{\mu\nu}$ differ from that in (14.14),

$$\delta U_{\mu\nu}(x) = J_\mu^\sigma \partial_\sigma(\delta U_\nu) - J_\nu^\sigma \partial_\sigma(\delta U_\mu)$$

$$= g[\phi_\mu^\sigma \partial_\sigma \partial_\nu \omega(x) - \phi_\nu^\sigma \partial_\sigma \partial_\mu \omega(x)] \neq 0. \quad (14.15)$$

Similarly, the SU_2 gauge transformation of the gauge curvature $W_a^{\mu\nu}$ also differs from that in (14.13),

$$\delta W_a^{\mu\nu} = J^{\mu\sigma}\partial_\sigma(\delta W_a^\nu) - J^{\nu\sigma}\partial_\sigma(\delta W_a^\mu) + f\epsilon_{abc}(\delta W_b^\mu)(\delta W_c^\nu)$$

$$\neq [\delta W_a^{\mu\nu}]_{g=0}, \quad (14.16)$$

where δW_a^μ is given by (14.13). Thus, the Lagrangian L_{WS} in (14.4), which involves $W_{a\mu\nu}W_a^{\mu\nu}$ and $U_{\mu\nu}U^{\mu\nu}$, has only approximate $SU_2 \times U_1$ symmetry due to the presence of a very small quantity $g\phi_\mu^\sigma$ in (14.15) and (14.16).

As a result of this violation of $SU_2 \times U_1$ gauge symmetry by gravity, we have modified wave equations (14.5) for the gauge bosons. This violation is interesting because in the classical limit we obtain the eikonal equation (14.6) for the

$W^{\pm}$ gauge boson, which involves an effective Riemann metric tensor $G_L^{\mu\nu}$ given by (14.7).

To see the effect of the violation of the U_1 gauge symmetry by gravity on the conservation of electric charge, consider the electromagnetic sector of the unified model. It suffices to concentrate on the action $\int d^4x L_{em}$ involving only the electromagnetic field A_μ and a charged fermion ψ, which are coupled to the gravitational field $\phi_{\mu\nu}$,[1,4]

$$L_{em} = -\frac{1}{4}F_{\mu\nu}F^{\mu\nu} + \frac{i}{2}\left[\,\overline{\psi}\gamma^\mu(\Delta_\mu - ieA_\mu)\psi - [(\Delta_\mu + ieA_\mu)\overline{\psi}]\gamma^\mu\psi\,\right] - m\overline{\psi}\psi,$$

$$F_{\mu\nu} = \Delta_\mu A_\nu - \Delta_\nu A_\mu, \quad \Delta_\mu = (\delta_\mu^\nu + g\phi_\mu^\nu)\partial_\nu \equiv J_\mu^\nu\partial_\nu, \quad e > 0, \tag{14.17}$$

in inertial frames with the metric tensors $\eta_{\mu\nu} = (1, -1, -1, -1)$ and $c = \hbar = 1$, where A_μ is assumed to satisfy the usual gauge condition $\partial^\mu A_\mu = 0$. The $W^\pm$ and Z^o gauge bosons and others are not considered here. The UGCD, $d_\mu = \Delta_\mu - ieA_\mu + \cdots = \partial_\mu + g\phi_\mu^\nu\partial_\nu - ieA_\mu + \cdots$, involves the generators of the space-time translational gauge symmetry, $p_\nu = i\partial_\nu$.

The generalized Maxwell's wave equations can be derived from (14.17). We have

$$\partial_\alpha(J_\mu^\alpha F^{\mu\nu}) = -e\overline{\psi}\gamma^\nu\psi, \quad J_\mu^\alpha = \delta_\mu^\alpha + g\phi_\mu^\alpha. \tag{14.18}$$

The wave equation (14.18) implies a modified continuity equation for electric current in the presence of gravity,

$$\partial_\nu J_{\text{total}}^\nu = 0, \quad J_{\text{tot}}^\nu = -e\overline{\psi}\gamma^\nu\psi - g\partial_\alpha(\phi_\mu^\alpha F^{\mu\nu}), \tag{14.19}$$

where we have used the identity $\partial_\mu\partial_\nu F^{\mu\nu} = 0$.

Thus, the usual current $e\overline{\psi}\gamma^\nu\psi$ in the electroweak theory and QED is no longer exactly conserved in the presence of gravity. Only a total current, composed of both the usual electromagnetic current and a new "gravity-em current" [corresponding to the second terms in (14.19)], is conserved. The total conserved charge in the gravity-electroweak model is given by

$$Q_{\text{tot}} = \int J_{\text{tot}}^0 d^3x = \int [-e\overline{\psi}\gamma^0\psi - g\partial_\alpha(\phi_\mu^\alpha F^{\mu 0})]d^3x. \tag{14.20}$$

The new effect involves a factor $|g\phi_0^0| \approx |g\phi_1^1| \approx Gm/r$, which is roughly 10^{-9} on the surface of Earth. Such an effect could be tested experimentally if the electric field **E** and magnetic induction **B** (i.e., the electromagnetic field strength $F_{\mu\nu}$) are very large.[d]

Another important effect of the violation of U_1 and SU_2 gauge symmetries by gravity is that the classical equations of motion of fermions, bosons and scalar particles now involve effective Riemann metric tensors, as shown in Section 14.3. Thus the perihelion shift of the planet Mercury could be considered a demonstration of the violation of U_1 and SU_2 gauge symmetry in the unified gravity-electroweak model. As discussed in Section 9.3, a crucial test of the unified model and Yang–Mills gravity would come from a more accurate measurement of the deflection of light with a frequency in the optical range.

References

1. J. P. Hsu, *Int. J. Mod. Phys. A* **21**, 5119 (2006); **24**, 5217 (2009); J. P. Hsu and L. Hsu, in *A Broader View of Relativity: General Implications of Lorentz and Poincaré Invariance* (World Scientific, 2006), Appendix D. Available online: "google books, a broader view of relativity, Hsu."
2. S. Weinberg, in *The Quantum Theory of Fields*, Vol. 2 (Cambridge University Press, 1996), p. 305; R. E. Marshak, in *Conceptual Foundations of Modern Particle Physics* (World Scientific, 1982), pp. 321–328.
3. J. P. Hsu, *Mod. Phys. Lett. A* **26**, 1707 (2011).
4. K. Huang, in *Quarks, Leptons and Gauge Field* (World Scientific, 1982), pp. 108–117, 152–156, 241–242; R. E. Marshak, in *Conceptual Foundations of Modern Particle Physics* (World Scientific, 1993), Chapter 2.
5. J. J. Sakurai, in *Advanced Quantum Mechanics* (Addison-Wesley, 1967), pp. 258–259.
6. L. Landau and E. Lifshitz, in *The Classical Theory of Fields*, Trans. by M. Hamermesh (Addison-Wesley, Cambridge, MA, 1951), pp. 136–137.

---

[d]The small new effect involves a term of the form $g\phi_\mu^\alpha (\partial_\alpha F^{\mu\nu})$, so if the field strength $F_{\mu\nu}$ has a very high frequency, one may be able to detect this new predicted effect.

Chapter 15

A Unified Gravity-Strong Force Model

15.1. Unified Gauge Covariant Derivatives and Gauge Curvatures

We now consider the unification of Yang–Mills gravity and quantum chromodynamics in the Yang–Mills–Utiyama–Weyl (YMUW) framework with two independent coupling constants. We call the resulting theory, which involves the gauge group $T_4 \times [SU_3]_{\text{color}}$, "quantum chromogravity" (QCG). Similar to the U_1 or SU_2 gauge symmetry in the electroweak theory, QCG involves T_4 gauge symmetry, so that the T_4 group generators appear in the Lagrangian through a gauge covariant derivative. In the YMUW framework, the T_4 gauge covariant derivatives are obtained through the following replacement[1]:

$$\partial_\mu \to \partial_\mu - ig\phi_\mu^\nu p_\nu = J_\mu^\nu \partial_\nu, \quad J_\mu^\nu = (\delta_\mu^\nu + g\phi_\mu^\nu), \tag{15.1}$$

in inertial frames with the metric tensor $\eta_{\mu\nu}$, where we use the representation $p_\mu = i\partial_\mu$. The operators p_ν denote the generators (in general) of the flat spacetime translation group and satisfy $[p_\mu, p_\nu] = 0$. This is similar to the minimal substitution in quantum electrodynamics for introducing the electromagnetic coupling. In (15.1), a mixed tensor field ϕ_μ^ν is naturally associated with the T_4 generator $p_\mu = i\partial_\mu = i\partial/\partial x^\mu$ and the coupling constant g has the dimension of length.

In a general frame of reference (which may be non-inertial) with the Poincaré metric tensors $P_{\mu\nu}$, the form of the gauge covariant derivative (15.1) still holds, provided we replace ∂_μ by the covariant derivative D_μ associated with $P_{\mu\nu}$.[1] Therefore, the unified gauge covariant derivatives $\mathbf{d}_\mu$ in QCG for general frames are defined as

$$\mathbf{d}_\mu \equiv D_\mu + g\phi_\mu^\nu D_\nu + ig_s G_\mu^a \frac{\lambda^a}{2}, \tag{15.2}$$

where λ^a ($a = 1, 2, 3, \ldots, 8$) is the fundamental representation of the eight generators of color SU_3. The fields G_μ^a are the color gauge vector potentials, i.e., the gluon fields.[2–4]

In order to see the gauge curvatures associated with the unified gauge covariant derivative $\mathbf{d}_\mu$ in inertial and non-inertial frames, we calculate the commutators $[\mathbf{d}_\mu, \mathbf{d}_\nu]$ using the general expression (15.2). The purpose is to show that the T_4 gauge invariant quantum chromogravity can be formulated in both inertial and non-inertial frames. After some straightforward calculations, we have[5]

$$[\mathbf{d}_\mu, \mathbf{d}_\nu] = C_{\mu\nu\sigma} D^\sigma + i g_s G_{\mu\nu a} \frac{\lambda_a}{2}, \tag{15.3}$$

where the T_4 gauge curvature $C_{\mu\nu\sigma}$ is given by (11.5) with $\partial_\mu \to D_\mu$ and $\eta_{\mu\nu} \to P_{\mu\nu}$. The new $[SU_3]_{color}$ gauge curvature (i.e., the color gauge field strength tensor) $G_{a\mu\nu}$ in the presence of the gravitational gauge potential is now given by

$$G_{a\mu\nu} = J_\mu^\sigma D_\sigma G_{a\nu} - J_\nu^\sigma D_\sigma G_{a\mu} - g_s f_{abc} G_{b\mu} G_{c\nu}$$
$$= \Delta_\mu G_{a\nu} - \Delta_\nu G_{a\mu} - g_s f_{abc} G_{b\mu} G_{c\nu}, \tag{15.4}$$

$$\Delta_\mu = J_\mu^\sigma D_\sigma, \quad J_{\mu\nu} = P_{\mu\nu} + g\phi_{\mu\nu}, \quad \phi_{\mu\nu} = \phi_{\nu\mu} = P_{\nu\lambda}\phi_\mu^\lambda, \tag{15.5}$$

where $a, b, c = 1, 2, \ldots, 8$, the structure constants f_{abc} are antisymmetric and Δ_μ is the T_4 gauge covariant derivative in arbitrary coordinates in flat space-time.

15.2. The Action of the Unified Model and Violations of Local SU_3 Gauge Symmetry by Gravity

Within the YMUW framework with the gauge curvatures (15.4) and (15.5), we can construct the Lagrangian L_G for gluon fields $G_{a\nu}(x)$ and the Lagrangian $L_{\phi q}$ for quarks fields $q(x)$ and gravitational fields $\phi_{\mu\nu}$, similar to those in quantum chromodynamics and Yang–Mills gravity.[2,5] The action functional S_{QCG} and the Lagrangian L_{QCG} of the unified quantum chromogravity (QCG) model in general frames are naturally assumed to take the form,[5]

$$S_{QCG} = \int L_{QCG} d^4x, \quad L_{QCG} = L_G + L_{\phi q}, \tag{15.6}$$

$$L_G = -\frac{1}{4}G^{a\mu\nu}G^a_{\mu\nu}\sqrt{-P}, \quad L_{\phi q} = [L_\phi + \bar{q}(i\Gamma^\mu d_\mu - M)q]\sqrt{-P},$$

$$L_\phi = \frac{1}{4g^2}(C_{\mu\nu\alpha}C^{\mu\nu\alpha} - 2C_{\mu\alpha}{}^\alpha C^{\mu\beta}{}_\beta), \quad \{\Gamma_\mu, \Gamma_\nu\} = 2P_{\mu\nu}, \tag{15.7}$$

$$\Gamma_\mu = \gamma_a e^a_\mu, \quad \{\gamma_a, \gamma_b\} = 2\eta_{ab}, \quad \eta_{ab}e^a_\mu e^b_\nu = P_{\mu\nu},$$

where $P = \det P_{\mu\nu}$ and q's denote collectively spinor quarks with three colors and six flavors.[2,3] There are two independent quadratic forms that one can construct with the T_4 gauge curvature $C_{\mu\nu\alpha}$. Their linear combination, as shown in L_ϕ, is consistent with gravitational experiments, as discussed in Chapter 9. Although there is a small difference in the field equations for quarks and antiquarks due to the presence of the gravitational gauge field $\phi_{\mu\nu}$ in (15.7), this difference can be eliminated by symmetrizing the fermion Lagrangian.[1] (See equation (15.13).) We ignore the difference here since it does not affect the observable physical implications of the theory.

The $[SU_3]_{color}$ gauge transformations in the unified model are the same as those in the usual quantum chromodynamics, provided the gravitational gauge fields $\phi_{\mu\nu}(x)$ are assumed to transform trivially under $[SU_3]_{color}$ because they are not affected by the strong force. Under the $[SU_3]_{color}$ gauge transformations, the gauge curvature $G_{a\mu\nu}$ in (15.4) transforms according to the usual relation in QCD,

$$\delta G_{a\mu\nu}(x) = f_{abc}\omega_b(x)G_{c\mu\nu}(x),$$

only when $g = 0$, i.e., in the absence of the gravitational field or when the gauge function $\omega_b(x)$ is a constant. Thus, the local color SU_3 gauge invariant Lagrangian L_G in (15.7) is violated by the presence of gravity due to the presence of $\Delta_\mu = J^\sigma_\mu D_\sigma$ rather than D_μ in (15.4).

Although the unified model does not have the exact local color SU_3 gauge symmetry, the action S_{QCG} in (15.6) still has T_4 gauge invariance. The T_4 gauge transformations are more involved because the infinitesimal and local space-time translations are given by

$$x^\mu \to x'^\mu = x^\mu + \Lambda^\mu(x) \tag{15.8}$$

with an arbitrarily infinitesimal vector function $\Lambda^\mu(x)$. Because these local translations (15.8) with an arbitrary $\Lambda^\mu(x)$ are also general infinitesimal coordinate transformations, the T_4 gauge transformations for vector or tensor fields (in a general frame of reference with arbitrary coordinates) are formally the same as the Lie variation of tensors.[1] For a space-time-dependent tensor $Q_{\alpha_1 \cdots \alpha_n}^{\mu_1 \cdots \mu_m}(x)$ in the action (15.6) including $P_{\mu\nu}$, the T_4 gauge transformations are assumed to be

$$
\begin{aligned}
Q_{\alpha_1 \cdots \alpha_n}^{\mu_1 \cdots \mu_m}(x) &\to \left(Q_{\alpha_1 \cdots \alpha_n}^{\mu_1 \cdots \mu_m}(x) \right)^{\$} \\
&= \frac{\partial x'^{\mu_1}}{\partial x^{\nu_1}} \cdots \frac{\partial x'^{\mu_m}}{\partial x^{\nu_m}} \frac{\partial x^{\beta_1}}{\partial x'^{\alpha_1}} \cdots \frac{\partial x^{\beta_n}}{\partial x'^{\alpha_n}} \left(1 - \Lambda^\lambda(x)\partial_\lambda \right) Q_{\beta_1 \cdots \beta_n}^{\nu_1 \cdots \nu_m}(x), \quad (15.9)
\end{aligned}
$$

where $\mu_1, \nu_1, \alpha_1, \beta_1$, etc. are flat space-time indices. Here, we have suppressed indices of internal groups because they do not change the external gauge transformations (15.9). As usual, both the (Lorentz) spinor field ψ and the (Lorentz) scalar field Φ are treated as "coordinate scalars" and have the same translational gauge transformation: $\psi \to \psi^{\$} = \psi - \Lambda^\lambda \partial_\lambda \psi$, and $\Phi \to \Phi^{\$} = \Phi - \Lambda^\lambda \partial_\lambda \Phi$. Moreover, the covariant derivative of a vector V_ν, i.e., $D_\mu V_\nu$, transforms like a covariant tensor $T_{\mu\nu}$.

In QCG model, the symmetry related to the transformation (15.8) is particularly interesting because symmetries usually imply a conservation law. In this case, the space-time translational symmetry implies the conservation of the energy–momentum tensor. This is important because otherwise, the QCG model would not be satisfactory because it would not be consistent with an established conservation law. As we have mentioned previously, the YMUW framework is just general enough to realize this symmetry-conservation-law connection in all reference frames[6] (see Appendix B). In a more general framework such as curved space-time, this important symmetry-conservation-law connection is, in general, lost according to Noether's Theorem II. (Cf. Section 0.1 in the Overview).

Based on the translational gauge transformations (15.9) or (8.4)–(8.6), we can show that the scalar Lagrangian $L_Q = L_{QCG}/\sqrt{-P}$ transforms as follows:

$$
L_Q \to (L_Q)^{\$} = L_Q - \Lambda^\lambda(\partial_\lambda L_Q). \quad (15.10)
$$

Since Λ^μ is an infinitesimal gauge vector function, the gauge transformation of $P_{\mu\nu}$ leads to

$$
\sqrt{-P} \to \sqrt{-P^{\$}} = [(1 - \Lambda^\sigma \partial_\sigma)\sqrt{-P}](1 - \partial_\lambda \Lambda^\lambda), \quad P = \det P_{\mu\nu}. \quad (15.11)
$$

As a result, the Lagrangian $\sqrt{-P}\,L_Q$ changes only by a divergence under the gauge transformation (15.9),

$$\int L_{QCG} d^4x = \int \sqrt{-P}\,L_Q d^4x \rightarrow \int [\sqrt{-P}L_Q - \partial_\lambda(\Lambda^\lambda L_Q \sqrt{-P})]d^4x$$

$$= \int \sqrt{-P}\,L_Q d^4x = \int L_{QCG} d^4x. \tag{15.12}$$

The divergence term in (15.12) does not contribute to the field equations because one can transform an integral over a four-dimensional volume into the integral of a vector over a hypersurface on the boundaries of the volume of integration where fields and their variations vanish. Thus, we have shown that, although the QCG Lagrangian is not explicitly gauge invariant, the action S_{QCG} given by (15.6) is invariant under the T_4 gauge transformations (15.9).

In order to see the big picture of gauge symmetries in all fundamental interactions in physics, consider both the unified gravity-electroweak and chromo-gravity models. In the gravity-electroweak model, the action $S_{gew} = \int d^4x L_{gew}$, where L_{gew} is given in (14.4), can be generalized to arbitrary coordinates, just like S_{QCG} in (15.6). One can follow the steps from (15.8) to (15.12) to show that the gravity-electroweak action S_{gew} is invariant under the T_4 gauge transformations in a general frame of reference. Thus, the only exact local gauge symmetry in the actions of (14.4) and (15.6) is the local space-time translation gauge symmetry. All other internal gauge symmetries such as $SU_2 \times U_1$ in (14.4) and color SU_3 in (15.6) are violated by the gravitational interaction, as discussed in Sections 14.4 and 15.2. These are only approximate symmetries in the physical world. The violation of all internal gauge symmetries by gravity is an important result in the unified models because it leads to an effective curved space-time for the motion of all classical objects. (Cf. Sections 14.5, Chapter 14 and Section 15.3 below.) Such violations of fundamental gauge symmetries are very difficult to detect in the high-energy laboratory however, unless a macroscopic object is involved.

15.3. Effective Curved Space-Time for Motions of Quarks and Gluons in the Classical Limit

As usual, for simplicity, we discuss classical equations of motion in chromogravity in inertial frames. We have, for example,

$$\Delta_\mu = J_\mu^\sigma \partial_\sigma, \quad J_{\mu\nu} = \eta_{\mu\nu} + g\phi_{\mu\nu}, \quad \phi_{\mu\nu} = \phi_{\nu\mu} = \eta_{\nu\lambda}\phi_\mu^\lambda,$$

instead of (15.5). As mentioned previously, the presence of the translation gauge symmetry of gravity in the QCG model leads to an effective curved space-time for the motion of all classical objects. The reason for this is the T_4 gauge symmetry dictates the presence of the T_4 gauge covariant derivatives $\Delta_\mu = J_{\mu\nu}\partial^\nu$ in the Lagrangians of all quark and gluon fields in the unified model. Since we are interested in the motion of classical objects, we set the strong coupling constant $g_s = 0$ in (15.4), (15.6) and (15.7) for simplicity. As shown in equation (15.7) for the presence of gravitational interaction, the ordinary partial derivative ∂_μ in the free Lagrangian for scalar and fermion fields will be replaced by the T_4 gauge covariant derivative Δ_μ:

$$\overline{\psi}i\gamma^\mu\partial_\mu\psi \to \overline{\psi}i\gamma^\mu\Delta_\mu\psi, \quad (\partial_\mu\Phi)\partial^\mu\Phi \to (\Delta_\mu\Phi)\Delta^\mu\Phi,$$

where we consider inertial frames with the metric tensor $\eta_{\mu\nu}$ and ignore mass terms, which have little to do with the gravitational interaction. The effects of this replacement can be seen in equations (15.2)–(15.7), where the gauge curvatures of gauge fields are also modified to be consistent with (15.1). This replacement implies that there is a universal coupling of gravity to all other fields through the unified gauge covariant derivatives (15.2), where $D_\mu = \partial_\mu$ for inertial frames, and the first two terms in (15.2) are the T_4 gauge covariant derivatives. Furthermore, all wave equations of quarks and gluons reduce to classical equations with "effective Riemann metric tensors" in the geometric-optics (or classical) limit. This limiting property is a key for the unified chromogravity theory to be consistent with all classical tests of gravity.

We now examine the classical limit of the quark and gluon wave equations. First, let us consider only quarks and a gravitational field for simplicity. In inertial frames, the Lagrangian for a quark q in the presence of the gravitational gauge potential takes the usual form

$$L_q = \frac{i}{2}[\overline{q}\gamma_\mu\Delta^\mu q - (\Delta^\mu\overline{q})\gamma_\mu q] - m\overline{q}q, \tag{15.13}$$

$$\gamma_\mu\gamma_\nu + \gamma_\nu\gamma_\mu = 2\eta_{\mu\nu}, \quad \Delta_\mu = J_{\mu\nu}\partial^\nu, \quad J_{\mu\nu} = J_{\nu\mu},$$

where we have symmetrized the quark Lagrangian so that both quarks and antiquarks have the same coupling to gravity. Although as far as we know, free quarks do not exist, the wave equation for such a quark can be derived from (15.13)

$$i\gamma_\mu\Delta^\mu q - mq + \frac{i}{2}\gamma_\mu[\partial_\nu(J^{\mu\nu})]q = 0. \tag{15.14}$$

Using the limiting expression[7] for the fermion field $q = q_o \exp{(iS)}$, where q_o is a constant spinor, and the properties that $g\phi_{\mu\nu}$ and $g\partial_\alpha\phi_{\mu\nu}$ are extremely small for gravity, we can write the quark equation (15.14) in the form

$$[\gamma_\mu J^{\mu\sigma}\partial_\sigma S + m]q_o = 0, \quad J^{\mu\sigma} = (\eta^{\mu\sigma} + g\phi^{\mu\sigma}). \tag{15.15}$$

In the classical limit, the momentum $\partial_\mu S$ and mass m are large quantities.[5,7] To eliminate the spin variables, we multiply (15.15) by a factor of $(\gamma_\sigma J^{\sigma\mu}\partial_\mu S - m)$. With the help of the anticommutation relation for γ_σ in (15.13), equation (15.15) leads to the Einstein–Grossmann equation with an effective Riemann metric $G_{\mu\nu}$,

$$G^{\mu\nu}(\partial_\mu S)(\partial_\nu S) - m^2 = 0, \quad G^{\mu\nu} = \eta_{\alpha\beta}J^{\alpha\mu}J^{\beta\nu}. \tag{15.16}$$

This is the equation of the motion of a classical particle in the presence of the gravitational tensor field $\phi^{\mu\nu}$.

For the classical limit of the gluon wave equation in the presence of a gravitational field, let us consider the wave equation of a gluon field coupled only to the gravitational tensor field $\phi_{\mu\nu}$. In inertial frames with $P_{\mu\nu} = \eta_{\mu\nu}$, the wave equation of the gluon can be derived from the Lagrangian L_G in (15.7),

$$\partial_\mu(J^\mu_\sigma G^{\sigma\nu}_a) - g_s f_{abc} G_{b\mu} G^{\mu\nu}_c = 0. \tag{15.17}$$

Using the limiting expression $G^\mu_a = b^\mu_a \exp{(iS')}$ and choosing a "Lorentz gauge" for the gluon field $\partial_\mu G^\mu_a = 0$, the wave equation (15.17) for the massless gluon reduces to

$$J^\mu_\sigma J^{\sigma\lambda}\partial_\mu\partial_\lambda(b^\mu_a e^{iS'}) - g\phi^\mu_\sigma J^{\nu\lambda}\partial_\lambda\partial_\mu(b^\sigma_a e^{iS'}) = 0. \tag{15.18}$$

In the geometric-optics limit, the wave 4-vector $\partial_\mu S'$ is large so that (15.18) can be written as

$$\overline{G}^{\mu\nu}(\partial_\mu S')(\partial_\nu S') = 0, \quad \overline{G}^{\mu\nu} = G^{\mu\nu} - \frac{g}{4}\phi^\mu_\sigma J^{\sigma\nu}, \tag{15.19}$$

where $G^{\mu\nu} = \eta_{\alpha\beta}J^{\alpha\mu}J^{\beta\nu}$. Since macroscopic objects are not made of gluons and, as far as we know, free gluons do not exist, the additional term $(g/4)\phi^\mu_\sigma J^{\sigma\nu}$ in (15.19) probably cannot be tested experimentally. The result (15.19) for gluon field is the same as the eikonal equation (8.36) for a light ray. The calculations in (15.13)–(15.19) can also be carried out in general frames of reference with arbitrary coordinates.

15.4.   Discussion

Just as in the usual quantum chromodynamics theory, the Faddeev–Popov method is used to quantize the gauge fields in the unified QCG model. The quantization of the strong-interaction sector (i.e., the chromodynamics sector) is well known,[2] while the quantization of the gravitational gauge field $\phi_{\mu\nu}$ has been discussed in Chapter 11, where the Feynman–Dyson rules for Yang–Mills gravity were also obtained.

The emergence of the effective metric tensor $G^{\mu\nu}$ in the geometric-optics limit is a general property of any field equation because all fields are coupled to gravity through the T_4 gauge covariant derivative $\Delta_\mu = J_{\mu\nu}\partial^\nu$ in equation (15.1). Thus, any classical object moves as if in a "curved space-time" with the metric tensor $G^{\mu\nu} = \eta_{\alpha\beta}J^{\alpha\mu}J^{\beta\nu}$, and behaves in a way consistent with the equivalence principle. However, the true underlying space-time of quantum chromogravity is flat. Consequently, one still has the true conservation of the energy–momentum tensor by Noether's theorem or by the Fock–Hilbert approach (see Appendix B). Furthermore, the usual quantization procedure for the gravitational field[5] can be carried out within QCG in inertial frames. The unified model suggests that the curvature of space-time revealed by the classical tests of gravity is a manifestation of the translational gauge symmetry in the classical limit within flat space-time.

Although (15.14) or (15.5) is, strictly speaking, only a formal wave equation for a single quark, one can consider the classical Einstein–Grossmann equation (15.16) as the equation for a bound system of a large number of quarks with a large mass. Physically, the large mass and large eikonal in the classical limit are actually due to the summation of the individual masses and eikonals for each particle in the system under consideration, as discussed in Chapter 8.

In light of previous discussions of the gravity-electroweak and QCG models, it seems fair to say that the gravitational, strong, and electroweak interactions are all dictated by gauge symmetries (in the YMUW framework), even though all internal gauge symmetries are subtly violated by the extremely weak gravitational coupling. That gauge symmetries dictate interactions were advocated by Utiyama *et al.*[8] and particularly stressed by Yang.[9] Dynamical symmetries such as U_1, SU_2, SU_3 and T_4 appear to be more profound (in the sense that they dictate fundamental interactions of nature) than the geometrical symmetries of the rotational group and Poincaré group, which are valid for all physical laws.[10] Furthermore, the unified model of chromogravity suggests that the fundamental role for our understanding of nature is played by the gauge curvatures of symmetry groups associated with conservation laws rather than by the space-time curvature of Riemannian geometry.

References

1. J. P. Hsu, *Int. J. Mod. Phys. A* **21**, 5119 (2006); J. P. Hsu and L. Hsu, in *A Broader View of Relativity: General Implications of Lorentz and Poincaré Invariance* (World Scientific, 2006), Appendix D.
2. K. Huang, in *Quarks, Leptons and Gauge Field* (World Scientific, 1982), pp. 241–242.
3. R. E. Marshak, in *Conceptual Foundations of Modern Particle Physics* (World Scientific, 1993), Chapter 5.
4. S. Weinberg, in *The Quantum Theory of Fields*, Vol. 2 (Cambridge University Press, 1996), pp. 152–157.
5. J. P. Hsu, *Chin. Phys. C* **36**, 403 (2012).
6. V. Fock, in *The Theory of Space Time and Gravitation* (Pergamon Press, 1959), pp. 163–166.
7. L. Landau and E. Lifshitz, in *The Classical Theory of Fields*, Trans. by M. Hamermesh (Addison-Wesley, Cambridge, MA, 1951), pp. 136–137.
8. R. Utiyama, in *100 Years of Gravity and Accelerated Frames: The Deepest Insights of Einstein and Yang–Mills*, eds. J. P. Hsu and D. Fine (World Scientific, 2005), p. 157; T. W. B. Kibble, *ibid*, p. 168.
9. C. N. Yang, in *100 Years of Gravity and Accelerated Frames: The Deepest Insights of Einstein and Yang–Mills*, eds. J. P. Hsu and D. Fine (World Scientific, 2005), pp. 527–538.
10. E. P. Wigner, *Proc. Natl. Acad. Sci.* **51**, 5 (1964).

Part III

Dynamics of Expansion and Cosmic Relativity with a Cosmic Time

Part III

Dynamics of Expansion and Cosmic Relativity
with a Cosmic Time

Chapter 16

New Dynamics of Cosmic Expansion

We explore the implications of quantum Yang–Mills gravity and cosmological principle for a new dynamics of cosmological expansion, including the geometry of an expanding space-time, the Hubble velocity, the cosmic redshift z and the age of the universe. Our approach is similar to that of the Friedmann–Lemaître–Robertson–Walker (FLRW) model, in which Einstein's gravitational field equation dictates the dynamics of the expansion of the universe.[1] To understand the motion and distribution of matter and energy across the universe, we employ an effective metric tensor, which emerges in quantum Yang–Mills gravity in the geometric-optics limit. This effective metric tensor causes macroscopic objects and light rays to move as if they were in a curved space-time, even though the real underlying space-time of quantum particles and gravitational fields is flat. Therefore, all our discussions can be based on inertial frames in which space-time coordinates have their usual well-defined operational meanings, as opposed to the space-time coordinates in the FLRW model based on general relativity.[2,a]

16.1. The Cosmological Principle and Simplified Effective Metric Tensor

Based on the requirements of homogeneity and isotropy on large scales, the FLRW model assumes

$$ds^2 = g_{\mu\nu}dx^\mu dx^\nu, \quad g_{\mu\nu} = \left(1, -a^2(t), -a^2(t), -a^2(t)\right),$$
$$T_{\mu\nu} = (\rho(t), a^2(t)p(t), a^2(t)p(t), a^2(t)p(t)), \quad c = \hbar = 1, \tag{16.1}$$

---

[a] cf. Section 0.1 and Ref. 3 in Overview.

where $T_{\mu\nu}$ is the energy–momentum tensor for macroscopic bodies.[1] The expansion of the universe, as described by the scale factor $a(t)$ in (16.1), is governed by the gravitational field equation in general relativity (GR). The Robertson–Walker scale factor $a = a(t)$ is a function of time only and denotes the change of distance between astronomical objects. Friedman (1922) was the first to find a solution of $a(t) \propto t^{2/3}$ for a matter-dominated universe.[1]

Since quantum Yang–Mills gravity is based on particle physics in flat space-time, it is natural to employ Yang–Mills gravity with translational gauge symmetry in inertial frames to discuss an alternative dynamics for the expanding universe, so we can determine the expansion rate and estimate the age of the universe.

In analogy with FLRW model for an expanding universe that is homogeneous and isotropic at large scales,[3] let us discuss the implications of quantum Yang–Mills gravity on the effective covariant metric tensor $G_{\mu\nu}(t)$, which is directly related to the line element or the metric ds^2 of the expanding space-time geometry,

$$ds^2 = G_{\mu\nu}(t)dx^\mu dx^\nu = B^2 dt^2 - A^2(dx^2 + dy^2 + dz^2),$$
$$G_{\mu\nu}(t) = (B^2(t), -A^2(t), -A^2(t), -A^2(t)), \quad \mu, \nu = 0, 1, 2, 3. \tag{16.2}$$

This is a basic assumption in the Hsu–Hsu–Katz (HHK) model[4] of particle-cosmology-based quantum Yang–Mills gravity. Note that the $G_{00}(t)$ component in HHK model is assumed to be a function of time, which differs from the assumption $g_{00} = 1$ given by (16.1) in the FLRW model. The Yang–Mills gravitational field equation with space-time translational gauge symmetry will determine this function $G_{00}(t)$ at large scales, where homogeneous and isotropic cosmos is a reasonable approximation rather than an exact property.

In the macroscopic world of our everyday experience, GR dictates that classical objects move in a curved space-time, obeying a relativistic Hamilton–Jacobi type equation with an effective Riemann metric tensor $G_{\mu\nu}(x)$. In the geometrical optics limit, this tensor involves ten functions of space-time.[5] However, at cosmological (or super-macroscopic) scales, the assumed homogeneity and isotropy on large Hubble scales simplify the metric tensor to only four functions of time, as shown in (16.2). In this sense, we may say that distant galaxies on the Hubble scale move as if they are in a super-macroscopic world with an effective expanding metric tensor given in (16.2).

In the HHK model of cosmic expansion,[4,6] quantum Yang–Mills gravity in inertial frames and the cosmological principle imply the following simple relations[b]

$$G_{\mu\nu}(t) = \eta^{\alpha\beta} J_{\alpha\mu}(t) J_{\beta\nu}(t), \tag{16.3}$$

$$J_{\mu\nu}(t) = (B(t), -A(t), -A(t), -A(t)), \quad \eta^{\alpha\beta} = (1, -1, -1, -1). \tag{16.4}$$

Originally, the T_4 gauge invariant gravitational field equations in inertial frames take the form,

$$g^2 S_{\mu\nu}(x) = \partial_\lambda (J_{\lambda'\rho'} C_{\rho\mu\nu} \eta^{\rho\rho'} - J_{\lambda'\alpha'} C^\beta_{\alpha\beta} \eta^{\alpha\alpha'} \eta_{\mu\nu} + C^\beta_{\mu\beta} J_{\nu\lambda'}) \eta^{\lambda\lambda'}$$

$$- C^\beta_{\mu\alpha} \partial_\nu J_{\alpha'\beta} \eta^{\alpha\alpha'} + C^\beta_{\mu\beta} \partial_\nu J_{\alpha\alpha'} \eta^{\alpha\alpha'} - C^\beta_{\lambda\beta} \partial_\nu J_{\mu\lambda'} \eta^{\lambda\lambda'}, \tag{16.5}$$

$$C_{\mu\nu\alpha}(x) = J_{\mu\lambda}(\partial_{\lambda'} J_{\nu\alpha}) \eta^{\lambda\lambda'} - J_{\nu\lambda}(\partial_{\lambda'} J_{\mu\alpha}) \eta^{\lambda\lambda'}, \quad C^\beta_{\mu\alpha} = C_{\mu\alpha\beta'} \eta^{\beta\beta'},$$

$$S_{\mu\nu}(x) = \overline{\psi} i \gamma_\mu \Delta_\nu \psi - i(\Delta_\nu \overline{\psi}) \gamma_\mu \psi, \quad \Delta_\nu = J_{\nu\lambda} \partial^\lambda, \quad J_{\nu\lambda}(x) = \eta_{\nu\lambda} + g\phi_{\nu\lambda}(x), \tag{16.6}$$

where $g^2 = 8\pi G$ and μ and ν should be made symmetric in the field equation (16.5).[7] In the original quantum Yang–Mills gravitational field equation, $C_{\mu\nu\alpha}(x)$ is the T_4 gauge curvature and $S_{\mu\nu}(x)$ is the energy–momentum of coupled fermions with $\partial^\mu S_{\mu\nu}(x) = 0$.

16.2. Super-Macroscopic World and Further Approximation of Yang–Mills Gravitational Equations

Based on the T_4 gauge curvature (16.6) in the super-macroscopic approximation (16.4), the non-vanishing components of $C_{\mu\nu\alpha}$ are

$$C_{0ik} = -C_{i0k} = B\dot{A}\eta_{ik}, \quad \dot{A} = dA(t)/dt. \tag{16.7}$$

All other components vanish at the super-macroscopic scale.

In analogy with (16.1) in the FLRW model, we assume that the universe is homogeneous and isotropic on Hubble scales, as observed in inertial frames. Thus,

---

[b]The $J_{ii}, i = 1, 2, 3$, components differ by a sign from those in Ref. 4 (i.e., $A \to -A$ in (16.4)) so that the spatial components of $T_{\mu\nu}(t)$ in (16.8) have the same conventional signs. The results (16.9)–(16.17) are not changed.

the energy–momentum tensor of the universe everywhere takes a simple diagonal form:

$$T_{\mu\nu}(t) = (\rho(t)B^2(t), P(t)A^2(t), P(t)A^2(t), P(t)A^2(t)),$$

$$T_{\mu}^{\nu}(t) = (\rho(t), -P(t), -P(t), -P(t)), \quad T_{\mu\nu}(t) = T_{\mu}^{\lambda}(t)G_{\lambda\nu}(t), \tag{16.8}$$

where $\rho(t)$ and $P(t)$ are the energy density and pressure of macroscopic bodies, respectively. Because of the universal gravitational interaction of the T_4 tensor fields for all matter, it is reasonable to approximate the energy–momentum $S_{\mu\nu}$ (16.6) of fields by the energy–momentum (16.8) of macroscopic bodies at the super-macroscopic scale.

Therefore, in the super-macroscopic limit with the cosmological principle, we approximate the Yang–Mills gravitational equation (16.5) by the replacements

$$S_{\mu\nu}(x) \to T_{\mu\nu}(t), \quad J_{\mu\nu}(x) \to J_{\mu\nu}(t).$$

Using these approximations, we now have the HHK model for particle cosmology in which the dynamics of expansion are dictated by Yang–Mills gravity. In other words, the T_4 gauge symmetry of quantum Yang–Mills gravity governs the scale factors $A(t)$ and $B(t)$ in (16.4) and, hence, the effective metric tensor $G_{\mu\nu}(t)$ in (16.2) and (16.3) for the expansion of the universe.

16.3.   Basic Equations for Expanding Universe in HHK Model

The Yang–Mills gravitational field equation (16.5), together with (16.3)–(16.4) and (16.6)–(16.8), leads to two independent equations

$$6B\dot{A}^2 = g^2 \rho B^2, \tag{16.9}$$

$$2(B^2\ddot{A} + 2B\dot{B}\dot{A}) = g^2 PA^2,$$

which relate the space-time scale factors $A(t)$ and $B(t)$. The first equation is derived from (16.5) in the super-macroscopic limit when $\mu = \nu = 0$. The second equation is obtained from (16.5) in the super-macroscopic limit when $\mu = \nu = 1$ or $\mu = \nu = 2$ or $\mu = \nu = 3$. These results arise from the special time dependence of $G_{\mu\nu}$ and $J_{\mu\nu}$ in (16.3)–(16.4), and the simplicity of the T_4 gauge curvature $C_{\mu\nu\alpha}$ as shown in (16.7).

Using the equation of state $P = \omega\rho$ with constant ω and the usual expanding relation of the energy density $\rho = \rho_0/A^3(t)$, (16.9) leads to the following equations:

$$(5A\ddot{A} + 6\dot{A}^2)A^6\dot{A}^4 = \frac{(g^2\rho_0)^3\omega}{72}, \quad B = \frac{6\dot{A}^2 A^3}{g^2\rho_0}. \tag{16.10}$$

These are the basic equations governing the expansion of the universe, according to quantum Yang–Mills gravity in flat space-time.

16.4. Solutions for Space-Time Expanding Factors $A(t)$ and $B(t)$ and the Age of the Universe

For a "matter-dominated" (md) cosmos, we use $A(t) = \alpha t^n$ with constant α and n in (16.10) to obtain the solutions for $A(t)$ and $B(t)$,

$$A(t)_{md} = \alpha t^{1/2}, \quad \alpha = \left(\frac{8g^6\omega\rho_0^3}{9}\right)^{1/12},$$

$$B(t)_{md} = \beta t^{1/2}, \quad \beta = \frac{3\alpha^5}{2g^2\rho_0}. \tag{16.11}$$

For a "radiation-dominated" (rd) cosmos, in which $\omega = 1/3$ and $\rho = \rho_0/A^4$,[1] (the extra factor of $1/A(t)$ in the energy density ρ is related to the redshift of radiation as the cosmos expands, cf. Okubo equation (16.35) with $m = 0$ and $\partial_\mu S = k_\mu$ below), the time-dependent scale factors $A(t)$ and $B(t)$ are found to be

$$A(t)_{rd} = \alpha' t^{2/5}, \quad \alpha' = \left(\frac{5^6 g^6\omega\rho_0^3}{2^6 \times 36}\right)^{1/15},$$

$$B(t)_{rd} = \beta' t^{2/5}, \quad \beta' = \frac{24\alpha'^6}{25g^2\rho_0}. \tag{16.12}$$

We interpret these results as implying that the effective covariant metric tensor $G_{\mu\nu}(t)$ in (16.2) and (16.3) and the space-time geometry of the expanding universe are dictated by the space-time translational gauge symmetry of the gravitational interaction.

For comparison, the scale factor $a(t)$ in the FLRW model is given by $a(t) \propto t^{2/3}$ for a matter-dominated cosmos and $a(t) \propto t^{1/2}$ for a radiation-dominated cosmos.[1] Thus, the results of Yang–Mills gravity in (16.11) and (16.12) lead to slower expansions than those based on GR, implying that the universe is older than estimates from, say, the Einstein–de Sitter model.[1,8]

Let us estimate the age of the universe t_0^{YM} based on Yang–Mills gravity in flat space-time with the scale factors in (16.11) for a matter-dominated cosmos. Since there is only one expansion rate $H(t) = \dot{A}(t)/A(t)$, it is natural to have the relation $\dot{X}(t) \propto H(t)$, which holds in any inertial frame with Cartesian coordinates

(X, Y, Z). Because of the isotropy, it suffices to consider only the X coordinate, without loss of generality. Furthermore, there is only one natural length scale involved in the problem, i.e., the present Hubble length $X_o = cH_o^{-1}$. Dimensional analysis suggests the following equation for the expanding universe:

$$\dot{X}(t) = H(t)X_o, \quad H(t) = \dot{A}(t)/A(t). \tag{16.13}$$

This relation enables us to compare the age of the universe for different models with different expansion rates $H(t)$. For a matter-dominated cosmos, let us write $A(t) = \alpha t^\sigma$. The solutions to (16.13) are then

$$\frac{1}{X_o} \int_{X_a}^{X_b} dX = \ln\left(\frac{t_b}{t_a}\right)^\sigma,$$

$$\frac{t_b}{t_a} = \frac{t_o}{t_a} = e^{1/\sigma}, \quad X_b - X_a = X_o = cH_o^{-1}, \quad e = 2.718, \tag{16.14}$$

where X_a and t_a are some suitably small numbers such that when the universe expands to the present size, i.e., $X_b - X_a = X_o = c/H_o$, the time t_b is naturally identified as the age of the universe, $t_o = t_b$.

For our calculations, the values of t_a and X_a are not important because the relationship between the ages of the universe calculated using two models with different scale functions is independent of them. It is the ratios of the values of $A(t)$ at different times that are important. For a matter-dominated universe, let us compare the age of the universe t_o^{GR} based on GR (in the Einstein–de Sitter model[1] with the scale factor $a(t) = a_o t^{2/3}$) and the age t_o^{YM} based on Yang–Mills gravity (in the present HHK model with the scale factor $A(t) = \alpha t^{1/2}$):

$$\frac{t_o^{YM}}{t_a} = e^2, \quad H^{YM}(t) = \frac{\dot{A}}{A} = \frac{1}{2t}, \quad \text{HHK model,} \tag{16.15}$$

$$\frac{t_o^{GR}}{t_a} = e^{3/2}, \quad H^{GR}(t) = \frac{\dot{a}}{a} = \frac{2}{3t}, \quad \text{EdS model.} \tag{16.16}$$

In the Einstein–de Sitter model, the age of the universe is $t_o^{GR} \approx 9.3 \times 10^9 years$, which appears to be too small and difficult to reconcile with the measured ages of old star clusters. [c] It follows from (16.15) and (16.16) that the "age of the universe"

---

[c]The ages of the oldest globular star clusters in the halo of our Milky Way galaxy are estimated to be $(16 \pm 2) \times 10^9 years$.[8]

t_o^{YM} according to Yang–Mills gravity is approximately

$$t_o^{\text{YM}} = \sqrt{e}\, t_o^{\text{GR}} \approx 15.3 \times 10^9\, years, \tag{16.17}$$

which is consistent with experiments. For a more accurate result, the calculation should take into account the different time dependences of the energy density ρ in the matter-dominated and radiation-dominated eras, rather than assuming that the universe is always dominated by non-relativistic matter. Nevertheless, it is reasonable to approximate that the universe has been dominated by non-relativistic matter for most of its existence. The value (16.17) appears to be reasonable because the cosmos expands slower in the HHK model than that in the Einstein–de Sitter or FLRW model,[9,10] so that it needs more time or a larger t_o^{YM} to reach its currents Hubble length $X_o = cH_o^{-1}$.

16.5. Cosmic Action and the Okubo Equation for Dynamics of Cosmic Expansion

Yang–Mills gravity and the cosmological principle lead to an effective metric tensor $G^{\mu\nu}(t)$ in (16.2) for the super-macroscopic world, as viewed from an inertial frame. As usual, the associated effective metric ds^2 can be used to construct a new cosmic action S_{cos} for the recessional motion of distant galaxies within the framework of flat space-time,

$$S_{\text{cos}}(r(t); t_1, t_2) = \int_1^2 (-mds) = \int_{t_1}^{t_2} \left(-mB(t)\sqrt{1 - \frac{\dot{r}^2}{C_o^2}} \right) dt, \tag{16.18}$$

where $\dot{r} = dr/dt$ and $C_o = B/A = $ constant. A distant galaxy is idealized as a point with mass m and radial coordinate $r(t)$. In view of the cosmological principle of homogeneity and isotropy of the visible cosmos, it suffices to consider the variation of the action (16.18) with the radial coordinate r and the time t in an inertial frame F. We postulate that this cosmic action is stationary, $\delta S_{\text{cos}} = 0$, and that the resultant Lagrange equation governs the motion of any distant galaxies in the super-macroscopic cosmos.

The effective metric tensor $G_{\mu\nu}(t)$ leads to an explicitly time-dependent cosmic Lagrangian L_{cos},

$$L_{\text{cos}}(r(t), \dot{r}(t), t) = -mB(t)\sqrt{1 - \frac{\dot{r}^2}{C_o^2}}. \tag{16.19}$$

Let us consider the variations of r and t. We note that when the trajectory is shifted,

$$r(t) \rightarrow r'(t) = r(t) + \delta r(t), \tag{16.20}$$

the time along the path changes differently at each point on the varied curve, including the endpoints.[11] That is, time changed locally,

$$t \rightarrow t'(t) = t + \delta t(t). \tag{16.21}$$

Thus we have

$$\delta(dt) = d(t + \delta t) - dt = dt \frac{d}{dt} \delta t(t), \tag{16.22}$$

where $\delta t(t_2) = \delta t_2$, and $\delta t(t_1) = \delta t_1$ at times t_1 and t_2. If time is not varied, we use δ_o instead of δ. Thus we have

$$\delta_o t = 0, \quad \text{or} \quad \delta_o \frac{d}{dt} = \frac{d}{dt} \delta_o. \tag{16.23}$$

As a result, the difference between δ and δ_o is given by the identity

$$\delta - \delta_o = \delta t \frac{d}{dt}. \tag{16.24}$$

We also have the following relations:

$$\delta L = \delta_o L + \det \frac{d}{dt} L, \quad \delta \dot{r}(t) = \frac{d}{dt} \delta r - r \frac{d}{dt} \delta t, \quad L \equiv L_{\text{cos}}. \tag{16.25}$$

Thus, the variation of the cosmic action S_{ca} is given by

$$\delta S_{\text{cos}} = \int_{t_1}^{t_2} [(\delta L) dt + L \delta(dt)] = \int_{t_1}^{t_2} \left[\frac{d}{dt} (L \delta t) dt + \delta_o L \right], \tag{16.26}$$

where we have used

$$L \frac{d}{dt} \delta(t) = \frac{d}{dt} (L \delta t) - \delta t \frac{dL}{dt}.$$

The total variation of the cosmic Lagrangian is

$$\delta L = \delta_o L + \delta t \frac{d}{dt} L = \frac{\partial L}{\partial r} \delta r + \frac{\partial L}{\partial \dot{r}} \delta \dot{r} + \frac{\partial L}{\partial t} \delta t$$

$$= \frac{\partial L}{\partial t} \delta t + \frac{\partial L}{\partial r} \delta r + m\dot{r} \frac{d}{dt} \delta r - m\dot{r}^2 \frac{d}{dt} \delta t, \tag{16.27}$$

where we have used $\delta\dot{r}(t)$ in (16.24). Thus $\delta S_{\cos}$ can be expressed as

$$\delta S_{\cos} = \int_{t_1}^{t_2} dt \left[-\left(\frac{d}{dt}\frac{\partial L}{\partial \dot{r}} - \frac{\partial L}{\partial r} \right) \delta r + \left(\frac{\partial L}{\partial t} - \frac{d}{dt}\left[L - \frac{\partial L}{\partial \dot{r}}\dot{r} \right] \right) \delta t + G_{\mathrm{sf}} \right]$$
(16.28)

$$G_{\mathrm{sf}} = \frac{d}{dt}\left[\frac{\partial L}{\partial \dot{r}}\delta r + L\delta t \right],$$

where $L \equiv L_{\cos}$ and G_{sf} is the surface term.

Since r is cyclic in (16.19), the two independent variations, in δr and δt, give two cosmic equations of motion,

$$\delta r : \frac{d}{dt}\frac{\partial L}{\partial \dot{r}} = \frac{\partial L}{\partial r} = 0, \quad p = \frac{\partial L}{\partial \dot{r}} = \frac{mB\dot{r}/C_o^2}{\sqrt{1 - \dot{r}^2/C_o^2}}, \quad (16.29)$$

$$\delta t : \frac{d}{dt}\left(\frac{\partial L}{\partial \dot{r}}\dot{r} - L \right) = -\frac{\partial L}{\partial t}, \quad E = \frac{\partial L}{\partial \dot{r}}\dot{r} - L = \frac{mB}{\sqrt{1 - \dot{r}^2/C_o^2}}. \quad (16.30)$$

Using $H = (\partial L/\partial \dot{r})\dot{r} - L$, one can obtain a time-dependent Hamiltonian[11] $H(p, r, t)$

$$H(p, r, t) = \sqrt{p^2 C_o^2 + m^2 B(t)^2}. \quad (16.31)$$

Hamiltonian's equation gives

$$p = \frac{\partial S}{\partial r}, \quad H(p, r, t) = \frac{\partial S}{\partial t}, \quad S \equiv S_{\cos}. \quad (16.32)$$

They lead to a Hamilton–Jacobi equation,

$$\frac{\partial S}{\partial t} = \sqrt{\left(\frac{\partial S}{\partial r} \right)^2 C_o^2 + m^2 B^2(t)} \quad (16.33)$$

$$\text{or} \quad \left(\frac{\partial S}{\partial t} \right)^2 = \left(\frac{\partial S}{\partial r} \right)^2 C_o^2 + m^2 B^2(t). \quad C_o = B/A.$$

This last equation in (16.33) can be expressed in terms of the effective (contravariant and diagonal) metric tensor $G^{\mu\nu}(t) = (G^{00}, G^{11}, G^{22}, G^{33}) = (B^{-2}, -A^{-2}, -A^{-2}, -A^{-2})$,

$$G^{\mu\nu}(t)\partial_\mu S \partial_\nu S - m^2 = 0, \quad G^{\mu\lambda}(t)G_{\lambda\nu}(t) = \delta^\mu_\nu, \quad (16.34)$$

or

$$G^{00}(t)\left(\frac{\partial S}{\partial t}\right)^2 - G^{ii}(t)\left(\frac{\partial S}{\partial r^i}\right)\left(\frac{\partial S}{\partial r^i}\right) = m^2, \quad m \geq 0 \quad r^i = (x, y, z). \quad (16.35)$$

We call this new cosmic equation the Okubo equation of motion and it governs the motion of a galaxy with mass m in the super-macroscopic world based on particle cosmology.[d]

As we shall discuss in Chapter 17, when the mass $m > 0$, the Okubo equation dictates the motion of distant galaxies and reveals the expansion of the universe because we have the scale factors $A \propto B \propto t^{1/2}$ for a matter-dominated universe. For the case, $m = 0$, the Okubo equation governs the propagation, frequency, and redshifts of light rays emitted from accelerated (or decelerated) distant galaxies.

The Okubo equation (16.34) is equivalent to the cosmic action S_{cos} in (16.18) or the time-dependent cosmic Lagrangian L_{cos} in (16.19). It is the basic equation or the cornerstone of the new dynamics of cosmic expansion based on quantum Yang–Mills gravity and the cosmological principle. One can see from, say, the time-dependent cosmic Lagrangian (16.19) that the energy E in (16.30) is not conserved because $\partial L/\partial t \neq 0$. This demonstrates that the conservation of energy in cosmic dynamics is violated by the approximation that the cosmos is homogeneous, which implies that the effective metric tensor can only depend on time in (16.2). Furthermore, the line element ds^2 in (16.2) also violate special relativity or Lorentz invariance. This violation of Lorentz invariance can be resolved when we introduce a cosmic time in the dynamics of expansion in Chapter 19.

16.6.   Time-Dependent Cosmic Force and the Cosmic Dynamics

From the cosmic Okubo equation (16.34) or (16.35) for the motion of distant galaxies, it is not clear what is the cosmic force acting on the distant galaxies. To see this force, let us consider the time-dependent Lagrangian (16.19). Since r is cyclic in the cosmic Lagrangian (16.19), the "generalized force" $\partial L/\partial r$ vanishes, so that

---

[d]We term this equation of motion for distant galaxies, derived from quantum Yang–Mills gravity and the cosmological principle, the "cosmic Okubo equation" of motion in memory of his endeavors in particle physics and his "departure ... to the black hole."

we have the conservation of "generalized momentum" p,

$$\frac{d}{dt}p = \frac{d}{dt}\left(\frac{mB\dot{r}/C_o^2}{\sqrt{1 - \dot{r}^2/C_o^2}}\right) = 0, \quad \dot{r} < C_o, \tag{16.36}$$

as shown in (16.29). However, one would like to know what is the usual force that corresponds to a change of the usual momentum, i.e., $m\dot{r}/\sqrt{1 - \dot{r}^2/C_o^2}$. For this purpose, we rewrite the Lagrange equation (16.36) in the following form involving the "usual force" $F_{\cos}$

$$\frac{d}{dt}\left(\frac{m\dot{r}}{\sqrt{1 - \dot{r}^2/C_o^2}}\right) = F_{\cos}, \tag{16.37}$$

$$F_{\cos} \equiv -\frac{dB/dt}{B}\left(\frac{m\dot{r}}{\sqrt{1 - \dot{r}^2/C_o^2}}\right), \tag{16.38}$$

where $B = B(t)$ is given by (16.11), $B = B(t)_{\mathrm{md}}$, for matter-dominated cosmos.

This cosmic force (16.38) is embedded in the Okubo equation (16.34). Such a force acts on all distant galaxies and is responsible for the Hubble recession velocity to be discussed in Chapter 17. The "mysterious force" related to the late-time acceleration of the universe will also be discussed later.

16.7. Space-Time Variations δx^v of the Cosmic Action $S_{\cos} = -m \int ds$ for the Okubo Equation

The Okubo equation (16.34) can also be derived more elegantly from the principle of least action with the cosmic action $S_{\cos} = -m \int ds$. Instead of varying space and time separately as shown in (16.20) and (16.21), we consider space-time variations δx^v of the effective cosmic metric ds^2 given by (16.2), i.e.,

$$\delta \int ds = 0, \quad ds^2 = G_{\mu v}(t)dx^\mu dx^v, \quad x^v = (t, x, y, z).$$

We have

$$\delta ds^2 = 2ds\delta ds = dx^\mu dx^v \delta G_{\mu v} + 2G_{\mu v}dx^\mu \delta dx^v$$

$$= dx^\mu dx^\lambda \frac{\partial G_{\mu\lambda}}{\partial x^v}\delta x^v + 2G_{\mu v}dx^\mu \delta dx^v, \quad G_{\mu v} = G_{\mu v}(t).$$

Suppose we consider the covariant space-time variation δx^v and the actual trajectory (i.e., those that satisfy the equations of motion) with a fixed initial point and a variable endpoint,[12] we have

$$\delta S_{\text{cos}} = m \int T_\nu \delta x^\nu \, ds + \left[mG_{\mu\nu} \frac{dx^\mu}{ds} \right] \delta x^\nu = p_\nu \delta x^\nu, \tag{16.39}$$

$$T_\nu = \left[\frac{1}{2} \frac{dx^\mu}{ds} \frac{dx^\lambda}{ds} \frac{\partial G_{\mu\lambda}}{\partial x^\nu} - \frac{d}{ds} \left(G_{\mu\nu} \frac{dx^\mu}{ds} \right) \right] = 0, \quad p_\nu = mG_{\mu\nu} \frac{dx^\mu}{ds}, \tag{16.40}$$

where the actual trajectory satisfies the equation $T_\nu = 0$, and the non-vanishing last term in (16.39) is due solely to the variable endpoint.[e] Based on (16.39), we define the generalized four-momentum of an object moving in the super-macroscopic world as the derivative $p_\nu = \partial S_{\text{cos}}/\partial x^\nu$,[12] we have

$$G^{\mu\nu}(t) p_\mu p_\nu = m^2. \tag{16.41}$$

Substituting $\partial S_{\text{cos}}/\partial x^\mu$ for p_μ in (16.41), we find the Okubo equation (16.34) for the motion of a distant galaxy with a mass m in the super-macroscopic world:

$$G^{\mu\nu}(t) \partial_\mu S \partial_\nu S - m^2 = 0, \quad S = S_{\text{cos}}.$$

One can verify that $T_0 = 0$ and $T_i = 0$ in (16.40) are, respectively, the same as (16.29) and (16.30)[f]:

$$T_0 = \frac{1}{2(1 - \dot{r}^2/C_o^2)} \left(\frac{1}{B^2} \frac{\partial B^2}{\partial t} - \frac{\dot{r}^2}{C_o^2 A^2} \frac{\partial A^2}{\partial t} \right) - \frac{1}{2B^2} \frac{d}{dt} \frac{B^2}{(1 - \dot{r}^2/C_o^2)}$$

$$= \frac{-1}{Bm\sqrt{1 - \dot{r}^2/C_o^2}} \left[\frac{\partial L_{\text{cos}}}{\partial t} + \frac{dE}{dt} \right] = 0, \quad \dot{r} = \frac{dr}{dt}, \quad \dot{r}^2 = \dot{x}^2 + \dot{y}^2 + \dot{z}^2, \tag{16.42}$$

$$T_i = \frac{1}{Bm\sqrt{1 - \dot{r}^2/C_o^2}} \frac{d}{dt} \left(\frac{Am\dot{x}^i}{C_o\sqrt{1 - \dot{r}^2/C_o^2}} \right) = 0, \quad x^i = (x, y, z), \tag{16.43}$$

$$L_{\text{cos}} = -mB\sqrt{1 - \dot{r}^2/C_o^2}, \quad E = \frac{mB}{\sqrt{1 - (\dot{r}^2/C_o^2)}}, \quad C_o = \frac{B}{A},$$

where $B^{-2}(\partial B^2/\partial t) = A^{-2}(\partial A^2/\partial t)$ because $A(t) \propto B(t)$ as shown in (16.11) and (16.12), while L_{cos} and E are the same as (16.19) and (16.30), respectively.

---

[e]The fixed initial point does not contribute to δS_{cos} in (16.39).

[f]At the end of calculations, we may choose the spatial coordinates to be $(x, y, z) = (r, 0, 0)$, so that (16.43) with $A(t) \propto B(t)$ is the same as (16.29).

16.8. Discussion

According to the cosmic Okubo equation of motion (16.34), F_{cos} in (16.38) is a universal cosmic force acting on all distant galaxies that slows down the expansion of the universe. In inertial frames, it appears that this cosmic force F_{cos} is a non-propagating force, which depends on time and the velocity of distant galaxies. This unusual cosmic force is embedded in the Okubo equation (16.34) and turns out to be responsible for causing distant galaxies to follow Hubble's law for small radial velocities.

We note that if one uses a slightly different relation $\dot{X}(t) = H(t)X(t)$,[8] instead of that in (16.13) to compare t_o^{YM} and t_o^{GR}, the result for the ratio $t_o^{\text{YM}}/t_o^{\text{GR}}$ then depends on the choice of the initial time t_a (the lower limit of the time integration) and hence, may not be reliable. Also, the singularity at $t_a = 0$ in (16.14) should not be taken seriously because (16.13) is only suggested by dimensional analysis in order to estimate the age of the universe. As long as the estimation does not depend on the choice of t_a, the result may be more reliable.

If the energy–momentum tensor $T_{\mu\nu}$ were dominated by the vacuum energy–momentum,[1] then $T_{\mu\nu} \approx T_{\mu\nu}^V = \rho_V G_{\mu\nu}$, where ρ_V is a constant independent of space-time position. Thus, if one uses $T_{\mu\nu}^V$ and equations (16.2)–(16.7), one obtains $A(t) \propto B(t) \propto t^2$. The possible physical origin of the vacuum energy–momentum and "dark energy" and their relations to the extremely small baryon–lepton charges will be discussed in Chapter 20.

Because the space-time framework of quantum Yang–Mills gravity is flat, the HHK model of cosmology has no flatness problem.[1] Moreover, the HHK model has interesting implications for the time-dependent cosmic force F_{cos} in (16.38), where the time-dependent scale factor $B(t)$ plays a double role in the HHK model: It slows down the expanding cosmos as shown in (16.38) and (16.37) and increases the scale of the time coordinate in the effective metric tensor $G_{\mu\nu}(t) = (B^2, -A^2, -A^2, -A^2)$ in (16.2).

We stress that both the resultant tensor $G_{\mu\nu}(x)$ in the Hamilton–Jacobi type equation (8.40) for the motion of macroscopic objects and the purely time-dependent $G_{\mu\nu}(t)$ in (16.2) for expanding space-time geometry are only effective metric tensors within the framework of quantum Yang–Mills gravity formulated in inertial frames. There are different physical reasons for their presence in Yang–Mills gravity. The effective metric tensor $G_{\mu\nu}(x)$ emerges in the equations of motion for classical objects after we take the geometric-optics limit of wave equations for quantum particles. However, the effective metric tensor $G_{\mu\nu}(t)$ emerges as a result of combining Yang–Mills gravity and the cosmological principle to explore alternative dynamics of cosmic expansion.

References

1. S. Weinberg, in *Cosmology* (Oxford University Press, 2008), pp. 1–14, 34–45.
2. E. P. Wigner, *Symmetries and Reflections Scientific Essays* (The MIT Press, 1967), pp. 52–53.
3. L. Landau and E. Lifshitz, in *The Classical Theory of Fields* (Addison-Wesley, 1951), p. 336.
4. J. P. Hsu, L. Hsu and D. Katz, *Mod. Phys. Lett. A* **36**, 1850116 (2018).
5. J. P. Hsu and S. H. Kim, *Eur. Phys. J. Plus* **127**, 146 (2012).
6. J. P. Hsu, *Int. J. Mod. Phys. A* **24**, 5217 (2009).
7. J. P. Hsu, *Int. J. Mod. Phys. A* **21**, 5119 (2006).
8. P. J. E. Peebles, *Principles of Physical Cosmology* (Princeton University Press, Princeton, NJ, 1993), pp. 70–130.
9. D. Katz, *Int. J. Mod. Phys. A* **30**, 1550119 (2015).
10. T. Buchert, A. A. Coley, H. Kleinert, B. F. Roukema, and D. L. Wiltshire, *Int. J. Mod. Phys. D* **25**, 1630007 (2016).
11. W. Dittrich and M. Reuter, in *Classical and Quantum Dynamics: From Classical Paths to Path Integrals*, 2nd ed. (Springer-Verlag, 2001), pp. 3–14.
12. L. Landau and E. Lifshitz, in *The Classical Theory of Fields* (Addison-Wesley, 1951), pp. 268–271.

Chapter 17

Finite General Hubble Velocity and
Exact Cosmic Redshift

17.1. Finite Hubble Recession Velocity and Cosmic Okubo Equation

Lorentz invariant Yang–Mills gravity with translational (T_4) gauge symmetry is formulated and quantized in inertial frames. The theory is consistent with experiments such as the perihelion shift of Mercury, the deflection of light by the sun, gravitational redshifts and gravitational quadrupole radiation.[1-3] We are also able to obtain the gravitational Feynman–Dyson rules and the S-matrix.[2,4] Yang–Mills gravity also brings gravity back to the arena of gauge field theory and quantum mechanics, and provides a solution to difficulties present in the conventional theory such as giving an operational meaning to space and time coordinates and developing a quantum theory of gravity.[5,6]

In the usual macroscopic (i.e., geometric–optics) limit, the wave equations of quantum particles with mass $m \geq 0$ in Yang–Mills gravity reduce to the Hamilton–Jacobi (HJ) type equation, $G^{\mu\nu}(x)(\partial_\mu S)(\partial_\nu S) - m^2 = 0$. Thus, the apparent curvature of macroscopic appears to be a manifestation of the flat translational gauge symmetry for the wave equations of quantum particles in the geometric-optics limit.[2,7] According to quantum Yang–Mills gravity, macroscopic objects move as if they were in a curved because their equations of motion involve an "effective metric tensor" $G^{\mu\nu}(x)$, which is actually a function of T_4 tensor gauge fields.

Conventionally, the FLRW model of cosmology,[8,9] based on general relativity, is used to model the expansion dynamics of the universe. In quantum Yang–Mills gravity, the effective metric tensor for the motion of macroscopic objects suggests that the Yang–Mills gravitational field equations can provide an alternative model for the dynamics of the expanding universe (i.e., a new HHK model of particle cosmology[10]).

In the super-macroscopic limit, the cosmological principle enables us to further simplify the dependent $G^{\mu\nu}(x)$ in the Hamilton–Jacobi equation. In this case, $G^{\mu\nu}(x)$ becomes a diagonal effective metric tensor $G^{\mu\nu}(t)$ dependent only on time in an inertial frame $F = F(t, x, y, z)$, where $c = \hbar = 1$,[10]

$$G_{\mu\nu}(t) = \left(B^2(t), -A^2(t), -A^2(t), -A^2(t)\right), \quad B = \beta t^{1/2}, \quad A = \alpha t^{1/2}. \quad (17.1)$$

This holds for a matter-dominated cosmos, where $\beta = 3\alpha^5/2g^2\rho_0$ and $\alpha = (8g^6\omega\rho_0^3/9)^{1/12}$.[a] One advantage of this HHK model is that because it is formulated in inertial frames, cosmological predictions are based on well-defined frames of reference, in which space and time coordinates have operational meanings.

Let us explore the physical and cosmological implications of the effective metric tensor $G_{\mu\nu}(t)$ in the super-macroscopic limit (with length scales roughly ≥ 300 million light years[8]). It is natural to use the following "Okubo equation" of motion for distant galaxies,

$$G^{\mu\nu}(t)\partial_\mu S\partial_\nu S - m^2 = 0, \quad G^{\mu\nu}(t)G_{\nu\lambda}(t) = \delta^\mu_\lambda, \quad (17.2)$$

where the effective $G^{\mu\nu}(t)$ is given by (17.1). Using spherical coordinates, we choose a specific radial direction with specific angles θ and ϕ to express (17.2) in the form,

$$B^{-2}(\partial_t S)^2 - A^{-2}(\partial_r S)^2 - m^2 = 0. \quad (17.3)$$

Since r is cyclic in the cosmic Okubo equation (17.2) or (17.3) (i.e., $G^{\mu\nu}(t)$ does not involve r[11]), we obtain the conservation of the "generalized momentum" $p = \partial_r S = \partial S/\partial r$. As usual, to solve equation (17.3), we look for an S in the form[3,11] $S = -f(t) + pr$. We have

$$f(t) = \int \sqrt{p^2 C_0^2 + m^2 B^2} \, dt, \quad C_0 = B/A = \text{constant}. \quad (17.4)$$

The trajectory of a distant galaxy is, as usual, determined by the equation $\partial S/\partial p = \text{constant}$.[3,11] Therefore, we have

$$r - \int \frac{pC_0^2}{\sqrt{p^2 C_0^2 + m^2 B^2}} dt = \text{constant}. \quad (17.5)$$

---

[a]In the conventional model of cosmology, one has $g^{\mu\nu}(t) = (1, -a^{-2}(t), -a^{-2}(t), -a^{-2}(t))$, where the scale factor is given by $a(t) = a_0 t^{2/3}$ for the matter-dominated universe.[8]

Since $B = \beta t^{1/2}$, equation (17.5) leads to the receding velocity $\dot{r}$ and acceleration $d\dot{r}/dt$ of a distant galaxy,

$$\dot{r} = \frac{dr}{dt} = \frac{pC_o^2}{\sqrt{p^2 C_o^2 + m^2 \beta^2 t}} = \frac{C_o}{\sqrt{1 + \Omega^2 t}}, \quad \Omega = \frac{m\beta}{pC_o},$$

$$\frac{d\dot{r}}{dt} = -\frac{C_o \Omega^2}{2(1 + \Omega^2 t)^{3/2}} \approx -\frac{C_o}{2\Omega t^{3/2}}, \quad m \gg p. \tag{17.6}$$

In order to discuss Hubble's law, we should also have the time-dependence of the radius $r = r(t)$, which can be derived from (17.6),

$$r = \frac{2pC_o^2}{m^2 \beta^2} \sqrt{p^2 C_o^2 + m^2 \beta^2 t} + r_a = \frac{2C_o}{\Omega^2} \sqrt{1 + \Omega^2 t} + r_a, \tag{17.7}$$

where r_a is a constant of integration. In the low velocity approximation, i.e., $m \gg p$, we have $f(t) \approx (2m\beta/3)t^{3/2}$, $r \approx t^{1/2}(2pC_o^2/m\beta)$. In this approximation, it is reasonable to ignore r_a because r is a super-macroscopic length scale. Thus, we obtain the usual Hubble laws as the low velocity approximation of the solution to Okubo equation (17.2),

$$\dot{r} = \frac{pC_o^2}{m\beta t^{1/2}} = H(t)r, \quad H(t) = \frac{\dot{A}}{A} = \frac{1}{2t}, \quad r \gg r_a, \tag{17.8}$$

which is approximately linear in r for $H(t) = H(t_o)$, where t_o is the present age of the universe.

However, suppose we do not take the low velocity approximation. We then have to solve for Ω^2 in terms of r and t by using (17.7),

$$\Omega^2 = \frac{2}{\sqrt{t^2 + r^2/C_o^2} - t}, \quad r \gg r_a. \tag{17.9}$$

It follows from (17.9) and (17.6) that we have the exact recession velocity $\dot{r}$ for a distant galaxy in terms of $rH(t)$,

$$\dot{r} = \frac{C_o}{\sqrt{1 + 2t[\sqrt{t^2 + r^2/C_o^2}) + t]/[r^2/C_o^2]}}$$

$$= \frac{rH}{\sqrt{(1/4) + [1/4 + r^2 H^2/C_o^2] + \sqrt{1/4 + r^2 H^2/C_o^2}}}$$

$$= \frac{rH}{1/2 + \sqrt{1/4 + r^2 H^2/C_o^2}} < C_o, \quad C_o = \frac{B}{A} = \text{constant}. \tag{17.10}$$

This is the generalized Hubble's law based on the HHK model. The upper limit C_o of the Hubble velocity in (17.10) can be seen as follows: When rH is very large, $\dot{r}$ approaches C_o, as shown in (17.10). We note that such a limiting velocity C_o is the recession velocity at time $t = 0$, as one can see in (17.6) with $t = 0$.

The HHK model, together with the cosmic Okubo equation, predicts that the exact recession velocity is given by (17.10) with the upper limit,

$$C_o = \frac{\beta}{\alpha} = (3\omega)^{1/3}, \quad c = \hbar = 1, \qquad (17.11)$$

where $\omega = P/\rho < 1/3$ is the ratio of pressure P to energy density ρ of macroscopic bodies.[8,10] There is little experimental data for the parameter ω. Presumably, ω is much smaller than $1/3$. For example, suppose (3ω) is 10^{-3}, the limiting speed C_o is still very large, $C_o = 0.1 (= 3 \times 10^7 \text{ m/s})$.

Thus, it is natural to interpret (17.10) as the exact "non-relative" Hubble velocity as measured in any inertial frame, according to quantum Yang–Mills gravity:

> *The recession velocity of a distant galaxy, as measured in any inertial frame, is dictated by the cosmic Okubo equation (17.2) with the effective metric tensor (17.1). It is exactly given by (17.10) at a given time and has an upper limit C_o in (17.11) for a matter-dominated cosmos.*

We may consider C_o in (17.11) as an "effective speed of light" because it is associated with (17.1) through the vanishing effective metric, $ds^2 = G_{\mu\nu}(t)dx^\mu dx^\nu = 0$.

To see the next correction term for $\dot{r}$ in (17.8), we expand the square-root in (17.10) for small rH/C_o. We obtain the approximate recession velocities,

$$\dot{r} \approx rH(t)\left[1 - \frac{r^2 H^2(t)}{C_o^2}\right], \quad \dot{r} \ll C_o;$$

$$\dot{r} \approx C_o\left[1 - \frac{C_o^2}{8r^2 H^2(t)}\right] < C_o, \quad \dot{r} \approx C_o. \qquad (17.12)$$

Results (17.6)–(17.12) are for a matter-dominated cosmos in the HHK model.

For a radiation-dominated cosmos, we obtain different scale factors $A_{\rm rd}(t)$ and $B_{\rm rd}(t)$,[10]

$$A_{\rm rd}(t) = \alpha' t^{2/5}, \quad \alpha' = \left(\frac{5^6 g^6 \omega \rho_o^3}{2^6 \times 6^2}\right)^{1/15}, \qquad (17.13)$$

$$B_{rd}(t) = \beta' t^{2/5}, \quad \beta' = \frac{24\alpha'^6}{25g^2 \rho_o}, \quad g^2 = 8\pi G, \tag{17.14}$$

where G is the Newtonian gravitational constant. Following the same calculational steps as before with $S = -f(t) + pr$, the recession velocity $\dot{r}_{rd}$ for a radiation-dominated cosmos is

$$\dot{r}_{rd} = \frac{pC_o'^2}{\sqrt{p^2 C_o'^2 + m^2 \beta'^2 t^{4/5}}} < C_o', \quad C_o' = B_{rd}/A_{rd} \tag{17.15}$$

$$\approx \frac{3Hr}{2}\left[1 - \frac{9H^2 r^2}{8C_o'^2}\right], \quad H = \frac{\dot{A}_{rd}}{A_{rd}} = \frac{2}{5t}, \quad m \gg p.$$

Based on (17.13)–(17.15) for the radiation-dominated cosmos, the HHK model with Yang–Mills gravity predicts the limiting recession speed C_o' to be

$$C_o' = \beta'/\alpha' = (6\omega)^{1/3}. \tag{17.16}$$

As shown in (17.15), C_o' is the maximum Hubble recession speed at time $t = 0$ for the radiation-dominated cosmos. For the electromagnetic radiation, one has the usual value, $\omega = 1/3$.

The limiting recession speed (17.16) for distant galaxies appears to be in harmony with the four-dimensional space-time symmetry of quantum Yang–Mills gravity. Both limiting speeds (17.11) and (17.17) are "effective speeds of light" given by a vanishing effective line element $ds^2 = G_{\mu\nu}(t)dx^\mu dx^\nu = 0$. The scale factors are given by (17.1) for a matter-dominated universe and by (17.13)–(17.14) for a radiation-dominated universe.

17.2. Massless Okubo Equation and Accelerated Cosmic Redshifts

The cosmic redshift of radiation emitted from distant galaxies can be treated similar to Doppler shifts or, to be more specific, the accelerated Wu–Doppler shift.[3] The exact Doppler shift can be obtained by the transformation between two wave 4-vectors k_μ' and k_μ for radiation observed from two different inertial frames. Those wave vectors are related by the invariant law $\eta^{\mu\nu} k_\mu' k_\nu' = \eta^{\mu\nu} k_\mu k_\nu$. Similarly, the wave 4-vector $k_{e\mu}$ for radiation emitted from a distant galaxy is associated with an eikonal equation or the massless Okubo equation

$$G^{\mu\nu}(t)\partial_\mu \psi_e \partial_\nu \psi_e = 0, \quad \partial_\mu \psi_e = k_{e\mu}.$$

This is analogous to the massive ($m > 0$) Okubo equation (17.2) for distant galaxies. The wave 4-vector of light, as measured by observers in the inertial frame F, satisfies the usual eikonal equation $\eta^{\mu\nu}(t)\partial_\mu\psi\partial_\nu\psi = 0$ with $\partial_\mu\psi = k_\mu$.[11] Since the distant galaxy undergoes accelerated motion and the observer is in an inertial frame F, it is natural to treat the observed cosmic redshift in analogy to the accelerated Wu–Doppler effect with the "covariant" eikonal equation

$$G^{\mu\nu}(t)\partial_\mu\psi_e\partial_\nu\psi_e = \eta^{\mu\nu}\partial_\mu\psi\partial_\nu\psi, \quad \partial_\mu\psi_e = k_{e\mu}, \quad \partial_\mu\psi = k_\mu. \quad (17.17)$$

This is formally consistent with the idea of limiting Lorentz–Poincaré invariance when the galaxy moves with a constant velocity and gravity is negligible.[3] For simplicity, we choose a specific radial direction in a spherical coordinate with $k_{e\mu} = (k_{e0}, k_e, 0, 0)$ and $k_\mu = (k_0, k, 0, 0)$, similar to (17.3). We have

$$B^{-2}(t)k_{e0}^2 - A^{-2}(t)k_e^2 = k_0^2 - k^2 = 0, \quad (17.18)$$

for a matter-dominated cosmos. The recession velocity of a distant light source is the Hubble velocity $V_r = \dot{r}/C_o$ and, hence, the observed frequency k_0 decreases. The relation (17.18) leads to "cosmic transformations" for these covariant wave 4-vectors:

$$\frac{k_{e0}}{B} = \Gamma\left[k_0 + V_r k\right], \quad \frac{k_e}{A} = \Gamma\left[k + V_r k_0\right], \quad \Gamma = \frac{1}{\sqrt{1 - V_r^2}}, \quad V_r = \frac{\dot{r}}{C_o}. \quad (17.19)$$

One can verify that the covariant law (17.18) is preserved by the transformations (17.19) with a velocity function V_r. Since a distant galaxy moves with a non-inertial velocity (17.10), it is natural to identify $V_r < 1$ with the recession velocity, $V_r = \dot{r}/C_o < 1$, where C_o is given by (17.11). Specifically, we choose $k = +k_0$ for the recession of the radiation source in the radial direction.

For convenience of explanation, let us introduce an "auxiliary expansion frame" F_e associated with the effective metric tensor $G^{\mu\nu}(t)$ such that all distant galaxies are, by definition, at rest in F_e. Thus, the frequency shift for a source at rest in a receding galaxy is given by

$$\left[\frac{k_{e0}}{B}\right]_{\text{at rest in } F_e} = \frac{k_0(1 + V_r)}{\sqrt{1 - V_r^2}}, \quad V_r = \frac{\dot{r}(t)}{C_o}, \quad (17.20)$$

where $k_0 = [k_0]_{\text{observed in } F}$, and $V_r = \dot{r}(t)/C_o$ is the non-relative recession velocity of the distant galaxy at the moment when the light was emitted.

We stress that the expansion frame F_e is only an "auxiliary frame," which does not have well-defined space and time coordinates over all space. Therefore, all experiments and observations must be carried out in the inertial frame $F = F(t, x, y, z)$, in which the space and time coordinates are operationally defined.

Thus, to obtain an expression for the redshift of distant galaxies based on (17.20), we cannot rely on k_{e0} measured in F_e. Furthermore, the F_e and F frames are not equivalent because F_e (with metric tensor $G_{\mu\nu}(t)$) is not an inertial frame in flat space-time. Consequently, we follow a similar procedure to what we have used in previous work calculating Wu–Doppler shifts, calculating the frequency shift of a radiation source at rest in an accelerated frame and observed from an inertial frame. Although F_e is a non-inertial frame, we assume a weak equivalence of F_e with an inertial frame on the basis of the principle of limiting continuation for physical laws.[3,12] Specifically, under weak equivalence, we have[12]

$$\left[\frac{k_{e0}}{\sqrt{G_{00}}}\right]_{\text{at rest in } F_e} = \left[\frac{k_0}{\sqrt{\eta_{00}}}\right]_{\text{at rest in } F'}, \quad \sqrt{G_{00}} = B, \tag{17.21}$$

where $[\ ]_{\text{at rest in } F'}$ should be understood as "the source is at rest in a moving frame F' and its emitted waves are observed from the F' frame." Such a weak equivalence has been supported by Davies–Jennison's two laser experiments involving orbiting laser sources observed in an inertial laboratory.[12,13]

Therefore, equations (17.20) and (17.21) with $\eta_{00} = 1$ lead to the following frequency redshift $k_0 = \omega$ as observed in the inertial frame F:

$$[k_0]_{\text{at rest in } F'} \equiv k_0(\text{emission}) = k_0(\text{observed})\frac{(1 + V_r)}{\sqrt{1 - V_r^2}}. \tag{17.22}$$

The cosmic redshift z is defined by[8,9]

$$\frac{k_0(\text{emission})}{k_0(\text{observed})} = 1 + z. \tag{17.23}$$

It follows from (17.22) and (17.23) that the exact law of the redshift z is related to the recession velocity V_r of a distant galaxy as follows:

$$z = \frac{1 + V_r}{\sqrt{1 - V_r^2}} - 1, \quad V_r = \dot{r}/C_o < 1. \tag{17.24}$$

This is the prediction of the HHK model with Yang–Mills gravity for the redshift of light emitted by a distant galaxy in terms of its recession velocity. Only for small recession velocities $V_r \ll 1$ do we have the usual approximate relation[9] with a second-order correction term,

$$z \approx V_r + \frac{1}{2} V_r^2. \qquad (17.25)$$

For a matter-dominated cosmos, the result (17.24) enables us to express the recession velocity $\dot{r}/C_o = V_r$ of a distant galaxy in terms of its directly measurable redshift z,

$$V_r = \frac{\dot{r}}{C_o} = \frac{(1+z)^2 - 1}{(1+z)^2 + 1} = \frac{2z + z^2}{2 + 2z + z^2} < 1, \qquad (17.26)$$

as measured from the inertial frame F. One can also verify that $z = 0$ and $z \to \infty$ correspond to $V_r = 0$ and $V_r \to 1$, respectively. If $V_r = \dot{r}/C_o$ is given by 0.5 and 0.8, one has $z = 0.8$ and 2, respectively.

Thus, according to the present HHK model[10] of particle cosmology based on Yang–Mills gravity, the Hubble recession velocity V_r of a distant galaxy, as measured in an inertial frame, can only take on values between 0 and C_o. However, the cosmic redshift z can take on values between 0 and infinity in a matter-dominated cosmos.

17.3.   A Sketch of the Total Cosmic History by the Okubo Equation

The Okubo equation of motion for a distant galaxy leads to exact solutions for $r(t)$ and $\dot{r}(t)$ in (17.6) and (17.7) for, say, the matter-dominated universe. These solutions sketch a total cosmic history from the beginning to the end of the universe:

(i) In the beginning $t = 0$, we have the following features:

      initial mass run away velocity $\dot{r} = C_o$,
      initial non-vanishing radius $r = 2p^2 C_o^3/(m^2 \beta^2) \equiv r_o$,
      initial Hubble recession velocity $V_r = \dot{r}/C_o = 1$,
      initial cosmic frequency redshift

$$z = \frac{k_o(\text{emission})}{k_o(\text{observed})} - 1 = \infty.$$

(ii) At the end $t \to \infty$, we have the following features:

      final velocity of galaxies $\dot{r} \to 0$,
      final radius $r \to \infty$,
      final Hubble recession velocity $V_r \to 0$,
      final cosmic frequency redshift $z \to 0$,

where we have used equations (17.1), (17.6), (17.7) and (17.24). Of course, these properties will be modified when the quantum nature of Yang–Mills gravity and other new long-range and short-range forces[b] in particle physics are taken into account. One can imagine that the universe has been doing extremely complicated multi-tasks during its evolution.[c]

These simple results in (i) and (ii) are obtained under the postulate that the properties of homogeneity and isotropy also hold in the early universe, $t \approx 0$. The initial Hubble velocity is C_o in (17.11), and the initial redshift z is infinity. It is intriguing that our results imply that the universe began with a finite radius ($r = r_o$ in (i)) rather than a singularity.[d]

Within the Big Jets model,[10] which is suggested by the fundamental CPT invariance in particle physics, all the previous results of HHK model hold in our matter "half-universe." Moreover, all these results (such as the effective metric tensor (17.1), Okubo equations of motions (17.2) for distant galaxies, light rays, and redshifts, etc.) should also hold in the antimatter "half-universe,"[e] because CPT invariance implies the maximum symmetry between particles and antiparticles regarding their masses, lifetimes and interactions.[14]

17.4. Discussions

It may be interesting to observe that we assume local metric tensor in (17.1) and (17.2) to derive some properties about the behavior of point-like galaxies at extremely large distances. This result may not be surprising because the local effective metric tensor (17.1) embodies the super-macroscopic properties of homogeneity and isotropy. Thus, such a treatment of the physical system of distant galaxies in the super-macroscopic limit seems to resemble "Riemann geometry in the large."[f] In this sense, Riemann geometry in the large may play a role as the

---

[b] For example, the extremely weak linear force of baryon–lepton charges with general Lee–Yang U_1 symmetry. Cf. M. Khan, Y. Hao, J.P. Hsu, https://doi.org/10.1051/epjconf/201816804004.

[c] These multi-tasks are presumably beyond our understanding based on the present-day particle physics and quantum field theory, let alone general relativity, which is incompatible with quantum mechanics and the existence of antiparticles.[5,6]

[d] Perhaps such an Okubo equation for the extremely early universe (or in the super-microscopic limit) may not be completely farfetched because it can be derived from the Dirac equation in Yang–Mills gravity in the geometric-optics limit,[3] where only a large mass or energy is required, rather than a large length scale.

[e] It is presumably located far away from our matter half-universe.

[f] A curved space which extends to far away, not just treated locally, is the subject of Riemannian geometry in the large (or Global Riemannian Geometry).[15]

mathematical base for the expanding cosmos in the HHK model with quantum Yang–Mills gravity.

In the conventional FLRW model with general relativity, one has the Hamilton–Jacobi equation

$$g^{\mu\nu}(t)\partial_\mu S \partial_\nu S - m^2 = 0, \quad g^{\mu\nu}(t) = (1, -a^{-2}(t), -a^{-2}(t), -a^{-2}(t)), \quad (17.27)$$

where the scale factor is given by $a(t) = a_o t^{2/3}$ for a matter-dominated universe.[8] Following similar calculations from (17.2) to (17.8), one obtains the recession velocity $\dot{r} \approx (\dot{a}/a)(3r/2)$ in the non-relativistic approximation ($m \gg p$). However, there is no constant upper limit for the recession velocity $\dot{r}$ at large momenta ($p \gg m$). Since these results are not obtained in an inertial frame, it is difficult to have a satisfactory comparison between these results and those obtained by Yang–Mills gravity for an inertial frame. Note that because Yang–Mills gravity is Lorentz invariant, the results in the HHK model hold in all inertial frames.[8]

Within Yang–Mills gravity, the effective cosmic metric tensor $G^{\mu\nu}(t) = (B^{-2}, -A^{-2}, -A^{-2}, -A^{-2})$ appears to play a more useful role than that of $g^{\mu\nu}(t) = (1, -a^{-2}, -a^{-2}, -a^{-2})$ in the conventional theory. The reason is that the cosmic Okubo equations,

$$G^{\mu\nu}(t)\partial_\mu S \partial_\nu S - m^2 = 0, \quad \text{with} \quad m > 0 \quad \text{and} \quad m = 0, \quad (17.28)$$

can lead to Hubble velocities for $0 \leq \dot{r}/C_o < 1$ and cosmic redshift z for $0 \leq z < \infty$ without making low velocity approximations.

It is interesting to note that since $r = t^{1/2}(2pC_o^2/m\beta)$ in the low velocity approximation $m \gg p$, as one can see from (17.7), $\dot{r}$ in (17.8) can also be expressed in terms of $1/r$ without involving the function $H(t)$,

$$\dot{r} = \left(\frac{2C_o^2}{\Omega^2}\right)\frac{1}{r}, \quad (17.29)$$

where the constant of proportionality $(2C_o^2/\Omega^2)$ does not involve the time t. At first glance, one might be puzzled by the apparently inconsistency of different pictures of the expanding universe, as shown by (17.8) and (17.29). Equation (17.8) suggests that at the present age of the universe $t = t_o \approx 10^{10}$ years, $H(t_o)$ may be considered to have been a constant during say, the past 10^6 years, so that the

---

[8]The observation in the F frame that "all distant galaxies recede with non-constant velocities" does not make the F frame privileged and unique because this F frame can be any inertial frame rather than a particular one. This is demonstrated in the HHK model based on Lorentz invariant quantum Yang–Mills gravity.

Table 17.1. Consistency pictures of $\dot{r} = rH(t) = r/(2t)$ and $\dot{r} = 2/r$ derived from Okubo's cosmic equation.

t	1	10	90	100	110
$H = 1/(2t)$	0.5	0.05	0.0055	0.0050	0.0045
$r = 2t^{1/2}$	2	6.32	18.97	20	20.97
$\dot{r} = r/(2t)$	1	0.316	0.105	0.100	0.0953
$\dot{r} = 2/r$	1	0.316	0.105	0.100	0.0954
**$[[\dot{r} = r/(200)$	....	.....	0.095	0.100	0.105]]

**If one approximates $H(t)$ by a constant, say, $H \approx 1/(200)$ for a rough comparison, one gets a different and inconsistent result, as one can see from the cases $t = 90, 100$, and 110.

velocity $\dot{r}$ of a distant galaxy is roughly proportional to its distance $r = r(t)$. In contrast, equation (17.29) indicates that the velocity $\dot{r}$ of a distant galaxy is proportional to the inverse of its distance $1/r$. However, both (17.8) and (17.29) are derived from the same solutions (i.e., (17.6)–(17.7)) of the cosmic Okubo equation (17.3) in the low velocity approximation $m \gg p$. Their consistency is demonstrated in Table 17.1, where the units of r and t are chosen to facilitate comparison. The recession velocity (17.29) should give a better picture than that of (17.8) throughout the cosmic evolution. In (17.29), different distant galaxies have different constants Ω given by (17.6).

The usual FLRW model also has result similar to (17.29). To wit, one considers (17.27) with $a(t) = a_0 t^{2/3}$. Following (17.2) to (17.8), one obtains

$$\dot{r}(t) = H(t)\frac{r(t)}{2} = \frac{9}{a_o^3}\frac{1}{r^2(t)}, \quad H(t) = \frac{\dot{a}}{a} = \frac{2}{3t}, \quad r(t) = \frac{3t^{1/3}}{a_o}, \quad (17.30)$$

where the condition $m \ll p$ is used for simplicity.

Furthermore, Yang–Mills gravity and cosmic Okubo equation suggest several new views and understandings of our universe:

(A) The linear Hubble law is the low-velocity approximate solution to the Okubo equation for distant galaxies. The generalized Hubble law for all velocities is given by (17.10).

(B) The universe began with the maximum recession velocity $\dot{r} = C_o = (3\omega)^{1/3}$, $r = 2p^2 C_o^3/(m^2\beta^2) \equiv r_o$, and cosmic redshift $= z = \infty$, at time $t = 0$ for mass-dominated universe, and the universe will end with $\dot{r} \to 0$, $r \to \infty$, $z \to 0$, as $t \to \infty$. Of course, a more realistic model will not be completely dominated by the mass. It may be dominated by a combination of mass, radiation and some sort of effective "vacuum energy."

(C) The Okubo equations (17.2) with $m \geq 0$ in the HHK model are the basic equations of motion for all distant galaxies in the super-macroscopic limit. Interestingly, it appears that the Okubo equation with $m > 0$ automatically initiates the mass runaway, which resembles some sort of detonation at time $t = 0$ with a maximum speed, as discussed in (B).

(D) The cosmic "force" F_{cos} associated with the motion of distant galaxies, as described by the Okubo equation (17.2) or by the cosmic action (16.18), is intriguing. As shown in (16.38), it is proportional to momentum (since $m\dot{r}/\sqrt{1 - \dot{r}^2/C_o^2}$ corresponds to the usual momentum) and is inversely proportional to time (since $(dB/dt)/B(t) = 1/(2t)$). The origin of this time-dependent cosmic force may be traced to the presence of the effective time-dependent metric tensor $G_{\mu\nu}(t)$ given by (17.1), in the Okubo equation (17.2).

References

1. J. P. Hsu, *Int. J. Mod. Phys. A*, **21**, 5119 (2006).
2. J. P. Hsu, *Eur. Phys. J. Plus*, **126**, 24 (2011).
3. L. Hsu and J. P. Hsu, in *Space-Time Symmetry and Quantum Yang–Mills Gravity* (World Scientific, 2013), Part I, pp. 23–30, 38–42, 109–123, 134–136.
4. S. H. Kim, Ph.D. Thesis, University of Massachusetts Dartmouth (2012).
5. E. P. Wigner, in *Symmetries and Reflections, Scientific Essays* (The MIT Press, 1967), pp. 52–53.
6. F. J. Dyson, in *100 Years of Gravity and Accelerated Frames: The Deepest Insight of Einstein and Yang–Mills*, eds. J.-P. Hsu and D. Fine (World Scientific, 2005), pp. 347–351.
7. S. H. Kim and J. P. Hsu, *Eur. Phys. J. Plus* **127**, 146 (2012).
8. S. Weinberg, in *Cosmology* (Oxford University Press, 2008), pp. 10–11, 40, 62.
9. P. J. E. Peebles, in *Principles of Physical Cosmology* (Princeton University Press, Princeton, NJ, 1993), pp. 70–108.
10. J. P. Hsu, L. Hsu and D. Katz, *Mod. Phys. Lett. A* **33**, 1850116 (2018).
11. L. Landau and E. Lifshitz, in *The Classical Theory of Fields* (Addison-Wesley, 1951), pp. 268–271, 312–313, 340–344, 350.
12. J. P. Hsu and L. Hsu, *Eur. Phys. J. Plus* **128**, 74 (2013). Appendix A.
13. L. Hsu and J. P. Hsu, *Nuovo Cimento B* **112**, 1147 (1997). Appendix.
14. T. D. Lee, in *Particle Physics and Introduction to Field Theory* (Hardwood Academic Publishers, New York, 1981), pp. 320–333.
15. W. H. Huang, in *Global Riemannian Geometry* (in Chinese, National Taiwan University Press, 2019, forthcoming) Sections 1, 1.3.

Chapter 18

Standard Cosmic Time and "Cosmic Relativity"

18.1. Can One Accommodate Both Relativity and Cosmic Time?

Since quantum Yang–Mills gravity leads to an effective metric tensor $G_{\mu\nu}(x)$ for the motion of macroscopic objects and satisfies Lorentz–Poincaré invariance, it is of interest to construct a model of particle cosmology that also satisfies Lorentz–Poincaré invariance. Any reasonable model of particle cosmology must accommodate the cosmological principle of homogeneity and isotropy. Homogeneity implies that the effective metric tensor $G_{\mu\nu}$ in the super-macroscopic limit cannot depend on spatial coordinates; it can only depend on time,[1]

$$G_{\mu\nu}(t) = (G_{00}(t), G_{11}(t), G_{22}(t), G_{33}(t)).$$

Moreover, isotropy implies that

$$G_{11}(t) = G_{22}(t) = G_{33}(t).$$

Thus, at super-macroscopic length scales, the time coordinate turns out to play a more important role than the spatial coordinates. However, in special relativity, space and time are equally important. Thus, a metric tensor $G_{\mu\nu}(t)$ that depends only on time and hence the cosmological principle, appear to be difficult to accommodate in special relativity. The key problem is how one can accommodate both the four-dimensional space-time framework with the principle of relativity and the cosmological principle of homogeneity and isotropy.

In this chapter, we demonstrate that it is possible to solve this non-trivial problem by introducing a new definition of the time t for all observers in different inertial frames. As Taylor and Wheeler[2] showed in their excellent textbook *Spacetime Physics*, the essential feature of Einstein's relativity is the principle of relativity, which states that the form of all physical laws must be the same in all inertial frames. Operationally, this is achieved by demanding that the four space-time coordinates, three spatial and one temporal, must transform as a 4-vector *when they are all expressed in the same units*. In *Spacetime Physics*, Taylor and Wheeler express all four space-time coordinates in terms of units of length to reflect the fact that all four are on the same footing. In effect, the proper space-time coordinates are (ct, x, y, z). As long as the vector (ct, x, y, z) transforms as a 4-vector, all the results and implications of special relativity will be preserved. For example, such a transformation implies that the speed of light, measured in units of length per length, is a universal constant, without the need for an additional postulate. This is the principle of relativity.

In this light, Einstein's second postulate, that the speed of light (measured in units of length over time) should be universal (i.e., isotropic and with the same value in every inertial frame), can be seen for what it really is, a particular definition of the unit of time (e.g., "second") in terms of the unit of length (e.g., "meter"). (In the SI definitions of the units, the unit "meter" is defined in terms of the unit "second" for practical reasons of accuracy, repeatability, etc., that have nothing to do with the conceptual basis of physics. The essential concept is that the units of length and time are not independent.[a]) Operationally, this is achieved by separating the single transformation for the quantity ct into one for c ($c' = c$) and one for t ($t' = [t - Vx/c^2]/\sqrt{1 - V^2/c^2}$) between the frames F and F' when F' moves with a constant speed V relative to F along the $+x$-direction. This choice is inextricably linked with the currently accepted definition of the meter: "The meter is the length of the path traveled by light in vacuum during a time interval of 1/299792458 second."[b] However, since such a definition is a human convention, there is no particular difficulty in adopting a different definition for convenience in attacking a particular problem.

Nevertheless, the principle of relativity allows us the freedom to choose other definitions of the units of time and length. Although the transformation between ct and $c't'$ (or $(ct)'$) is fixed by the principle of relativity, the transformations between c and c' (or t and t') are not. We may choose those to be any that (1) preserve the transformation between ct and $c't'$ and (2) are convenient for the problem we wish to solve. This freedom is similar to the way in which we are free

---

[a] See Section 1.2 in Chapter 1.

[b] NIST. *Special Publication* 330, 2008 Edition, eds. B. N. Taylor and A. Thompson.

to choose a convenient coordinate system (Cartesian, polar, hyperbolic, etc.) when analyzing a problem.

As we shall see, making the choice $t' = t$[3,4] and

$$c' = \frac{(c \mp V)}{\sqrt{1 - V^2/c^2}}$$

between those same two F and F' frames enables us to accommodate both relativity and cosmic time and both the principle of relativity and the cosmological principle of homogeneity and isotropy.

18.2.   Why a New Time?

Before getting involved in the technical details of a new transformation between t' and t, and c' and c, we first discuss more fully the answer to the question "For what reason would one want to choose a different transformation?"

In the conventional space-time framework, there are infinitely many inertial frames and infinitely many relativistic times. This makes the comparison of cosmological quantities such as the age of the universe and other time-dependent quantities such as measurements of recession velocities and cosmic redshifts $z(t)$ of distant galaxies very complex. Outside of a cosmological context, the same difficulty arises when tackling problems involving many bodies, such as the problem of describing the canonical evolution[5,6] of a system of many particles in statistical mechanics. If $N \approx 10^{24}$ particles in a system travel with up to N different velocities, one is forced to describe such a situation using a system of $\approx 10^{24}$ coupled differential equations,[6] each with a different time t. Even a supercomputer cannot handle such a calculational nightmare and this inconvenience has implications in developing a practical definition for the temperature of a system in a relativistic framework.[c] One obvious way around these difficulties is to define a new unit of time t in which observers in all inertial frames use the same clock system, $t' = t$. We will call this particular definition "cosmic time" and the accompanying framework "cosmic relativity" in order to distinguish it from the conventional framework of "special relativity." As we have stressed, both are correct and useful in the same sense that both Cartesian coordinates and spherical coordinates are correct and useful.

---

[c]L. Hsu and J.-P. Hsu, Experiments on the CMB spectrum, Big Jets model and their implications for the missing half of the Universe, *EPJ Web Conf.* **168**, 01012, Published online: 09 January 2018, DOI 10.1051/epjconf/201816801012.

In fact, such a synchronization has already been discussed in our previous work, in which we termed it "common time" and showed that such a definition of time continued to satisfy all experimental tests of special relativity (as it must, since the product ct continues to transform as the zeroth component of a 4-vector). Here, we term it "cosmic time" and use the quantity "t_c" to denote it in order to better match the cosmological problems to which we apply it here and to distinguish it from the conventional definition of the time.

18.3.   Cosmic Time

We first describe how one might operationally set up clocks in such a way as to read cosmic time before discussing in more depth the differences between cosmic time and Galilean time. We begin by choosing any one inertial frame, which we label the F frame, in which to construct a system of synchronized clocks. One can imagine this as a collection of clocks arranged on a three-dimensional lattice, and the clocks are synchronized in such a way so that the speed of light is constant and isotropic in F. This clock system then becomes the standard of time for all observers in all inertia frames. Now, regardless of in which inertial frame observers happen to be located, they all use the nearest F clock (the one sharing their position in space) to record time. All observers then have, by definition, a common or cosmic time.

Although such a synchronization seems similar to Galilean time, there is one important distinction, besides the fact that cosmic time adjusts the transformation between c and c' to remain consistent with the principle of relativity. In Galilean time, it is assumed that nature specifies one preferred inertial frame in the universe in which the speed of light is truly constant and isotropic. The procedure of synchronizing clocks such that the speed of light is constant and isotropic may only be correctly carried out in that one inertial frame. In cosmic relativity however, nature has no such preference. That synchronization procedure, being merely a human construct relating the definitions of "meter" and "second," may be carried out in any inertial frame. As a result, the speed of light, when measured in units of meter per second (x/t) may no longer be constant or isotropic except in one frame.[d] However, the speed of light, as measured in units of meter per meter $(x/(ct))$ is

---

[d]Or even in no frames. One can imagine a case where after such a synchronization procedure, the F frame is removed from the universe, leaving no frames in which the speed of light (measured in meters/second) is constant and isotropic.

most definitely constant and isotropic in all inertial frames, as demanded by nature.

18.4.   Coordinate Transformations with the Cosmic Time

Although the derivation of the coordinate transformations of cosmic relativity, along with explicit verifications of its consistency with experimental results have been carried out elsewhere,[e] we briefly recapitulate some of the most important results here for convenience, beginning with the derivation of the coordinate transformations.

We consider two inertial frames F and F', whose relative motion is along parallel x- and x'-axes, with the usual simplifying assumptions that the origins of the frames coincide at $t_c = 0$. Suppose that we have synchronized the clocks in F according to the procedure described previously and that F' moves with a constant velocity $\mathbf{V} = (V, 0, 0)$ (measured in meter per cosmic-second) as measured by observers in F. The coordinate 4-vectors of an event as recorded by observers in F and F' can in general be denoted by

$$x^\mu = (bt_c, x, y, z), \quad x'^\mu = (b't_c, x', y', z'), \tag{18.1}$$

in F and F', respectively. Because the cosmic time t_c is a scalar, it does not transform as a component of a 4-vector. In order to obtain a quantity with the correct transformation properties, we must pair the cosmic time t_c with a function b such that $bt_c = w$ and $b't_c = w'$, since we know from four-dimensional taiji relativity[7,8,f] that bt does have the correct transformation properties.[g] In the F frame, where the speed of light (in meter per cosmic-second) is isotropic and a constant, $b = c$. In F', where the speed of light (in meter per cosmic-second) is not isotropic, b' is a function in the transformation.

From the invariant law for the motion of particles in F and F', we have $s^2 = x_\mu x^\mu = x'_\mu x'^\mu = s'^2$. The four-dimensional space-lightime transformation in cosmic relativity is found to be

$$b't_c = \gamma(ct_c - \beta x), \quad x' = \gamma(x - \beta ct_c), \quad y' = y, \quad z' = z,$$

$$\gamma = \frac{1}{\sqrt{1 - \beta^2}}, \qquad \beta = V/c, \tag{18.2}$$

---

[e] Cf. Refs. 9–11.

[f] It provides a convenient framework to discuss accelerated frames and spacetime transformations.
[g] We use the quantity b instead of c here because the relationship between b and the speed of light is not necessarily straightforward.

with a corresponding inverse transformation

$$ct_c = \gamma(b't_c + \beta x'), \quad x = \gamma(x' + \beta b't_c), \quad y = y', \quad z = z', \qquad (18.3)$$

where β is a dimensionless constant characterizing the magnitude of the relative motion between F and F'. As one expects, this is precisely the result obtained by replacing w with ct_c and w' with $b't_c$ in the taiji transformations (7.4) in Chapter 7.[7]

The "ligh function" b is peculiar to cosmic relativity and does not always have a simple physical interpretation. We examine this function more closely in the following section.

18.5.　Physical Meaning of the "Ligh Function" b

Taking the derivatives of the first equation in both (18.2) and (18.3) with respect to the cosmic time t_c gives

$$\frac{d(b't_c)}{dt_c} = \gamma \left(c - \beta \frac{dx}{dt_c} \right), \qquad (18.4)$$

$$c = \gamma \left(\frac{d(b't_c)}{dt_c} + \beta \frac{dx'}{dt_c} \right). \qquad (18.5)$$

Because (18.2) and (18.3) are inverses of each other, it is natural to associate b' with the speed of light as measured by F' observers. However, the presence of the dx/dt_c term can be confusing. How is it possible that the speed of light depends on the velocity of some object that has not even been specified?

To answer this question, first let us consider the case where the equations describe the propagation of a light signal in a vacuum, so that

$$\Delta s^2 = 0. \qquad (18.6)$$

In this situation, since $\Delta(b't_c) = \Delta r'$ in every frame, it is clear that the lighttime interval $\Delta(b't_c)$ in the F' frame can be interpreted as the vacuum optical path of the light, i.e., the distance traveled by the light signal. For a light signal propagating along the $+x/x'$-axes, the ligh function b is given by

$$b = c \quad \text{in} \quad F(ct_c, x, y, z),$$
$$\frac{d(b't_c)}{dt_c} = b' = \gamma \left(c - \beta \frac{dx}{dt_c} \right) = \gamma(c - \beta c) \quad \text{in} \quad F'(b't_c, x', y', z'). \qquad (18.7)$$

We know that $d(b't_c)/dt_c = b'$ because $\gamma(c - \beta c)$ is independent of t_c. In this case, b' can be interpreted as the one-way speed of light in F'. As a check, one can verify that $d(b't_c)/dt_c = dx'/dt_c$. Furthermore, in the more general case of a light signal propagating in an arbitrary direction, $b'^2 = v_x'^2 + v_y'^2 + v_z'^2$.

In a situation when the equations are used to describe the motion of a particle with non-zero mass so that $\Delta s^2 \neq 0$ and $dx/dt_c \neq c$, the lightime interval $\Delta(b't_c)$ does not correspond to the distance traveled by any actual light signal. In this case, the ligh function b' is the average speed, as measured by F' observers, of a light signal that travels a closed path in an inertial frame moving with a velocity dx/dt_c relative to F. Thus, the ligh function does, in a way, refer to the speed of light as measured by F' observers. However, the interpretation is somewhat complex, because of the way the unit cosmic-second has been defined.

To see that the given interpretation is correct, consider the simple case $dx'/dt_c = 0$, so that $dx/dt_c = \beta c$. We can think of this relation $dx'/dt_c = 0$ as corresponding to the motion of an object at rest relative to F'. In this situation,

$$\frac{d(b't_c)}{dt_c} = b' = \gamma\left(c - \beta\frac{dx}{dt_c}\right) = \gamma(c - \beta^2 c) = \frac{c}{\gamma}. \qquad (18.8)$$

We see that b' in this case is the average speed, measured by F' observers, of a light signal that travels a closed path in the F' frame. Thus, the interpretation that b' is the average speed of a light signal as measured by F' observers that travels a closed path in a frame that moves at a velocity dx/dt_c relative to F holds. The purpose of the above discussion is to serve as a caution to readers that the "ligh function" b', although related to the speed of light in F', is not simply the speed of a light signal as measured by F' observers.

18.6.   Implications of Cosmic Time

Using the cosmic time, many of the effects and properties ordinarily associated with relativity, such as the universality of the speed of light (measured in units of meter per second), the concept of simultaneity (of time measured in seconds), and the relativistic length contraction and time dilation are lost. However, it is important to keep in mind that these properties are lost only when one uses the human-conceived unit "cosmic-second" rather than the conventional "second" to measure time and that the loss in that case is due to the particular relationship between the meter and the cosmic-second. Because cosmic relativity retains the principle of relativity, its consistency with experimental results is retained.[7]

Some readers might still object that cosmic time, even as a human convention, is untenable because the speed of light is isotropic in one frame only and that such a definition violates the principle of relativity by singling out that frame in particular. However, we note that although our scheme for clock synchronization defines the speed of light (measured in meter per cosmic-second) to be isotropic in only the F frame, the frame chosen to be F could be any inertial frame. Cosmic time differs from the absolute time of Galilean relativity in that Galilean relativity assumes that there exists somewhere in the universe a frame of absolute rest, that this would be the only frame in which the speed of light would be isotropic, and that one could measure the velocity of Earth relative to that frame. In cosmic time, on the other hand, any frame could be arbitrarily chosen to be the one in which clocks are synchronized such that the speed of light in that frame (measured in meter per cosmic-second) is isotropic. Thus, all inertial frames are equivalent in the sense that any of them could be chosen (by human definition) to be this frame.

One might also claim that the requirement that the speed of light be isotropic in one frame constitutes a third postulate or assumption. However, the existence of such a frame is not necessary to cosmic relativity. One could imagine synchronizing clocks in F as specified above, placing identical clocks at the vertices of a grid in F', synchronizing the F' clocks by adjusting their readings to match that of the nearest F clocks at a particular time, and then destroying all the F clocks along with any objects in the universe that were at rest with respect to those F clocks. In effect, all physical traces of the F frame would have been eliminated from the universe. However, no physics would change. The existence of an F frame in which the speed of light (measured in meter per cosmic-second) is merely a technical assumption for simplifying our calculations.

Finally, one might conclude that since the definition of the cosmic second renders the one-way speed of light non-isotropic, as well as losing conventional concepts of simultaneity, length contraction, and time dilation, such a definition of time cannot possibly be consistent with experimental results. However, we would remind readers that cosmic relativity still satisfies the principles of relativity and that, when re-cast in terms of space-time coordinates in which all four are expressed in the same units, satisfies the same coordinate transformation as the conventional special relativity. Explicit demonstrations of the consistency of cosmic (common) relativity with experimental results such as the Michelson–Morley experiment, the Kennedy–Thorndike experiment, the Fizeau experiment, the Ives–Stilwell experiment, the lifetime dilation of cosmic-ray muon, and the decay-length dilation in high energy particle experiments can be found in our previous work.[9–11]

As an example, the Michelson–Morley experiment is conventionally interpreted as an experimental test and proof of the universality of the speed of light. Strictly speaking, it shows only that the two-way (or round-trip) speed of light is universal, which remains true in cosmic relativity. Similarly, experiments of the lifetime dilation of unstable particles in flight in fact measure the distances traveled by such particles before they decay (their decay length). Cast in this light, these experimental results are still consistent with the predictions of cosmic relativity.[10,h]

References

1. J. P. Hsu, L. Hsu and D. Katz, *Mod. Phys. Lett. A* **36**, 1850116 (2018).
2. E. F. Taylor and J. A. Wheeler, in *Spacetime Physics*, 2nd ed. (W. H. Freeman, New York, 1992), pp. 1–5.
3. J. P. Hsu, *Found. Phys.* **8**, 371 (1978); **6**, 317 (1976); J. P. Hsu and T. N. Sherry, *Found. Phys.* **10**, 57 (1980); J. P. Hsu, *Il Nuovo Cimento B* **74**, 67 (1983).
4. Editorial, *Nature* **303**, 129 (1983).
5. J. P. Hsu, *Nuovo Cimento B* **80**, 201 (1984); J. P. Hsu and T. Y. Shi, *Phys. Rev. D* **26**, 2745 (1982).
6. R. Hakim, *J. Math. Phys. (N.Y.)* **8**, 1315 (1967).
7. J. P. Hsu and L. Hsu, in *A Broader View of Relativity* (World Scientific, 2006), Chapters 7, 9 and 10.
8. J. P. Hsu and L. Hsu, *Phys. Lett. A* **196**, 1 (1994).
9. J. P. Hsu and L. Hsu, in *A Broader View of Relativity* (World Scientific, 2006), pp. 114–142.
10. J. P. Hsu and Y. Z. Zhang, in *Lorentz and Poincaré Invariance, 100 Years of Relativity* (World Scientific, 2001), pp. 509–561.
11. Y. Z. Zhang, in *Special Relativity and Its Experimental Foundations* (World Scientific, 1998).

---

[h] The decay-length dilation of particle decaying in flight can be calculated by using quantum field theory formulated on the basis of cosmic/common relativity.

Chapter 19

Dynamics of Expansion with Cosmic Time

19.1. Yang–Mills Gravity with Cosmic Time and Four-Dimensional Symmetry of Cosmic Relativity

Within a four-dimensional symmetry framework with cosmic time t_c, the four-dimensional space-time coordinates for the inertial frames $F = F(x^\mu)$ and $F'(x'^\mu)$ can be denoted by

$$x^\mu = (w, x, y, z) = (bt_c, x, y, z), \quad \text{in } F, \tag{19.1}$$

$$x'^\mu = (w', x', y', z') = (b't_c, x', y', z'), \quad \text{in } F', \tag{19.2}$$

where $\mu = 0, 1, 2, 3$. As usual, the four-dimensional coordinate transformations between F and F' form the Lorentz group.[1,a] Suppose, as described in Chapter 18, a synchronized clock system corresponding to cosmic time is set up in the inertial frame F, with the F clocks used by all observers in all inertial frames. As a result, the speed of light in (and only in) the F frame is constant and isotropic, i.e., $b = c$, by definition.

As discussed in Chapter 18, although the cosmic time t_c transforms like a scalar, all equations in quantum Yang–Mills gravity and quantum electrodynamics remain unchanged because they can be expressed in terms of the space-time coordinate 4-vectors in (19.1) and (19.2), which transform exactly as in the conventional framework of special relativity. Let us verify this explicitly for a few basic equations in Yang–Mills gravity.

---

[a] See also Sections 2.4 and 2.5 in Chapter 2 of the present volume.

The invariant action $S_{\phi\psi}$ for fermions and T_4 gauge fields in inertial frames can be expressed in terms of the coordinate 4-vectors with cosmic time t_c,[b]

$$S_{\phi\psi} = \int L_{\phi\psi} d^4 x, \quad x \equiv x^\mu = (bt_c, x, y, z), \quad b = c, \tag{19.3}$$

$$L_{\phi\psi} = \frac{1}{2g^2}(C_{\mu\alpha\beta}(x)C^{\mu\beta\alpha}(x) - C_{\mu\alpha}{}^\alpha(x)C^{\mu\beta}{}_\beta(x)) + L_\psi, \tag{19.4}$$

$$L_\psi = \frac{i}{2}[\overline{\psi}(x)\Gamma_\mu \Delta^\mu \psi(x) - (\Delta^\mu \overline{\psi}(x))\Gamma_\mu \psi(x)] - m\overline{\psi}(x)\psi(x), \tag{19.5}$$

$$\Delta^\mu \psi(x) = J^{\mu\nu}(x)\partial_\nu \psi(x), \quad J^{\mu\nu}(x) = \eta^{\mu\nu} + g\phi^{\mu\nu}(x) = J^{\nu\mu}(x), \tag{19.6}$$

$$C^{\mu\nu\alpha}(x) = J^{\mu\lambda}(x)(\partial_\lambda J^{\nu\alpha}(x)) - J^{\nu\lambda}(x)(\partial_\lambda J^{\mu\alpha}(x)), \tag{19.7}$$

where

$$C_{\mu\alpha\beta}(x)C^{\mu\alpha\beta}(x) = 2C_{\mu\alpha\beta}(x)C^{\mu\beta\alpha}(x),$$

$$\partial_\mu = (\partial/\partial(bt_c), \partial/\partial x, \partial/\partial y, \partial/\partial z), \quad \eta_{\mu\nu} = (1, -1, -1, -1).$$

To see the T_4 invariance of the action $S_{\phi\psi}$, let us consider the local space-time translations with an arbitrary infinitesimal vector functions $\Lambda^\mu(x)$. Under the T_4 gauge transformations, scalar function $Q(x)$ and vector function $\Gamma(x)$, etc., transform as follows:[c]

$$x^\mu \to x^\mu + \Lambda^\mu(x),$$

$$Q(x) \to (Q(x))^\$ = Q(x) - \Lambda^\lambda \partial_\lambda Q(x), \quad Q(x) = \psi, \overline{\psi}, \Phi, \tag{19.8}$$

$$\Gamma^\mu(x) \to (\Gamma^\mu(x))^\$ = \Gamma^\mu(x) - \Lambda^\lambda(x)\partial_\lambda \Gamma^\mu(x) + \Gamma^\lambda(x)\partial_\lambda \Lambda^\mu(x), \quad \text{etc.}$$

We have

$$L_{\phi\psi} \to (L_{\phi\psi})^\$ = L_{\phi\psi} - \Lambda^\lambda(\partial_\lambda L_{\phi\psi}). \tag{19.9}$$

If the action (19.3) is formulated in a general frame (inertial or non-inertial frame) with a Poincaré metric tensor $P_{\mu\nu}(x)$, the gravitational action in a general

---

[b] See Section 8.3 in Chapter 8.
[c] See Section 8.1 in Chapter 8; Section 4.2 in Chapter 4.

frame takes the form

$$S_{\phi\psi} = \int L_{\phi\psi}\sqrt{-P}d^4x, \quad P = \det P_{\mu\nu}, \quad x \equiv x^\mu = (bt_c, x, y, z), \quad (19.10)$$

which reduces to (19.3) in the limit of zero acceleration, where $P_{\mu\nu} \to \eta_{\mu\nu}$ and $P \to -1$.[2,d]

Under the T_4 gauge transformation, we have

$$S_{\phi\psi} \to S_{\phi\psi}^\$ = \int (L_{\phi\psi}Wd^4x)^\$, \quad W \equiv \sqrt{-P}. \quad (19.11)$$

The T_4 gauge transformation is the same as the Lie derivative, e.g., $(\Gamma^\mu(x))^\$ - \Gamma^\mu(x) = -(\mathcal{L}_\Lambda\Gamma)^\mu$, as shown in Appendix A. We can use Cartan's formula to prove the invariance of the gravitational action (19.11):

$$S^\$ = \int (LWd^4x)^\$ = \int [LWd^4x - \Lambda^\mu(\partial_\mu L)Wd^4x - L\Lambda^\mu\partial_\mu(Wd^4x)]$$

$$= S - \int \left[\Lambda^\mu(\partial_\mu L)W + L\Lambda^\mu(\partial_\mu W) + L(\partial_\mu\Lambda^\mu)W\right]d^4x$$

$$= S - \int \partial_\mu(\Lambda^\mu LW)d^4x = S. \quad (19.12)$$

(See Appendix A for detailed calculations of $\Lambda^\mu\partial_\mu(Wd^4x)$.)

The Lorentz invariant Yang–Mills gravitational field equation can be derived from $S_{\phi\psi}$ in inertial frames

$$H_{\mu\nu}(x) - g^2 S_{\mu\nu}(x) = 0, \quad \text{in} \quad F(w, x, y, z), \quad w = ct_c,$$
$$H'_{\mu\nu}(x') - g^2 S'_{\mu\nu}(x') = 0, \quad \text{in} \quad F'(w', x', y', z'), \quad w' = b't_c, \quad (19.13)$$

which is formally the same as that in (16.5), as expected.

In the following discussions, which will be based on the cosmic time t_c, because the speed of light is, by definition, no longer a universal constant, we shall not use $c = \hbar = 1$.

19.2. Relativistic Speed of Light and Maximum Speed in a Given Direction in F'

Within the framework of cosmic relativity, although the speed of light is not isotropic in F', the speed of light in a given direction remains the maximum speed

---

[d] See also Section 8.3 in Chapter 8.

of a physical object in that direction in F'. This can be seen by considering the motion of a particle in the absence of gravity. The (invariant) action with cosmic time is[1]

$$S_c = \int (-mds) = \int L_c dt_c, \quad L_c = -mv\sqrt{c'^2 - \mathbf{v}'^2}, \quad c' = \frac{d(b't_c)}{dt_c}, \quad (19.14)$$

where $ds^2 = d(b't_c)^2 - d\mathbf{r}'^2$ in the F' frame. The energy and momentum of a particle in F' frame can be derived from (19.14),

$$p_0' = \frac{m}{\sqrt{1 - v'^2/c'^2}}, \quad \mathbf{p}' = \frac{m\mathbf{v}'/c'}{\sqrt{1 - v'^2/c'^2}}, \quad p'^\mu p_\mu' = m^2, \quad (19.15)$$

where $\mathbf{p}' = \partial L_c/\partial \mathbf{v}'$ and $H' = (\partial L_c/\partial \mathbf{v}') \cdot \mathbf{v}' - L_c = p_0'$. This result shows that the velocity of a particle $\mathbf{v}' = d\mathbf{r}'/dt_c$ in a given direction must be smaller than the speed of light in that direction in F'.

19.3. A Relativistic-Cosmology Model with a Cosmic Time t_c

The HHK model of particle cosmology with the cosmic Okubo equation and the effective cosmic metric tensor $G_{\mu\nu}(t)$ does not have a proper and simple transformation within the framework of special relativity.[3] It appears to be non-trivial to formulate it as a special relativistic theory. However, the situation is different from the viewpoint of cosmic relativity.

Let us consider a simple relativistic-cosmology (RC) model based on cosmic relativity. We assume that the visible universe is homogeneous and isotropic on large or Hubble scales at any given cosmic time t_c. The "effective" metric tensor $G_{\mu\nu}(t_c)$ for the super-macroscopic world is assumed to be the same everywhere at any given moment in time t_c and must have the diagonal form

$$G_{\mu\nu}(t_c) = (B^2, -A^2, -A^2, -A^2) = \eta_{\mu\nu}U(t_c), \quad F,$$
$$G_{\mu\nu}'(t_c) \equiv G_{\mu\nu}(t_c) = \eta_{\mu\nu}U(t_c), \quad F', \qquad (19.16)$$

where $U(t_c)$ is the universal four-dimensional space-time scale factor and the Minkowski metric tensor is $\eta_{\mu\nu} = (1, -1, -1, -1)$. Comparing (19.16) and (17.1), we have the following invariant relations under the Lorentz transformation with a cosmic time (18.2), in which t_c is an invariant and transforms like a scalar

$$U(t_c) = B(t_c) = A(t_c) = \alpha t_c^{1/2}, \tag{19.17}$$

$$\alpha = (8g^6\omega\rho_0^3/9)^{1/12} = \beta = 3\alpha^5/2g^2\rho_0. \tag{19.18}$$

Thus, in the RC model, the result (19.18) suggests that the equation of state in the super-macroscopic limit may not be the same as the usual equation of state for a universe completely dominated by non-relativistic matter or for one dominated by relativistic matter and radiation.

The effective line element ds^2 is an invariant in cosmic relativity because $U(t_c)$ is a scalar function,

$$
\begin{aligned}
ds^2 &= G_{\mu\nu}(t_c)dx^\mu dx^\nu \\
&= U^2(t_c)\eta_{\mu\nu}dx^\mu dx^\nu = U^2(t_c)\eta_{\mu\nu}dx'^\mu dx'^\nu.
\end{aligned} \tag{19.19}
$$

The equation of motion of galaxies in the super-macroscopic world can be derived from an invariant action S_{cos}

$$S_{cos} = \int (-mds), \quad ds^2 = U^2(t_c)[dw^2 - dr^2]. \tag{19.20}$$

It also leads to the following invariant Okubo equation in RC model:

$$U^2(t_c)\eta^{\mu\nu}\partial_\mu S\partial_\nu S - m^2 = 0. \tag{19.21}$$

19.4. Generalized Hubble's Law in Inertial Frames with the Standard Cosmic Time

Hubble recession velocities in an inertial frame $F(ct_c, x, y, z)$ are exactly the same as those discussed in Chapter 17 with the time t replaced by cosmic time t_c. Let us now discuss Hubble's law in a different inertial frame F' with a cosmic time t_c. We use the Okubo equation (19.21) in the $F'(w', x', y', z')$ frame with the cosmic time t_c and cosmic relativity,

$$U^2(t_c)\partial'_\mu S\partial'_\nu S - m^2 = 0, \quad x'^\mu = (w', x', y', z'), \quad w' = b't_c. \tag{19.22}$$

Following the same steps as from (17.3) to (17.7), we obtain

$$r' - \int \frac{p'}{\sqrt{p'^2 + m^2 U^2}} dw' = \text{constant}, \quad U(t_c) = \alpha t_c^{1/2}, \qquad (19.23)$$

$$\frac{dr'}{dw'} = \frac{1}{\sqrt{1 + \Omega'^2 t_c}}, \quad \Omega' = \frac{m\alpha}{p'}, \qquad (19.24)$$

$$r'(t_c) = \frac{2b'}{\Omega'^2} \sqrt{1 + \Omega'^2 t_c}, \qquad (19.25)$$

where p' is the constant generalized momentum of a distant galaxy in the F' frame.

When we are dealing with space and time coordinates of a light signal or an object, the quantity b' in w' is independent of the cosmic time t_c. (See discussions in Section 18.5.) Thus we have the relation

$$\frac{d}{dw'} = \frac{1}{(dw'/dt_c)} \frac{d}{dt_c} = \frac{d}{b' dt_c}, \qquad (19.26)$$

to express the recession velocity of a distant galaxy in terms of the standard cosmic time t_c.

From equation (19.24), we have

$$\frac{dr'(t_c)}{dt_c} = \frac{b'}{\sqrt{1 + \Omega'^2 t_c}}. \qquad (19.27)$$

In terms of $H' = \dot{U}(t_c)/U(t_c)$, we then have the generalized Hubble's law,

$$\dot{r}'(t_c) = \frac{r' H'}{1/2 + \sqrt{1/4 + r'^2 H'^2/b'^2}} < b', \qquad (19.28)$$

where b' in (19.25)–(19.28) may be interpreted as the speed of light c' in the direction of the velocity of the distant galaxy under consideration. In the low velocity approximation, $r'^2 H'^2/b'^2 \ll 1$, we have the linear Hubble's law with an additional correction term that is $\propto r'^3$,

$$\dot{r}'(t_c) \approx r' H' + r'^3 H'^3/b'^2. \qquad (19.29)$$

We hope that the generalized Hubble's law (19.28) and the correction term in (19.29), which are predictions of the RC model based on quantum Yang–Mills gravity and cosmological principle, can be tested in the future.

To derive the exact relation for the recession velocity of any distant galaxy and its distance, as measured in an inertial frame with the standard cosmic time, let us consider the inertial frame F with the coordinate $x^\mu = (bt_c, x, y, z)$, $b = c$ in (19.3). From (19.24)–(19.27) with $r' \to r, b' \to c, \Omega' \to \Omega = m\alpha/p$, we obtain the exact result,

$$\dot{r}(t_c) = \frac{dr(t_c)}{dt_c} = \frac{c}{\sqrt{1 + \Omega^2 t_c}} = \frac{2c^2}{\Omega^2} \frac{1}{r(t_c)}, \tag{19.30}$$

for matter-dominated universe with $U(t_c) = \alpha t_c^{1/2}$. Thus, the RC model with a cosmic time t_c predicts that the recession velocity of any distant galaxy is exactly proportional to the inverse of its distance, i.e., $\dot{r}(t_c) \propto 1/r(t_c)$, for all time and for all values of $\Omega = m\alpha/p$, as measured in the inertial frame F. Note that the constant Ω in (19.30) involves the mass m and the constant generalized momentum p and, hence, it takes different constant values for different distant galaxies. The time-dependent recession velocity (19.30) leads to the deceleration of a distant galaxy,

$$\frac{d\dot{r}(t_c)}{dt_c} = -\frac{4c^4}{\Omega^4} \frac{1}{r^3(t_c)} = -\frac{\Omega^2}{2c^2} \dot{r}^3(t_c). \tag{19.31}$$

It leads to the deceleration parameter $q_o \equiv -\ddot{U}U/\dot{U}^2 = 1$, where the cosmic scale factor $U(t_c)$ is given by (19.17) and (19.16) for matter-dominated universe.

However, in the non-relativistic approximation $m \gg p$ (or large Ω), the above equation (19.30) can be written in the usual form, involving the Hubble function $H(t_c) = \dot{U}/U = 1/(2t_c)$, for any distant galaxies,

$$\dot{r}(t_c) \approx \frac{c}{\Omega t_c^{1/2}} \approx H(t_c) \frac{2ct_c^{1/2}}{\Omega} = H(t_c)r(t_c), \tag{19.32}$$

because $r(t_c) \approx (2ct_c^{1/2})/\Omega$, which can be obtained from (19.25) for the inertial frame F.

References

1. J. P. Hsu and L. Hsu, in *A Broader View of Relativity: General Implications of Lorentz and Poincaré Invariance*, 2nd ed. (World Scientific, 2006), Chapter 11 and pp. 159–160; J. P. Hsu, *Nuovo Cimento B* **80**, 201 (1984); J. P. Hsu and T. Y. Shi, *Phys. Rev. D* **26**, 2745 (1982).
2. J. P. Hsu, *Int. J. Mod. Phys. A* **21**, 5119 (2006).
3. J. P. Hsu, L. Hsu and D. Katz, *Mod. Phys. Lett. A* **33**, 1850116 (2018).

Chapter 20

Late-Time Accelerated Cosmic Expansion:
Conserved Baryon–Lepton Charges
with General U_1 Symmetry and "Dark Energy"

20.1. General Baryon–Lepton U_1 Symmetry — A Generalization of Lee–Yang's Idea for Conserved Baryon–Lepton Charges

To illustrate generalized gauge symmetry, let us consider new U_1 gauge transformations, which can be applied to both baryon and lepton charges. For simplicity, we consider only baryon charges here. The same discussions and equations can also be applied to lepton charges. The original proposal for understanding the conservation of baryon number (or baryon charge) was to associate it with U_1 gauge symmetry, just like the electromagnetic U_1 symmetry. The idea was first proposed and discussed by Lee and Yang in 1955.[1] Their formulation of dynamics for baryon charges was identical to that of quantum electrodynamics (QED). Based on Eötvös' experiments, they estimated that the coupling strength of such an inverse-square baryon force would be about 10^{-5} smaller than that of the gravitational force. According to Lee–Yang's formulation, this extremely weak baryon force should not have any observable effects in the physical world. Indeed, there have been no physical effects of Lee–Yang's U_1 fields or bosons detected in the high energy laboratory, thus far. As a result, some do not consider the U_1 symmetries corresponding to conserved baryon and lepton numbers to be local gauge symmetries, contrary to Lee–Yang's original idea.[1,a]

However, in view of the success of the unified electroweak theory and quantum chromodynamics, it is more natural to assume that gauge symmetry

---

[a] The baryon conservation law was first postulated by Stueckelberg.

is a universal symmetry principle,[2,b] so that baryon and lepton charges must correspond to local U_1 gauge symmetries. The reason that these massless gauge fields are not observed in the laboratory could be due to their extremely small coupling strength. However, this property turns out to be just right for them to play an important role in the late-time cosmological evolution.[3,c] We shall demonstrate that their physical effects can be related to the late-time accelerated cosmic expansion, provided that their usual U_1 symmetry with a scalar gauge function and the usual phase function is generalized to a new U_1 symmetry with vector functions and with Hamilton's characteristic phase function.

To demonstrate the cosmological effects of baryon–lepton charges, we first generalize Lee–Yang's idea and demonstrate that the conservation of baryon and lepton charges can lead to the observed late-time accelerated cosmic expansion[3] and hence, could effectively play the role of "dark energy." In other words, the driving force behind the late-time accelerated cosmic expansion is due to ubiquitous baryon (and lepton) charges, rather than an ad hoc "dark energy" everywhere in the universe.

In the following discussions, we shall often use "phase symmetry" instead of the usual term "gauge symmetry" in order to emphasize that phase symmetries involve the Lorentz vector gauge functions $\Lambda_\mu(x)$, rather than the usual scalar gauge function $\Lambda(x)$. In addition, phase symmetries are dynamical symmetries associated with exact conservation laws for color, baryon and lepton charges. Furthermore, phase fields satisfy fourth-order field equations rather than the usual second-order gauge field equations. Such an exact phase symmetry is in contrast to the electroweak $SU_2 \times U_1$ gauge symmetries, which involve spontaneous symmetry breaking in the original Lagrangian,[d] and the usual Lorentz scalar gauge functions.

Let us use U_{1b} to denote the generalized U_1 gauge (or phase) symmetry associated with conserved baryon charges. The new baryon U_{1b} phase transformations for a quark field $q(x)$, which carries the conserved baryon charge[3,4] g_b, are assumed to take the form,

$$\psi'(x) = e^{-iP_1}\psi(x), \quad \overline{\psi}'(x) = \overline{\psi}(x)e^{+iP_1}, \quad \psi(x) = q(x), \tag{20.1}$$

---

[b] The fundamental interactions of particles are determined by gauge invariance. This was stressed by Utiyama, Yang and others.

[c] Our understanding of the universe as a whole is very limited.

[d] After spontaneous symmetry breaking, the original Lagrangian leads to an electromagnetic Lagrangian with exact U_1 gauge symmetry, which still involves a scalar gauge function. In this sense, the electromagnetic U_1 gauge symmetry differs from phase dynamics with vector gauge functions.

where P_1 is an "action integral" with a variable endpoint $x'_e = x$ and a fixed initial point x'_o

$$P_1 = \left(g_b \int_{x'_o}^{x'_e=x} dx'^{\mu} \Lambda_{\mu}(x') \right)_{Le1}, \quad c = \hbar = 1, \tag{20.2}$$

which is Hamilton's characteristic "phase function."[5,6] New U_{1b} transformations for the "phase fields," $B_{\mu}(x)$, are given by

$$B'_{\mu}(x) = B_{\mu}(x) + \Lambda_{\mu}(x), \tag{20.3}$$

which involve a vector phase function $\Lambda_{\mu}(x)$ rather than the usual scalar function. To see the meaning of the subscript $Le1$ in (20.2), let us consider the variation of the phase function P_1. We have

$$\delta P_1 = g_b \Lambda_{\mu}(x') \delta x'^{\mu} |_{x'_o}^{x'_e=x} + g_b \int_{x'_o}^{x} \left(\frac{\partial \Lambda_{\mu}(x')}{\partial x'^{\lambda}} dx'^{\mu} - d\Lambda_{\lambda}(x') \right)_{Le1} \delta x'^{\lambda}. \tag{20.4}$$

We can determine the variation of the action functional P_1 as a function of the coordinates because the initial point x_o is fixed, so that $\delta x'(x'_o) = 0$ and the endpoint is variable, $x'_e = x$. Furthermore, the path in (20.4) must satisfy[7] the "Lagrange equation $Le1$," as indicated in (20.2), i.e.,

$$Le1: \quad d\Lambda_{\lambda}(x') - \frac{\partial \Lambda_{\mu}(x')}{\partial x'^{\lambda}} dx'^{\mu} = 0, \tag{20.5}$$

$$\text{or} \quad (\partial_{\mu} \Lambda_{\lambda}(x) - \partial_{\lambda} \Lambda_{\mu}(x)) dx^{\mu} = 0,$$

where $x' = x$. This requirement is necessary for the integral in (20.4) to vanish so that we have[6]

$$\delta P_1 = g_b \Lambda_{\mu}(x) \delta x^{\mu} \quad \text{or} \quad \partial_{\mu} P_1 = g_b \Lambda_{\mu}(x). \tag{20.6}$$

This equation for the U_{1b} phase "characteristic function" P_1 is crucial for an interaction Lagrangian to have phase symmetry.

One can verify that $\overline{\psi}(\partial_{\mu} + i g_b) \psi$ is invariant under the general U_{1b} transformations (20.1) and (20.3),

$$\overline{\psi}'(\partial_{\mu} + i g_b B'_{\mu}) \psi' = \overline{\psi}(\partial_{\mu} + i g_b B_{\mu}) \psi, \tag{20.7}$$

where we have used equations (20.1), (20.2), (20.3) and (20.6).

The U_{1b} "phase covariant derivative," $\partial_\mu + ig_b B_\mu$, in (20.7) is the same as that in the usual U_1 gauge theory QED. The U_{1b} "phase curvature" is, as usual, determined by the commutator of the U_{1b} covariant derivative,

$$[\partial_\mu + ig_b B_\mu, \partial_\nu + ig_b B_\nu] = ig_b B_{\mu\nu},$$
$$B_{\mu\nu} = \partial_\mu B_\nu - \partial_\nu B_\mu. \tag{20.8}$$

Under the U_{1b} phase transformation (20.3), the U_{1b} "phase curvature" $B_{\mu\nu}$ is not invariant,

$$B'_{\mu\nu} = B_{\mu\nu} + \partial_\mu \Lambda_\nu - \partial_\nu \Lambda_\mu \neq B_{\mu\nu}.$$

However, the divergence of the phase curvature $\partial^\mu B_{\mu\nu}$ is phase invariant, provided a constraint for $\Lambda_\mu(x)$ is satisfied,

$$\partial^\mu B'_{\mu\nu} = \partial^\mu B_{\mu\nu},$$
$$\partial^\mu[\partial_\mu \Lambda_\nu(x) - \partial_\nu \Lambda_\mu(x)] = 0. \tag{20.9}$$

We may remark that the constraints in (20.9) and (20.5) are compatible and that one still has infinitely many vector functions that satisfy these two constraints.

20.2.　Phase Invariant Lagrangians

The U_{1b} phase invariant Lagrangian is

$$L_b = \frac{L_s^2}{2}[\partial^\mu B_{\mu\lambda} \partial_\nu B^{\nu\lambda}] + \overline{\psi}[i(\partial_\mu + ig_b B_\mu) - m]\psi, \tag{20.10}$$

and the phase field equations are

$$\partial^2 \partial^2 B_\mu = \frac{g_b}{L_s^2}\overline{\psi}\gamma_\mu\psi, \quad \partial_\mu B^\mu = 0,$$

where $\psi = \psi(x)$ stands for the quark field $q(x)$ with the baryon charge g_b, and the usual "gauge condition" $\partial_\mu B^\mu = 0$ has been imposed for simplicity. In the phase invariant Lagrangian (20.10), the constant length L_s appears in the quadratic terms of the phase curvatures of U_{1b} symmetry. Baryon–lepton symmetry suggests that the length L_s should be a universal constant for baryon–lepton dynamics.[7]

If the vector function $\Lambda_\mu(x)$ can be expressed as the space-time derivative of an arbitrary scalar function $\Lambda(x)$, i.e., if $\Lambda_\mu = \partial_\mu \Lambda(x)$, then equation (20.5) and the constraint equation (20.9) become identities for the arbitrary scalar function $\Lambda(x)$. Also, the phase transformations (20.1), (20.2) and (20.3) become the same as the usual U_1 gauge transformations. However, in U_{1b} phase transformations (20.1) and (20.3), the vector function $\Lambda_\mu(x)$ does not take the special form $\partial_\mu \Lambda(x)$.

Our previous discussions, as well as equations (20.1)–(20.10) can similarly be applied to the conserved lepton charges g_ℓ with $U_{1\ell}$ phase symmetry. The $U_{1\ell}$ phase invariant Lagrangian for lepton phase fields $L_\mu(x)$ and the electron field $e(x)$ takes the form

$$L_\ell = \frac{L_s^2}{2}[\partial^\mu L_{\mu\lambda} \partial_\nu L^{\nu\lambda}] + \overline{e}(x)[i(\partial_\mu + ig_\ell L_\mu) - m_e]e(x), \tag{20.11}$$

$$L_{\mu\nu} = \partial_\mu L_\nu(x) - \partial_\nu L_\mu(x),$$

which is formally identical to the baryon Lagrangian L_b in (20.10). Thus we have assumed a baryon–lepton symmetry[e] for baryon–lepton dynamics with the $U_{1b} \times U_{1\ell}$ phase symmetries, and the constant L_s is assumed to be a universal length for the Lagrangian with phase symmetry (such as internal "phase groups" U_1 and color SU_3). To wit, the phase invariant Lagrangian of the color SU_3 phase fields $H_\mu^a(x)$ takes the form

$$L_c = \frac{L_s^2}{2}[\partial^\mu H_{\mu\lambda} \partial_\nu H^{\nu\lambda}] + \overline{q}(x)\left[i(\partial_\mu + ig_c H_\mu^a)\frac{\lambda^a}{2} - m_q\right]q(x), \tag{20.12}$$

where λ^a is the Gell-Mann matrix. The assumption that (20.12), (20.11) and (20.10) have the same universal length L_s is consistent with the idea of quark–lepton symmetry and baryon–lepton symmetry.[7,8]

Let us consider general U_{1b} group properties. For U_{1b} phase transformations (20.1) and (20.3), it is straightforward to verify that the group operations have the inverse, identity (i.e., $\Lambda_\mu = 0$) and associativity properties. To see the closure property, let us consider two consecutive transformations $\exp(-iP_a)$ and $\exp(-iP_b)$. We have

$$\exp(-i[P_a + P_b]) = \exp(-iP_c),$$

$$P_c = \left(g_b \int_{x_0'}^{x_e'=x} dx'^\mu \left[\Lambda_{c\mu}(x')\right]\right)_{L_{ec}} \tag{20.13}$$

where $\Lambda_{c\mu} = \Lambda_{a\mu} + \Lambda_{b\mu}$ and the Lagrange equation L_{ec} is given by $[\partial_\mu \Lambda_{c\lambda}(x) - \partial_\lambda \Lambda_{c\mu}(x)]dx^\mu = 0$. One can verify that the results in (20.4)–(20.9) also hold for the consecutive transformations when P_1 and Λ_μ are replaced by P_c and $\Lambda_{c\mu}$.

---

[e]For an early discussion of baryon–lepton symmetry, see Ref. 8.

Thus, the phase transformations (20.1), (20.2) and (20.3) together with the constraint in (20.9) can be considered to be a generalization of Lee–Yang's U_1 gauge transformations for baryon gauge fields.[1]

20.3. Linear Okubo Force Due to Baryon Charges on a Supernova or Galaxy

Now let us consider a simple model of particle-cosmology to investigate the cosmic baryon force in detail. The static equation and the solution for the potential $B_0(r)$ can be obtained from (20.10) with a point charge at the origin, i.e.,

$$\nabla^2 \nabla^2 B_0 = \frac{g_b}{L_s^2} \delta^3(\mathbf{r}),$$

$$B_0(r) = -\frac{g_b r}{8\pi L_s^2}.$$

$$(20.14)$$

Thus, the repulsive force $\mathbf{F}_{bb}$ between two point-like baryon charges is

$$\mathbf{F}_{bb} = -g_b \nabla B_0 = \frac{g_b^2}{8\pi L_s^2} \frac{\mathbf{r}}{r}, \qquad (20.15)$$

which is independent of distance rather than inverse-square as with the gravitational or electric forces.[f]

Let us consider the Okubo force on a supernova inside an extremely large sphere of galaxies due to the baryon charges.[9] Such a cosmic Okubo force can be calculated in two steps and the result could provide a way to test the cosmic baryon force by measuring the dependence of the accelerations of supernovae on their distance from Earth.

(A) First, we model the universe as a spherically symmetric distribution of uniform baryon charge with a very large radius R_o[10,8] and we approximate the supernova as a point inside that sphere.

(B) Second, we replace the point-like supernova by a sphere of baryon charge with a radius $R_s \lll R_o$.

We model a supernova as an object of mass m and radius R_s located inside a sphere consisting of approximately 100 billion galaxies. These baryonic galaxies

---

[f] We will call this force the Okubo force in recognition of Prof. Susumu Okubo's endeavors in particle physics.
[g] Here, $R_o \gg$ Hubble distance $= c/H_0 \approx 14$ billion light years. For cosmological parameters in Planck 2018 results, see the paper by N. Aghanim *et al.*

can be idealized as points that are uniformly distributed throughout the sphere with a baryon mass density of ρ. The mass of this sphere is then $M = \rho 4\pi R_0^3/3$. Each nucleon has three quarks, i.e., $3g_b$ of baryonic charges. We now calculate the cosmic force F_{CS1} between the uniform sphere of baryon charges and the point-like supernova inside the sphere.[7]

If m_p stands for the mass of a baryon (proton or neutron), then the supernova with mass m has a baryonic charge $3g_b m/m_p$. We divide the universe with radius R_0 into thin uniform shells with radius R, where R is less than R_0. The thickness of each shell is dR. Figure 20.1 shows the case for a supernova located at a distance r from the center of the sphere and a shell with a radius R where $r > R$. Taking a point P on the shell, the angle SCP is θ. The mass dM_r of the ring is

$$dM_r = \rho 2\pi R^2 dR \sin\theta \, d\theta, \tag{20.16}$$

where $2\pi R \sin\theta$ is the circumference of the ring centered on the line segment SC. At point P, let us consider an infinitesimal element. The effective repulsive force between this infinitesimal element and the point-like supernova is dF_p. The component of the force dF_p parallel to line segment SC is given by $dF_p \cos\varphi$. When we integrate this infinitesimal element over the ring, the components of the force from the elements dF_p perpendicular to SC cancel because of the symmetry of the ring around line segment SC. Calculating the total force dF_{CS} due to the ring on the point-like supernova S using the equation for the effective force between two

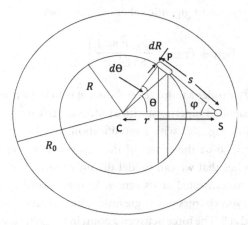

Fig. 20.1. A schematic diagram for calculations of the cosmic baryon (or Okubo) force F_{CS1} between a gigantic uniform sphere of baryon galaxies and a point-like baryon supernova.

point-like baryonic charges (20.23), we find

$$dF_{CS} = A \int_0^\pi \left[dR\, R^2 \cos \varphi \right] \sin \theta \, d\theta = A \int_{R-r}^{R+r} [dF] \frac{s}{Rr} ds, \qquad (20.17)$$

$$A = \frac{(3g_b)^2 m\rho}{4L_s^2 m_p^2}, \quad dF \equiv dR\, R^2 \left[\frac{s^2 + r^2 - R^2}{2sr} \right], \qquad (20.18)$$

where each baryon carries a charge of $3g_b$ and we have used the relations (see Fig. 20.1)

$$s^2 = R^2 + r^2 - 2Rr \cos \theta, \quad R^2 = s^2 + r^2 - 2rs \cos \varphi. \qquad (20.19)$$

The magnitude of the total repulsive force can be obtained[7] from the integration of ds for two cases:

$$F_{CS1} = A \int_0^r \frac{R\,dR}{2r^2} \int_{r-R}^{r+R} ds(s^2 + r^2 - R^2), \quad r > R,$$

$$F_{CS2} = A \int_r^{R_o} \frac{R'\,dR'}{2r^2} \int_{R'-r}^{R'+r} ds(s^2 + r^2 - R'^2), \quad R_o > R' > r. \qquad (20.20)$$

For the calculation of F_{CS2}, we modify Fig. 20.1 such that the distance SC is smaller than R' ($=R$). We obtain the distance-dependence of the baryon force between a point baryonic charge and a gigantic spherical charge with radius R_o,

$$F_{CS1} + F_{CS2} = \frac{(3g_b)^2 mM}{8\pi L_s^2 m_p^2} \left[\frac{r}{R_o} - \frac{r^3}{5R_o^3} \right]. \qquad (20.21)$$

Let us now consider a more general case, in which the point-like supernova is replaced by an extended supernova with a radius R_s. To carry out this calculation, we may use the same Fig. 20.1 with a modification, i.e., the interchange of C and S. Let us consider C to be the center of the supernova, so we replace R_o by R_s. Result (20.21) implies that we can model the universe as a sphere with all of its baryonic charges concentrated at its center. In other words, in this calculation, we consider all baryon charges of the gigantic sphere representing the universe as being concentrated at S. The force between a point in the volume element within the supernova and the total baryon charge of the universe concentrated at S depends on the distance s, i.e., $[s/R_o - s^3/5R_o^3]$, as shown in (20.21). Thus, we should make the replacement,

$$[dF]\frac{s}{Rr}ds \rightarrow [dF]\frac{s}{Rr}\left[\frac{s}{R_o} - \frac{s^3}{5R_o^3}\right]ds, \qquad (20.22)$$

for the two baryonic systems (supernova and universe) with masses m and M. After some tedious but straightforward calculations, the total cosmic baryon force exerted on a supernova with radius R_s by the universe (modeled as a gigantic uniform sphere of baryons charge with radius R_o) is

$$F^{Ok} = \left(\frac{g_1 g_2}{8\pi L_s^2}\frac{r}{R_o}\right)\left[1 - \frac{r^2}{5R_o^2} - \frac{R_s^2}{5R_o^2}\right] \approx \frac{g_1 g_2}{8\pi L_s^2}\frac{r}{R_o}, \quad r < R_o,$$

$$g_1 = \frac{3g_b m_1}{m_p}, \quad m_1 = m, \quad g_2 = \frac{3g_b m_2}{m_p}, \quad m_2 = M, \qquad (20.23)$$

where $r < R_o$ corresponds to the case where the baryonic system with radius R_s (the supernova) is inside the gigantic sphere (the universe) with radius R_o.

Thus, approximating the universe as a sphere of uniform baryonic charge of radius R_o, we find that the cosmic repulsive baryon force on a spherical baryonic system due to the other baryons in the rest of the universe is approximately proportional to the distance of the spherical system from the center of the universe sphere, as shown in (20.23) when $r \ll R_o$. In the limit $R_s \rightarrow 0$, the expression for the force in (20.23) simplifies to the expression for $F_{CS1} + F_{CS2}$ in (20.21), as expected.

The dominant linear term of the Okubo force in (20.23) predicts that the measured acceleration of the cosmic expansion should be linear in distance r. This prediction could be tested in the future by measuring the frequency shifts[12,h] of accelerating supernovae at different distances.

The previous discussions and results (20.16)–(20.23) can also be applied to lepton charges, with the baryon charge $3g_b$ replaced by the lepton charge g_ℓ.

20.4. Dynamics of Expansion and "Dark Energy" Due to Linear Force of Baryon Charges and Quantum Yang–Mills Gravity in a Super-Macroscopic Limit

To see the implications of the baryon–baryon and lepton–lepton interactions for the expansion of the universe, let us consider a system interacting only via the

---

[h]The Doppler effect for accelerated systems can be calculated from the Wu transformations for frames with a constant linear acceleration based on limiting Lorentz–Poincaré invariance.

gravitational and baryon forces. We ignore lepton charges here for simplicity because they can be treated in the same way as the baryon charges. The Lagrangian of the system in an inertial frame is given by

$$L_{\phi\psi} = L_\phi + L_\psi,$$

$$L_\phi = \frac{1}{2g^2}(C_{\mu\alpha\beta}C^{\mu\beta\alpha} - C_{\mu\alpha}{}^\alpha C^{\mu\beta}{}_\beta),$$

$$L_\psi = \frac{i}{2}[\overline{\psi}\gamma_\mu\Delta^\mu\psi - (\Delta^\mu\overline{\psi})\gamma_\mu\psi] - m\overline{\psi}\psi,$$

$$\Delta^\mu\psi = J^{\mu\nu}\partial_\nu\psi, \quad J^{\mu\nu} = \eta^{\mu\nu} + g\phi^{\mu\nu} = J^{\nu\mu},$$

(20.24)

where $L_{\phi\psi}$ is the same as that given by (19.4) and (19.5).[i] The Yang–Mills gravitational equation can be derived from (20.24),

$$g^{-2}H_{\mu\nu} = S_{\mu\nu}(\psi),$$

(20.25)

where $g^{-2}H_{\mu\nu}$ is derived from the T_4 gauge invariant Lagrangian L_ϕ. Based on previous discussions in (8.26)–(8.29), the source tensor $S_{\mu\nu}(\psi)$ comes from the baryon mass source for gravity

$$S_{\mu\nu}(\psi) = \frac{i}{2}\left[\overline{\psi}\gamma_\mu\partial_\nu\psi - (\partial_\nu\overline{\psi})\gamma_\mu\psi\right].$$

(20.26)

We note that the source tensor $S_{\mu\nu}(\psi)$ of the massive fermion field in (20.26) is approximately the same as the energy–momentum tensor of the fermion field in the presence of Yang–Mills gravity, i.e.,

$$T_{\mu\nu}^{YM}(x) = -\eta_{\mu\nu}L_\psi + \frac{\partial L_\psi}{\partial(\partial^\mu\psi)}\partial_\nu\psi + \partial_\nu\overline{\psi}\frac{\partial L_\psi}{\partial(\partial^\mu\overline{\psi})}$$

$$\approx \frac{i}{2}[\overline{\psi}\gamma_\mu\partial_\nu\psi - (\partial_\nu\overline{\psi})\gamma_\mu\psi] = S_{\mu\nu}(\psi), \quad J_{\lambda\nu} \approx \eta_{\lambda\nu}, \quad (20.27)$$

where $L_\psi = 0$ due to the fermion equations (8.30) in Chapter 8.[j] Thus, in the macroscopic approximation (the geometric-optics limit of quantum wave equations), it is reasonable to make the approximation $S_{\mu\nu}(\psi) \approx T_{\mu\nu}^{YM}(x)$ in

---

[i]The usual "gauge-fixing" terms are not included because we are not dealing with quantization of $\phi_{\mu\nu}$ here.

[j]In (20.27), $J_{\lambda\nu}$ contains another term involving the gravitational constant, which is extremely small compared with $\eta_{\lambda\nu}$.

(20.25) and then to replace that term with the energy–momentum tensor of macroscopic bodies,[6] i.e., $T_\mu^\lambda = (\rho, -P, -P, -P)$ in inertial frames,

$$S_{\mu\nu}(\psi) \approx T_{\mu\nu}^{YM}(x) \to T_\mu^\lambda \eta_{\lambda\nu} = (\rho, P, P, P),$$

because the wave equation of quantum fermions with mass m will reduce to a classical Hamilton–Jacobi type equation with the same mass m in the geometric-optics limit, as discussed in Section 8.6 of Chapter 8.

For completeness, we stress that the linear Okubo force (20.23) due to the extremely small baryonic charge that exists in-between galaxies should also be taken into account in the dynamics of cosmic expansion. The reason is that this linear repulsive force could be comparable or even more important than the attractive inverse-square gravitational force on Hubble scales and beyond, since it could be stronger than the gravitational force for a sufficiently large distance between two galaxies. This property of an extremely small and linear force turns out to be crucial for explaining the observed the cosmic acceleration, which shows up during the late-time cosmic evolution, about 5 billion years ago.[13,14]

The Einstein–Grossmann approach to gravity suggests an elegant and convenient way to investigate macroscopic phenomena based on the metric tensors $g_{\mu\nu}$. Interestingly enough, quantum Yang–Mills gravity has an effective metric tensor $G_{\mu\nu}$ in the macroscopic limit. Originally, we had a Lorentz invariant gravitational field equation and the Dirac equation for fermions such as quarks and leptons in inertial frames with well-defined space-time coordinates. They both lead to an effective metric tensor $G_{\mu\nu}$, which emerges in the Hamilton–Jacobi type equations for the motion of classical objects, in the geometric optics limit of quantum particle wave equations in inertial frames.

Thus, quantum Yang–Mills gravity suggests a new approach to the dynamics of cosmic expansion based on the following three observations:

(i) Based on the cosmological principle of homogeneity and isotropy and the macroscopic relation $G_{\mu\nu} = \eta^{\alpha\beta} J_{\alpha\mu} J_{\beta\nu}$, the effective metric tensor $G_{\mu\nu}(x)$ and the fields $J_{\mu\nu}(x)$ can be consistently simplified to a diagonal form and to a function of time only. That is, $G_{\mu\nu}(t) = (B^2(t), -A^2(t), -A^2(t), -A^2(t))$ and $J_{\mu\nu}(t) = (B(t), -A(t), -A(t), -A(t))$. We see that at a given instant of time, $G_{\mu\nu}(t)$ is the same everywhere in the universe, as required by homogeneity. The diagonal form of $G_{\mu\nu}$ is directly related to the requirement of the isotropy of matter in the universe.

(ii) To investigate the motions of distant galaxies, which are made of stable baryons and leptons, all long-range forces that can act on galaxies should be taken into account, including the gravitational force and linear Okubo force due to baryon and lepton charges.

(iii) In order to see the quantitative effect of baryon charges on the metric tensor, we embed the linear Okubo force in the source tensor terms in the field equations (20.25). In this way, the effective metric tensor $G_{\mu\nu}(t) = \eta^{\alpha\beta} J_{\alpha\mu}(t) J_{\beta\nu}(t)$ is also generated from a new source of baryon charges, in addition to the usual source for the gravitational interaction.

Based on these properties, it is natural to take a more general viewpoint regarding the T_4 gauge fields in the super-macroscopic world. Those fields are now generated by all relevant long-range forces in the universe. They are effectively super-macroscopic "time-dependent gauge fields," which govern the geometry of cosmic space-time and the motion of distant galaxies. In this sense, we have generalized the original T_4 gauge fields in equation (20.25) to include a new source with baryon charges. As discussed in Chapter 19, within the relativistic-cosmology model based on cosmic relativity, the effective metric tensor $G_{\mu\nu}(t) = U^2(t)\eta_{\mu\nu}$ involves a universal space-time scale factor $U(t)(=B(t) = A(t))$. (See (19.16)–(19.21).) Thus, we postulate the "cosmological T_4 gauge equation" with new effective source terms and the effective super-macroscopic tensors $G_{\lambda\nu}(t)$ and $T^\lambda_\mu = (\rho, -P, -P, -P)$:

$$g^{-2} H_{\mu\nu} = S_{\mu\nu}(\text{baryon mass}) + t_{\mu\nu}(\text{baryon charge}),$$

$$S_{\mu\nu} = T^\lambda_\mu G_{\lambda\nu}(t) = (\rho(t)U^2, P(t)U^2, P(t)U^2, P(t)U^2), \qquad (20.28)$$

$$t_{\mu\nu} = \rho^b G_{\mu\nu}(t), \quad \rho^b = \text{constant}.$$

This is the basic equation for the dynamics of the cosmic expansion of the baryon–matter universe within the framework of quantum Yang–Mills gravity in the super-macroscopic limit. It is gratifying that the linear Okubo force (20.23) can be effectively realized if $t_{\mu\nu}$ due to baryon charges is a constant of the form given by (20.28), which resembles the "vacuum energy."[10] This result leads to a new understanding of the physical origin of the so-called "dark energy," i.e., it is a long-range force due to baryon and lepton charges.

To investigate potentials and forces produced by baryon masses and baryon charges, we may linearize the cosmological equation (20.28) for simplicity:[k]

$$g^{-2}\partial^\alpha \partial_\alpha J_{\mu\nu} \approx \left(S_{\mu\nu} - \frac{1}{2}\eta_{\mu\nu}S_\lambda^\lambda\right) + \left(t_{\mu\nu} - \frac{1}{2}\eta_{\mu\nu}t_\lambda^\lambda\right), \tag{20.29}$$

where we have imposed the gauge condition $\partial_\lambda \phi^{\lambda\nu} = (1/2)\partial^\nu \phi_\lambda^\lambda$ and μ and ν should be made symmetric, just like in (8.26).

The appearance of energy density and pressure in (20.28) or (20.29) is reasonable because a moving baryon mass can have energy and momentum. Thus, we can approximate the quantum source term (20.26) in the microscopic world by the energy density ρ and pressure P of macroscopic bodies.[l] On the other hand, the baryon numbers (or charges) do not involve the energy and momentum of material baryons, so it is natural to include them as an effective constant energy density in (20.28).

In order to treat the forces arising from baryon charges in the same way as the gravitational force in Yang–Mills gravity, let us consider the gravitational and baryonic forces in the Newtonian static limit. From (20.28) and (20.29), we have the following results:

(A) In the usual macroscopic limit, the gravitational force produced by a system of N_1 baryons acting on a body of N_2 baryons is as follows:

The static field $\phi^m = g\phi_{00}$ produced by the energy density ρ due to a system of N_1 baryons (or protons) can be derived from (20.28) and (20.29) in inertial frames,

$$\nabla^2 \phi^m = \frac{g^2}{2}\rho, \quad \phi^m = -\frac{g^2(N_1 m_p)}{8\pi r}, \quad \phi^m = g\phi_{00},$$
$$\rho = N_1 m_p \delta^3(\mathbf{r}), \quad \rho^b = 0, \tag{20.30}$$

where $N_1 m_p$ is assumed to be concentrated at the origin for simplicity. It implies that the gravitational force F^m between two bodies with $N_1 m_p = m_1$

---

[k] Cf. equations (8.32) and (8.33) in Chapter 8.

[l] In contrast, the appearance of the energy–momentum tensor of macroscopic bodies in the conventional field equation of gravity is taken for granted.

and $N_2 m_p = m_2$ is

$$F^m = -(N_2 m_p)\frac{\partial \phi^m}{\partial r} = -\frac{g^2 m_1 m_2}{8\pi r^2}. \tag{20.31}$$

(B) The linear Okubo force produced by a system of N_1 baryons^m with baryon charges $N_1 3g_b$ acting on a body of N_2 baryons is as follows:

The static potential ϕ^b produced by N_1 baryon can also be obtained from (20.28) and (20.29) in inertial frames (with $\eta_{\mu\nu}$)

$$\nabla^2 \phi^b = -\frac{g}{2}\rho^b, \quad \phi^b = -\frac{g\rho^b}{12}r^2, \quad \phi^b = \phi_{00}, \quad S_{\mu\nu} = 0. \tag{20.32}$$

In analogy to (20.31) where the gravitational potential ϕ^m contains g^2, the baryonic potential ϕ^b or ρ^b should contain g_b^2 to be consistent within the framework of Yang–Mills gravity. (Cf. (20.35) below.) Accordingly, the baryonic force F^b should be related to ϕ^b by the relation,

$$F^b = -N_2 \frac{\partial \phi^b}{\partial r} = \frac{N_2 g \rho^b}{6}r. \tag{20.33}$$

20.5.   An Estimation of Dark Energy Density

The crucial idea to link an effective constant energy density of dark energy and the Okubo force due to baryon charges is as follows: For cosmological dynamics related to baryon charges, this linear force F^b is naturally identified with the linear Okubo force (20.23), so that we have

$$F^b = \left(\frac{gN_2\rho^b}{6}\right)r = \frac{g_1 g_2}{8\pi L_s^2}\frac{r}{R_0} = F^{Ok}, \tag{20.34}$$

where $g_1 = 3g_b N_1$ and $g_2 = 3g_b N_2$. This new relation (20.34) enables us to relate the effective constant energy density ρ^b to other physical quantities

$$\rho^b = \frac{3(3g_b)^2 N_1}{4\pi g L_s^2 R_0}. \tag{20.35}$$

Thus, quantum Yang–Mills gravity and the cosmological principle in the super-macroscopic limit enable us to estimate roughly the "effective dark energy density" ρ^b contributed from the ubiquitous baryon charge in the matter universe.

---

^{m} Each baryon contains three quarks and three baryon charges, i.e., $3g_b$.

Let us first consider the baryon charge g_b. It can be estimated on the basis of a critical distance R_c at which the repulsive linear Okubo force (20.23) with $r < R_o$ exactly balances the gravitational attractive inverse-square force in (20.31). Since $F^{Ok} = F^m$ for $r = R_c$,

$$(3g_b)^2 = \frac{8\pi G L_s^2 m_p^2 R_o}{R_c^3}, \qquad (20.36)$$

where we approximate the entire matter universe as a gigantic sphere with a radius R_o. It follows from (20.36) and (20.35) that

$$\rho^b = \frac{N_1 g m_p^2}{4 R_c^3} = \left(\frac{R_o}{R_c}\right)^3 \frac{g \pi N_b m_p^2}{3}, \qquad (20.37)$$

where N_1 is the total number of baryons (or protons) in the whole universe and N_b is the baryon density. We have used the relation $N_1 = 4\pi R_o^3 N_b/3$ in (20.37).

Data suggest that the cosmic expansion began to accelerate roughly 5 billions year ago.[13,14] They also suggest that the critical distance R_c is about 9 billion lightyears. Other constants such as $g^2 = 8\pi G$ are known. The baryon density N_b of the matter universe is obtained using data from the Planck space telescope.[10] Thus, we have

$$R_c \approx 8 \times 10^{26}\,\mathrm{m}, \quad g = \sqrt{8\pi G} \approx 10^{34}\,\mathrm{m}, \quad N_1 \approx 10^{80},$$
$$N_b \approx 9.9 \times 10^{-30}\,\mathrm{g/cm^3} \approx 5.9\,\mathrm{protons/m^3}. \qquad (20.38)$$

The biggest uncertainty is due to the size R_o of the whole matter universe (not necessarily just the visible universe), where R_o was given in (20.21). It could be anywhere from the Hubble length to infinity. In order to get a feeling of the value of the "effective dark energy density" ρ^b, let us assume a range of values for R_o,

$$R_o \approx (10^{26} - 10^{32})\,\mathrm{m}. \qquad (20.39)$$

We may then roughly estimate the effective dark energy density as

$$\rho^b \approx (10^{-66} - 10^{-48})\,\mathrm{GeV^4}, \quad c = \hbar = 1, \qquad (20.40)$$

where the experimental value is $\approx 10^{-47}\,\mathrm{GeV^4}$.[10] The ranges of numerical values for ρ^b in (20.39) and (20.40) are roughly estimated to get a feeling of their magnitudes and uncertainties.

It appears that phase symmetry and quantum Yang–Mills gravity naturally suggest a new physical understanding of the late-time cosmic acceleration based on the experimentally established conserved baryon (and lepton) charges, rather than a mysterious "dark energy." In other words, the effective cosmic repulsive force due to dark energy can be interpreted as the long-ranged linear repulsive force due to baryon (and lepton) charges. This is motivated by the fundamental principle of gauge symmetries, i.e., the general U_1 symmetry and the space-time translational gauge symmetry.

In summary, we have the following results:

(i) The theoretical estimate of the "effective dark energy density" (20.40) is roughly consistent with the experimental value $\approx 10^{-47}$ GeV^4.[10] This substantiates our understanding of the late-time accelerated cosmic expansion and "dark energy," based on baryon charges associated with general U_1 symmetry.

(ii) The result in (20.36) or the effective energy density (20.37) suggests that the physical universe is finite rather than infinite. Based on the experimental value $\rho^b \approx 10^{-47}$ GeV^4, quantum Yang–Mills gravity and conserved baryon charges suggest that the whole matter universe is finite with the radius

$$R_o \approx 10^{16} \text{ light years.} \qquad (20.41)$$

This estimated value implies a size of the matter universe that is much larger than the visible universe at present.

References

1. T. D. Lee and C. N. Yang, in *100 Years of Gravity and Accelerated Frames: The Deepest Insights of Einstein and Yang–Mills*, eds. J. P. Hsu and D. Fine (World Scientific, 2005), p. 155; E. C. G. Stueckelberg, *Helv. Phys. Acta* **11**, 299 (1938); T. D. Lee, in *Particle Physics and Introduction to Field Theory* (Hardwood Academic, 1981), Chapters 10 and 14, pp. 746–752.

2. Y. Utiyama, *Phys. Rev.* **101**, 1597 (1956); C. N. Yang, *Phys. Today* **33**, 42 (1980).

3. J. P. Hsu, *Chin. Phys. C* **41**, 015101 (2017); J. P. Hsu, *Int. J. Mod. Phys. A* **24**, 5217 (2009); T. Rothman and E. C. G. Sudarshan, in *Doubt and Certainty* (Helix Book, Reading, 1998), pp. 214–264.

4. M. Khan, Y. Hao and J. P. Hsu, *EPJ Web Conf.* **168**, 04004, (2018); J. P. Hsu and K. O. Cottrell, *Chin. Phys. C* **39**, 105101 (2015).

5. W. Yourgrau and S. Mandelstam, in *Variational Principles in Dynamics and Quantum Theory*, 3rd ed. (Dover, 1979), p. 50.

6. L. Landau and E. Lifshitz, in *The Classical Theory of Fields* (Addison-Wesley, 1951), p. 29 and pp. 91–93.

7. J. P. Hsu. *Chin. Phys. C* **41**, 015101 (2017).

8. A. Gamba, R. E. Marshak and S. Okubo, *Proc. Natl. Acad. Sci.* **45**, 881 (1951).

9. G. R. Fowles and G. L. Cassiday, in *Analytical Mechanics* (Thomson, 2005), pp. 223–225.

10. S. Weinberg, in *Cosmology* (Oxford University Press, 2008), p. 39, p. 57, p. 510; P. J. E. Peebles and B. Ratra, in *100 Years of Gravity and Accelerated Frames: The Deepest Insights of Einstein and Yang–Mills*, eds. J. P. Hsu and D. Fine (World Scientific, 2005), p. 592; N. Aghanim *et al.* (Planck Collaboration), arXiv:1807.06209.

11. M. Khan and Y. Hao, Private correspondences (2018); M. Khan, Y. Hao and J. P. Hsu, *Proc. 2017 ICGAC-XIII and IK15.* DOI:https://doi.org/10.1051/epjconf/201816801012.

12. L. Hsu and J. P. Hsu, in *Space-time Symmetry and Quantum Yang–Mills Gravity* (World Scientific, 2013), pp. 25–42, pp. 129–142.

13. S. Perlmutter *et al.*, *Astrophys. J.* **517**, 565 (1999).

14. A. G. Riess *et al.*, *Astron. J.* **116**, 1009 (1998).

Chapter 21

A Big Picture of the Universe:
Big Jets Model and Physical Laws in Microscopic,
Macroscopic and Super-Macroscopic Worlds

21.1.   Big Jets Model Based on CPT Invariance

With our limited knowledge of the universe at present, it would be premature to conclude that we have observed the largest portion of the universe.[1] In particular, the Big Bang model of the universe has difficulties with the established laws of particle physics, such as the fundamental CPT invariance[2] and the maximal matter–antimatter symmetry. The fact that, so far, we have not detected any antimatter stars or antimatter galaxies in the visible universe could be considered observational evidence for an alternative model of the universe: Namely that the CPT theorem suggests that there are antimatter galaxies separated from our visible universe composed of matter galaxies. To be more specific, the fundamental CPT invariance suggests a Big Jets model for the birth of the universe, leading to two gigantic blackbodies, one composed mainly of matter and the other of antimatter, which are now presumably separated by a great distance.

When particles collide in high energy experiments, many jets can be produced, with their production and decay processes dictated by basic dynamic and geometric symmetry principles of the space-time framework. The evolution processes of the particles in each jet are determined by the conservation laws of electric charge, baryon number, energy–momentum.

Based on particle cosmology, we propose that the universe began similarly with two Big Jets. One may roughly picture the Big Jets model as two gigantic fireballs being created and then moving away in opposite directions, each with an unimaginable number of particles and antiparticles. We assume that the ensuing evolution of each fireball was dictated by fundamental dynamic and geometric symmetry principles in particle physics, such as color SU_3 and four-dimensional space-time symmetry. After various annihilation and decay processes, one fireball happened to be dominated by stable particles (baryons, electrons, antineutrinos and dark matter), forming the "matter half-universe," while the other was dominated by stable antiparticles (antibaryons, positrons, neutrinos and dark antimatter[a]) and formed the "antimatter half-universe."[3] To observers in each half-universe, the evolution processes would have looked similar to that of a hot Big Bang model, as implied by the fundamental CPT theorem. Thus, the Big Jets model preserves matter–antimatter symmetry according to the CPT theorem and explains the absence of antimatter galaxies in our matter half-universe.

In particle physics, the combination of local quantum fields, Lorentz invariance and the spin-statistics relations for quantizations of fields implies the exact CPT invariance,[2] where C (charge conjugate) denotes changing the sign of a charge, P (Parity, $\mathbf{r} \rightarrow -\mathbf{r}$) denotes space inversion and T denotes time reversal ($t \rightarrow -t$). This exact CPT invariance assures exact lifetime and mass equalities between particles and antiparticles. It also implies opposite electroweak (and chromo-) interaction properties between particles and antiparticles, and the same gravitational interactions between particles and antiparticles (in the sense of local quantum Yang–Mills gravity based on flat space-time). Thus, the fundamental CPT invariance leads to a maximum symmetry between matter and antimatter in the whole universe, suggesting a Big Jets Model.

In the Big Jets model, our observable universe is the original matter fireball that has expanded and cooled to become a "matter blackbody" at about 2.7 K. Thus, the antimatter fireball should also have expanded and cooled to become an "antimatter blackbody" at about 2.7 K, according to CPT invariance. Presumably, these two blackbodies are separated by an extremely large distance (probably much more than 10 billion light years) so that electromagnetic radiations emitted from bodies in the antimatter half of the universe are too weak to be detected by us. Nevertheless, the microwaves emitted from the antimatter blackbody as a whole may be detectable. If the matter blackbody can be modeled as a spherical blackbody, then the Big Jets model predicts that the side that faces the antimatter blackbody should be slightly warmer than the opposite hemisphere. This prediction of a "semi-dipole anisotropy," on the order of 10^{-5} or smaller,[3,4] could be tested

---

[a]We assume the existence of both dark matter and dark antimatter based on the CPT theorem in particle physics.

by satellite CMB data.[4] The usual Big Bang model would not result in such a "semi-dipole asymmetry" of temperature in the CMB data.

The Okubo force for two baryon systems separated by a distance r that is larger than the size of either system, i.e., solving for the case $r > R_o$ in (20.23). This can be calculated by modifying the situation shown in Fig. 20.1 from Chapter 20 with the "supernova" S replaced by a big spherical baryon system and located outside the gigantic sphere. And the expression in (20.22) is now replaced by

$$[dF]\frac{s}{Rr}ds \rightarrow [dF](s/Rr)[1 - R_o^2/(5s^2)]ds.$$

We obtain

$$F^{Ok} = \left(\frac{g_1 g_2}{8\pi L_s^2}\right)\left[1 - \frac{R_o^2}{5r^2} - \frac{R_s^2}{5r^2}\right], \quad r > (R_o + R_s).$$

This Okubo force can then be applied to the Big Jets model, in which the two spheres represent a matter half-universe and an antimatter half-universe, rather than a supernova and the universe. If the two half-universes form a bound system and rotate around their common mass center, they could form a stable system. The Okubo force F^{Ok} between these two approximately spherical half-universes would be attractive and given by

$$F^{Ok} \approx \frac{-|g_1 g_2|}{8\pi L_s^2}, \quad |g_1| = |g_2| \approx \frac{3g_b M_{tot}}{m_p},$$

where the dominant term is independent of distance, and M_{tot} is the mass of the matter half-universe. According to CPT invariance, the masses and the absolute values of the baryon charges $|g_1|$ of these two half-universes are the same. We may remark that the trajectories of two rotating matter and antimatter half-universes have been numerically calculated.[5] For the case corresponding to the elliptical orbit in the Kepler problem, one has a new "distorted elliptical orbit."[b]

21.2. Laws of Physics in the Microscopic, Macroscopic and Super-Macroscopic Regimes

Following our work, we assume that the underlying space-time of all physics in both the "matter half-universe" and the "antimatter half-universe" in the Big Jets model is flat. In this section, we summarize the picture that quantum Yang–Mills

---

[b]The elliptical orbit is slightly distorted to be slightly closer to a rectangle of similar size.

gravity gives us at three different length scales, the microscopic, macroscopic, and super-macroscopic. As we have seen, Yang–Mills gravity is formulated in a flat space-time using inertial frames, and is consistent with experimental results as discussed in Chapter 9. The space-time coordinates of inertial frames exhibit four-dimensional symmetry (or Lorentz–Poincaré invariance) and may be either the (ct, x, y, z) of special relativity with relativistic time or the (bt_c, x, y, z) of cosmic relativity with the cosmic time t_c for all observers in inertial frames.

Since only long-range forces play important roles at all three length scales, we use the gravitational interaction to illustrate the new big picture of nature given to us by quantum Yang–Mills gravity. For simplicity, we consider the case of the matter half-universe.

(A) Microscopic World

In the microscopic world, the space-time of all physics is characterized by the Minkowski metric tensor $\eta_{\mu\nu} = (1, -1, -1, -1)$. Interactions between all known elementary particles such as leptons, quarks, gauge bosons and their antiparticles, are described by quantum field theories, namely quantum electroweak theory, quantum chromodynamics, quantum Yang–Mills gravity, etc. The space-time T_4 gauge fields $\phi_{\mu\nu}$ and fermion fields are quantized. In inertial frames, the basic equations of quantum Yang–Mills gravity are:

A1. Space-time T_4 gauge field equations for gravity

$$H^{\mu\nu} = g^2 S^{\mu\nu}, \quad g^2 = 8\pi G, \tag{21.1}$$

where

$$H^{\mu\nu} = \partial_\lambda (J^\lambda_\rho C^{\rho\mu\nu} - J^\lambda_\alpha C^{\alpha\beta}_\beta P^{\mu\nu} + C^{\mu\beta}_\beta J^{\nu\lambda})$$

$$- C^{\mu\alpha\beta}\partial^\nu J_{\alpha\beta} + C^{\mu\beta}_\beta \partial^\nu J^\alpha_\alpha - C^{\lambda\beta}_\beta \partial^\nu J^\mu_\lambda, \tag{21.2}$$

$$S^{\mu\nu} = \frac{1}{2}\left[\overline{\psi} i\gamma^\mu \partial^\nu \psi - i(\partial^\nu \overline{\psi})\gamma^\mu \psi \right], \tag{21.3}$$

$$C^{\mu\nu\alpha} = J^{\mu\lambda}(\partial_\lambda J^{\nu\alpha}) - J^{\nu\lambda}(\partial_\lambda J^{\mu\alpha}), \quad J^{\alpha\mu}(x) = \eta^{\alpha\mu} + g\phi^{\alpha\mu}(x).$$

A2. Dirac equation in the presence of Yang–Mills gravity

$$i\gamma_\mu (\eta^{\mu\nu} + g\phi^{\mu\nu})\partial_\nu \psi - m\psi + \frac{i}{2}[\partial_\nu(J^{\mu\nu}\gamma_\mu)]\psi = 0. \tag{21.4}$$

In a general frame (either inertial or non-inertial) with the Poincaré metric $P_{\mu\nu}$, these basic equations are given in (8.26)–(8.30). As discussed in Chapters 10–13, one obtains a set of new gravitational Feynman–Dyson rules for Feynman diagrams to calculate the graviton self-energy, space-time gauge identities, the S-matrix, etc.

(B) Macroscopic World

In the macroscopic world, space-time is characterized by an effective metric tensor $G^{\mu\nu}(x)$ rather than the Minkowski metric tensor $\eta^{\mu\nu}$. This effective metric tensor $G^{\mu\nu}(x)$ in the classical equations of motion is derived from the wave equations for quantum particles in the geometric-optics limit, as discussed in Chapter 8.

The Dirac equation for quantum particles and quantized space-time T_4 gauge fields $\phi_{\mu\nu}$ are not directly useful in the macroscopic world with classical objects. Instead, we use modified classical space-time gauge fields, in which the original source tensor $S^{\mu\nu}$ of the quantum field in (21.3) is approximated by the energy–momentum tensor of usual macroscopic bodies $T_m^{\mu\nu}(x) = (\rho(x), P(x), P(x), P(x))$.

B1. The effective space-time gauge field equation with a classical gravitational source $T_m^{\mu\nu}$ consisting of macroscopic bodies is

$$g^{-2} H^{\mu\nu}(x) = T_m^{\mu\nu}(x), \tag{21.5}$$

$$T_m^{\mu\nu}(x) = (\rho(x), P(x), P(x), P(x)). \tag{21.6}$$

B2. The basic equation of motion for classical objects in the Hamilton–Jacobi form is

$$G^{\mu\nu}(x)(\partial_\mu S)(\partial_\nu S) - m^2 = 0,$$
$$G^{\mu\nu}(x) = \eta_{\alpha\beta} J^{\alpha\mu}(x) J^{\beta\nu}(x). \tag{21.7}$$

The emergence of the effective metric tensor $G_{\mu\nu}(x)$ in (21.7) suggests an interesting and new macroscopic picture of gravity. It is that macroscopic objects and light rays behave as if they were in a curved space-time with an effective metric tensor $G^{\mu\nu}(x)$, which may be associated with a non-vanishing Riemann–Christoffel curvature tensor, even though the true underlying space-time for quantum fields and elementary particles is flat. Thus, quantum Yang–Mills gravity implies that the apparent curvature of space-time is a manifestation of the flat space-time translational gauge symmetry for the motion of quantum particles in the macroscopic or classical limit.

(C) Super-Macroscopic World with Cosmic Relativity

In the super-macroscopic world, the cosmological principle (a homogeneous and isotropic cosmos) dictates that the effective metric tensor is further simplified. The cosmic metric tensor $G_{\mu\nu}$ cannot depend on spatial coordinates $\mathbf{r} = (x, y, z)$ (because of the assumption of homogeneity) and can have only the diagonal (or isotropic) form within the framework of cosmic relativity,

$$G^{\mu\nu}(t_c) = \left(G^{00}(t_c), G^{11}(t_c), G^{22}(t_c), G^{33}(t_c)\right) = \eta^{\mu\nu} U^{-2}(t_c).$$

The cosmic time t_c and one single space-time expansion factor $U(t_c)$ are particularly useful in the super-macroscopic limit since measurements by all observers in all inertial frames then have a single, universal, cosmic time with which their data of recession velocities, expansion rate, distance, the ages of star clusters, and the age of the universe are expressed.

The universal space-time scale factor $U(t)$ in the relativistic-cosmology model (RC model) is determined by further modified space-time T_4 gauge fields, which are generated by the sources of all long-range cosmic forces acting on galaxies, including gravitational and baryon–lepton forces. The novel repulsive linear forces between distant galaxies produced by the extremely weak baryon–lepton charges[6,c] could provide a new quantitative understanding of the mysterious "dark energy" related to the late-time accelerated cosmic expansion, as discussed in Chapter 20.

C1. The cosmic space-time gauge field equations with gravitational source $T^{\mu\nu}$ and baryon–lepton-charge sources $t^{\mu\nu}$ are as follows:

$$g^{-2}H_{\mu\nu}(J_{\alpha\beta}) = T^m_{\mu\nu}(t_c) + t_{\mu\nu}, \tag{21.8}$$

$$T^m_{\mu\nu}(t_c) U^{-2}(t_c) = (\rho(t_c), P(t_c), P(t_c), P(t_c)), \tag{21.9}$$

$$J_{\alpha\mu}(t_c) = \eta_{\alpha\mu} U(t_c), \tag{21.10}$$

$$G_{\mu\nu}(t_c) = \eta_{\mu\nu} U^2(t_c), \qquad U(t_c) \propto t_c^{1/2}. \tag{21.11}$$

C2. The cosmic Okubo equation for the expanding cosmos and redshifts is

$$G^{\mu\nu}(t_c)(\partial_\mu S)(\partial_\nu S) - m^2 = 0, \quad m \geq 0,$$
$$G^{\mu\nu}(t_c) G_{\nu\lambda}(t_c) = \delta^\mu_\lambda. \tag{21.12}$$

---

cThe baryon conservation law was first postulated by Stueckelberg.

The cosmic Okubo equation governs the motion of distant galaxies and their cosmic redshifts. It leads to the generalized Hubble law for all recession velocities and exact redshifts in RC model. For small velocities, the recession velocity of a distant galaxy at a given time is proportional to its distance from us, observers in an inertial frame, as shown in (17.12).

Thus, it appears that the laws governing the motion of distant galaxies as a whole on large scales is quite different from those governing the motion of classical objects in the usual macroscopic world. Furthermore, the Hamilton–Jacobi type equation (21.7), which is equivalent to Newton's law $F = ma$ in the non-relativistic limit, becomes powerless to describe and to predict the exact Hubble recession velocity of distant galaxies in the super-macroscopic limit. Nevertheless, quantum Yang–Mills gravity and the cosmological principle in flat space-time suggest that the cosmic expansion of all distant galaxies can be described by a new and simpler effective time-dependent metric tensor (21.11). Assuming the principle of least action with a new cosmic action S_{cos} given by (16.18) or (16.9), we can derive the Lagrange equation (16.29) or equivalently, the Okubo equation (16.34) or (21.12). These equations immediately lead to the solution (16.36) for the motion of distant galaxies: They all move in such a way that the "generalized momentum" p is constant in the F frame (in which $t = t_c$ and $C_o = c$ in RC model):

$$\frac{d}{dt}p = \frac{d}{dt}\left(\frac{mU\dot{r}/c^2}{\sqrt{1 - \dot{r}^2/c^2}}\right) = 0, \quad \dot{r} < c, \tag{21.13}$$

as shown in (16.29).[d] To have a more conventional and intuitive understanding of this motion, we rewrite (21.13) in the form involving the "usual momentum" $m\dot{r}/\sqrt{1 - \dot{r}^2/c^2}$ and "usual cosmic force" F_{cos},

$$\frac{d}{dt}\left(\frac{m\dot{r}}{\sqrt{1 - \dot{r}^2/c^2}}\right) = F_{cos}, \tag{21.14}$$

$$F_{cos} \equiv -\frac{dU/dt}{U}\left(\frac{m\dot{r}}{\sqrt{1 - \dot{r}^2/c^2}}\right) = \frac{-1}{2t}\left(\frac{m\dot{r}}{\sqrt{1 - \dot{r}^2/c^2}}\right), \tag{21.15}$$

where U is given by (21.11). This force acts on all distant galaxies to slow the rate of their expansion and is inversely proportional to time as well as proportional to

---

[d]In the RC model with cosmic relativity, in which $c \neq 1$, we should replace m by mc in the cosmic action S_{cos} and Lagrangian L_{cos} in (16.18) and (16.19).

the momentum of a galaxy, i.e., $\propto (1/t)$ and $\propto (m\dot{r}/\sqrt{1 - \dot{r}^2/c^2})$. However, as we have seen in Chapter 20, this gravitational interaction may be overcome by the baryon–baryon and lepton–lepton repulsive interaction, leading to the observed accelerated cosmic expansion.

In the Newtonian dynamics, we have in general the conservation of energy, but the momentum is not conserved due to the presence of forces and accelerations. For comparison, in the dynamics of expansion based on Yang–Mills gravity and the cosmological principle, we have a new conservation of law of as generalized momentum, as shown in (21.13) or (16.29). However, we do not have the conservation of energy, as shown in (16.30), because the cosmic Lagrangian L_{cos} in (16.19) depends on time. These properties are intimately related to the cosmological principle. In other words, the assumption of homogeneity is incompatible with the conservation of energy in the Lagrangian formulation involving time-dependent $G^{\mu\nu}$.

It is gratifying that the dynamics of the recession of distant galaxies can be described by the space-time gauge field equation with source tensors involving baryon masses and baryon–lepton charges. This interaction produces long-range forces that act on all visible matter and antimatter, as implied by the fundamental CPT invariance. Such a space-time gauge field in an inertial frame leads to the effective cosmic metric tensor $G_{\mu\nu}(t_c)$, where cosmic time t_c is operationally defined by the synchronization procedure for the clock system in an arbitrarily chosen frame F. Thus, Yang–Mills gravity provides a cosmic dynamics to understand the general Hubble's law, the cosmic redshift, and the late-time accelerated cosmic expansion, as discussed in Chapter 20.

Strictly speaking, forces between particles contribute to the dynamics of the cosmic expansion of distant galaxies. However, these dynamics are dominated by the gravitational force and the baryon (and lepton) forces, since atoms are overwhelmingly neutral and the strong and weak forces are negligible outside atoms. Since all visible matter in the universe is made of baryons and leptons (i.e., protons, neutrons and electrons and neutrinos), it is interesting to note that the sources of these two forces, the "gravitational charge" (or mass) and the "baryon–lepton charge," have identical distributions in the universe. In other words, the cosmological principle of homogeneity and isotropy can be applied to both mass and baryon–lepton charge distributions.

References

1. T. Rothman and E. C. G. Sudarshan, in *Doubt and Certainty* (Helix Book, Reading, 1998), pp. 214–264.

2. T. D. Lee, in *Particle Physics and Introduction to Field Theory* (Harwood Academic, 1981), Chapters 10 and 14, pp. 746–752.

3. J. P. Hsu, L. Hsu and D. Katz, *Mod. Phys. Lett. A* **33**, 1850116 (2018).

4. L. Hsu and J. P. Hsu, *EPJ Web Conf.* **168**, 01012 (2018).

5. M. Khan, Thesis, Physics Department, UMass Dartmouth (2017)

6. A. Gamba, R. E. Marshak and S. Okubo. *Proc. Natl. Acad. Sci.* **45**, 881 (1951); T. D. Lee and C. N. Yang, in *100 Years of Gravity and Accelerated Frames: The Deepest Insights of Einstein and Yang–Mills*, eds. J. P. Hsu and D. Fine (World Scientific, 2005), p. 155; *Helv. Phys. Acta* **11**, 299 (1938).

Appendix A

Gauge Invariance of Yang–Mills Gravity
and the Vanishing of the Lie Derivative
of the Gravitational Action

A.1.   Gravitational Action with T_4 Gauge Invariance

The action functional S of quantum Yang–Mills gravity based on flat space-time takes the general form

$$S = \int L\sqrt{-P}d^4x \equiv \int Wd^4x, \quad W \equiv L\sqrt{-P}, \quad P = \det P_{\mu\nu}, \quad (A.1)$$

where $c = \hbar = 1$, $S = S_{\phi\psi}$ and $L = L_{\phi\psi}$ are given by equations (8.16) and (8.17). The action S is postulated in a general accelerated frame with the Poincaré metric tensor $P_{\mu\nu}(x)$. In the limit of zero acceleration, $P_{\mu\nu}(x)$ reduces to the Minkowski metric tensor $\eta_{\mu\nu} = (1,-1,-1,-1)$. We have carried out the quantization of Yang–Mills gravity in inertial frames to obtain Feynman–Dyson rules, to calculate the self-energy of the graviton and to discuss gauge identities, as discussed in Chapters 11–13.

Under the external space-time translational (T_4) gauge transformations (8.3)–(8.6), we have

$$S \rightarrow S^\$ = \int (Wd^4x)^\$. \quad (A.2)$$

To give a general proof of the invariance of action S in quantum Yang–Mills gravity under local space-time translational (T_4) gauge transformations, we first stress that the T_4 transformation with infinitesimal arbitrary vector function $\Lambda^\mu(x)$

$$x'^\mu = x^\mu + \Lambda^\mu(x) \quad (A.3)$$

is postulated to be interpreted simultaneously as follows:

(i) a local space-time translation with the displacement group generator $p_\mu = i\partial_\mu$ and an arbitrary infinitesimal vector function $\Lambda^\mu(x)$ and

(ii) an arbitrary coordinate transformation with the same arbitrary function $\Lambda^\mu(x)$.

For example, we have the space-time T_4 gauge transformations of vector functions $\Gamma^\mu(x)$ and $\Gamma_\mu(x)$,

$$\begin{aligned}
(\Gamma^\mu(x))^\$ &= \Gamma^\mu(x) - \Lambda^\lambda(x)\partial_\lambda\Gamma^\mu(x) + \Gamma^\lambda(x)\partial_\lambda\Lambda^\mu(x), \\
(\Gamma_\mu(x))^\$ &= \Gamma_\mu(x) - \Lambda^\lambda(x)\partial_\lambda\Gamma_\mu(x) - \Gamma_\lambda(x)\partial_\mu\Lambda^\lambda(x),
\end{aligned} \tag{A.4}$$

as given in (8.5). We have seen in Chapter 7 that the T_4 gauge transformations with infinitesimal vector gauge function $\Lambda^\mu(x)$ in Yang–Mills gravity turn out to be exactly the same as the Lie derivatives (in the coordinate expression in flat space-time) and the Pauli–Ślebodziński variation.[1,2] The Lie derivative in coordinate expression or the Pauli–Ślebodziński variation can be obtained for an arbitrary tensor density $Q^{\mu_1\cdots\mu_m}{}_{\alpha_1\cdots\alpha_n}(x)$ with weight w,[3,4]

$$\begin{aligned}
(Q^{\mu_1\cdots\mu_m}&{}_{\alpha_1\cdots\alpha_n}(x))^\$ - Q^{\mu_1\cdots\mu_m}{}_{\alpha_1\cdots\alpha_n}(x) \\
&= \delta^* Q^{\mu_1\cdots\mu_m}{}_{\alpha_1\cdots\alpha_n} = -(\mathcal{L}_\Lambda Q)^{\mu_1\cdots\mu_m}{}_{\alpha_1\cdots\alpha_n} \\
&= -\Lambda^\lambda\partial_\lambda Q^{\mu_1\cdots\mu_m}{}_{\alpha_1\cdots\alpha_n} + Q^{\lambda\mu_2\cdots\mu_m}{}_{\alpha_1\cdots\alpha_n}\partial_\lambda\Lambda^{\mu_1} \\
&\quad + Q^{\mu_1\lambda\mu_2\cdots\mu_m}{}_{\alpha_1\cdots\alpha_n}\partial_\lambda\Lambda^{\mu_2} + \cdots - Q^{\mu_1\cdots\mu_m}{}_{\lambda\alpha_2\cdots\alpha_n}\partial_{\alpha_1}\Lambda^\lambda \\
&\quad - Q^{\mu_1\cdots\mu_m}{}_{\alpha_1\lambda\alpha_3\cdots\alpha_n}\partial_{\alpha_2}\Lambda^\lambda - \cdots - w(\partial_\mu\Lambda^\mu)Q^{\mu_1\cdots\mu_m}{}_{\alpha_1\cdots\alpha_n}.
\end{aligned} \tag{A.5}$$

When $w = 0$, (A.5) is the same as the T_4 gauge transformation (8.3) with arbitrary infinitesimal gauge function $\Lambda^\mu(x)$. We may remark that, in contrast to the usual partial derivative, the Lie derivative acting on a tensor does not change the rank and the weight of the tensor.

To prove the invariance of the action (A.1) in Yang–Mills gravity, Cartan's formula[a] facilitates the calculation of the change of the volume $L\sqrt{-P}d^4x \equiv Wd^4x$ under the T_4 gauge transformations.[1] According to (A.5), this change of volume

---

[a] In notation more familiar to mathematician, Cartan's formula gives $\mathcal{L}_\Lambda(Wd^4x) = d\iota_\Lambda(Wd^4x) + \iota_\Lambda d(Wd^4x)$. The second term vanishes for dimensional reasons, given $S^\$ = S - \int d(\iota_\Lambda Wd^4x)$.

Wd^4x should be just a Lie derivative, $\mathcal{L}_\Lambda(Wd^4x) = \mathcal{L}_\Lambda(Wdx^0 \wedge dx^1 \wedge dx^2 \wedge dx^3)$, where $\wedge$ denotes the wedge product. We have

$$\mathcal{L}_\Lambda(Wd^4x) = \mathcal{L}_\Lambda(W)d^4x + W\mathcal{L}_\Lambda(d^4x), \quad W = L\sqrt{-P}. \tag{A.6}$$

Since L and $\sqrt{-P}$ are scalar densities with weight $w = 0$ and $w = +1$ respectively, (A.5) gives

$$\mathcal{L}_\Lambda(W) = \mathcal{L}_\Lambda(L\sqrt{-P}) = (\mathcal{L}_\Lambda L)\sqrt{-P} + L(\mathcal{L}_\Lambda\sqrt{-P})$$
$$= (\Lambda^\lambda \partial_\lambda L)\sqrt{-P} + L[\Lambda^\lambda \partial_\lambda \sqrt{-P} + (\partial_\lambda \Lambda^\lambda)\sqrt{-P}] = \partial_\lambda(L\Lambda^\lambda\sqrt{-P}). \tag{A.7}$$

The volume element d^4x is a scalar density with weight $w = -1$, (A.5) given

$$\mathcal{L}_\Lambda d^4x = \Lambda^\lambda \partial_\lambda d^4x + [w(\partial_\lambda \Lambda^\lambda)d^4x]_{w=-1}. \tag{A.8}$$

With the help of the Cartan formula, we have

$$\Lambda^\lambda \partial_\lambda d^4x = d(\Lambda^0 dx^1 \wedge dx^2 \wedge dx^3 - \Lambda^1 dx^0 \wedge dx^2 \wedge dx^3$$
$$+ \Lambda^2 dx^0 \wedge dx^1 \wedge dx^3 - \Lambda^3 dx^0 \wedge dx^1 \wedge dx^2)$$
$$= (\partial_0 \Lambda^0 dx^0 \wedge dx^1 \wedge dx^2 \wedge dx^3 + \partial_1 \Lambda^1 dx^0 \wedge dx^1 \wedge dx^2 \wedge dx^3$$
$$+ \partial_2 \Lambda^2 dx^0 \wedge dx^1 \wedge dx^2 \wedge dx^3 + \partial_3 \Lambda^3 dx^0 \wedge dx^1 \wedge dx^2 \wedge dx^3)$$
$$= (\partial_\mu \Lambda^\mu)d^4x. \tag{A.9}$$

It follows from (A.8) and (A.9) that

$$\mathcal{L}_\Lambda d^4x = 0. \tag{A.10}$$

From (A.6)–(A.9), we obtain

$$\mathcal{L}_\Lambda(L\sqrt{-P}\, d^4x) = \partial_\lambda(\Lambda^\lambda L\sqrt{-P})d^4x.$$

Therefore, the Lie derivative of the gravitational action S in (A.1) vanishes

$$\mathcal{L}_\Lambda S = \int \mathcal{L}_\Lambda(L\sqrt{-P}d^4x) = \int \partial_\lambda(\Lambda^\lambda L\sqrt{-P})d^4x = 0, \tag{A.11}$$

or

$$S^\$ = S - \mathcal{L}_\Lambda S = S.$$

The last term involving $\partial_\lambda(\Lambda^\lambda L\sqrt{-P}\,)d^4x$ in (A.11) vanishes because it can be transformed into the integral of a vector over a hypersurface on the boundary of the volume of integration, where fields and their variations vanish. The result (A.11) shows that the action S vanishes under the Lie derivative (or under the Pauli–Ślebodziński variation). Or equivalently, the action is invariant under the T_4 gauge transformations, $S^\$ = S$.

A.2.   Gravitational Lagrangians with Tensor and Spinor Fields

Let us demonstrate explicitly that L_ϕ and L_ψ,

$$L_\phi = \frac{1}{2g^2}\left[\frac{1}{2}C_{\mu\nu\alpha}C^{\mu\nu\alpha} - C_{\mu\alpha}{}^\alpha C^{\mu\beta}{}_\beta\right], \tag{A.12}$$

$$L_\psi = \frac{i}{2}[\overline{\psi}\gamma_\mu J^{\mu\nu}\partial_\nu\psi - J^{\mu\nu}\partial_\nu\overline{\psi}\gamma_\mu\psi] - m\overline{\psi}\psi, \tag{A.13}$$

transform like a scalar under the infinitesimal T_4 gauge transformations in inertial frames, for simplicity. We have

$$(\gamma^\mu)^\$(\partial_\mu\psi)^\$ = (\gamma^\mu - \Lambda^\lambda\partial_\lambda\gamma^\mu + \gamma^\lambda\partial_\lambda\Lambda^\mu)(\partial_\mu\psi - \Lambda^\sigma\partial_\sigma\partial_\mu\psi - (\partial_\sigma\psi)\partial_\mu\Lambda^\lambda)$$
$$= \gamma^\mu\partial_\mu\psi - \Lambda^\sigma\partial_\sigma(\gamma^\mu\partial_\mu\psi) + O(\Lambda^2) = (\gamma^\mu\partial_\mu\psi)^\$, \partial_\lambda\gamma^\mu = 0, \tag{A.14}$$

where γ^μ is a constant vector. The spinor ψ is a scalar density with weight $w = 0$, as defined in (8.4) in Chapter 8. Hence, $\partial_\mu\psi$ is a vector density with weight $w = 0$. Using (A.14), one can also verify the following relations:

$$\overline{\psi}^\$(\gamma^\mu\partial_\mu\psi)^\$ = (\overline{\psi} - \Lambda^\sigma\partial_\sigma\overline{\psi})(\gamma^\mu\partial_\mu\psi - \Lambda^\sigma\partial_\sigma(\gamma^\mu\partial_\mu\psi))$$
$$= \overline{\psi}\gamma^\mu\partial_\mu\psi - \Lambda^\sigma\partial_\sigma(\overline{\psi}\gamma^\mu\partial_\mu\psi) = (\overline{\psi}\gamma^\mu\partial_\mu\psi)^\$, \tag{A.15}$$

$$\overline{\psi}^\$(\gamma_\mu J^{\mu\nu})^\$(\partial_\mu\psi)^\$ = (\overline{\psi} - \Lambda^\sigma\partial_\sigma\overline{\psi})(\gamma_\mu J^{\mu\nu} - \Lambda^\lambda\partial_\lambda(\gamma_\mu J^{\mu\nu}) + \gamma_\mu J^{\mu\lambda}\partial_\lambda\Lambda^\nu)$$
$$\times (\partial_\nu\psi - \Lambda^\sigma\partial_\sigma\partial_\nu\psi - (\partial_\sigma\psi)\partial_\nu\Lambda^\lambda)$$
$$= \overline{\psi}\gamma_\mu J^{\mu\nu}\partial_\mu\psi - \Lambda^\lambda\partial_\lambda(\overline{\psi}\gamma_\mu J^{\mu\nu}\partial_\mu\psi) = (\overline{\psi}\gamma_\mu J^{\mu\nu}\partial_\mu\psi)^\$, \tag{A.16}$$

$$(\overline{\psi}\psi)^\$ = (\overline{\psi} - \Lambda^\sigma\partial_\sigma\overline{\psi})(\psi - \Lambda^\sigma\partial_\sigma\psi) = \overline{\psi}\psi - \Lambda^\lambda\partial_\lambda(\overline{\psi}\psi). \tag{A.17}$$

Similar relations can be obtained for $J^{\mu\nu}(\partial_\nu\overline{\psi})\gamma_\mu\psi$, i.e., it transforms like a scalar density with weight $w = 0$. Based on these calculations and (A.13), the fermion

Lagrangian L_ψ transforms like a scalar with weight $w = 0$,

$$L_\psi^\$ = L_\psi - \Lambda^\lambda \partial_\lambda L_\psi. \tag{A.18}$$

The T_4 curvature tensors $C_{\mu\nu\alpha}$ and $C^{\mu\nu\alpha}$ transform as follows:

$$(C_{\mu\nu\alpha})^\$ = C_{\mu\nu\alpha} - \Lambda^\lambda \partial_\lambda C_{\mu\nu\alpha} - C_{\rho\nu\alpha}\partial_\mu \Lambda^\rho$$
$$- C_{\mu\rho\alpha}\partial_\nu \Lambda^\rho - C_{\mu\nu\rho}\partial_\alpha \Lambda^\rho, \tag{A.19}$$

$$(C^{\mu\nu\alpha})^\$ = C^{\mu\nu\alpha} - \Lambda^\lambda \partial_\lambda C^{\mu\nu\alpha} + C^{\rho\nu\alpha}\partial_\rho \Lambda^\mu$$
$$+ C^{\mu\rho\alpha}\partial_\rho \Lambda^\nu + C_{\mu\nu\rho}\partial_\rho \Lambda^\alpha. \tag{A.20}$$

Tensor fields $\phi_{\mu\nu}$ and the gauge curvature $C_{\mu\nu\alpha}$ have zero weight, $w = 0$. Thus, we have the scalar T_4 transformation for the Lagrangian L_ϕ

$$(L_{\phi 1})^\$ = (C_{\mu\nu\alpha}C^{\mu\nu\alpha})^\$$$
$$= (C_{\mu\nu\alpha} - \Lambda^\lambda \partial_\lambda C_{\mu\nu\alpha} - C_{\rho\nu\alpha}\partial_\mu \Lambda^\rho - C_{\mu\rho\alpha}\partial_\nu \Lambda^\rho - C_{\mu\nu\rho}\partial_\alpha \Lambda^\rho)$$
$$\times (C^{\mu\nu\alpha} - \Lambda^\lambda \partial_\lambda C^{\mu\nu\alpha} + C^{\rho\nu\alpha}\partial_\rho \Lambda^\mu + C^{\mu\rho\alpha}\partial_\rho \Lambda^\nu + C_{\mu\nu\rho}\partial_\rho \Lambda^\alpha)$$
$$= C_{\mu\nu\alpha}C^{\mu\nu\alpha} - C_{\mu\nu\alpha}\Lambda^\lambda \partial_\lambda C^{\mu\nu\alpha} - \Lambda^\lambda(\partial_\lambda C_{\mu\nu\alpha})C^{\mu\nu\alpha} = L_{\phi 1} - \Lambda^\lambda \partial_\lambda L_{\phi 1}. \tag{A.21}$$

Similar relations can be obtained for $C^{\mu\alpha}{}_\alpha$ and $C_{\mu\beta}{}^\beta$,

$$(C^{\mu\alpha}{}_\alpha)^\$ = C^{\mu\alpha}{}_\alpha - \Lambda^\lambda \partial_\lambda C^{\mu\alpha}{}_\alpha + C^{\rho\alpha}{}_\alpha\partial_\rho \Lambda^\mu + C^{\mu\rho}{}_\alpha\partial_\rho \Lambda^\alpha - C^{\mu\alpha}{}_\rho\partial_\alpha \Lambda^\rho, \tag{A.22}$$

$$(C_{\mu\beta}{}^\beta)^\$ = C_{\mu\beta}{}^\beta - \Lambda^\lambda \partial_\lambda C_{\mu\beta}{}^\beta + C_{\mu\beta}{}^\lambda\partial_\lambda \Lambda^\beta - C_{\lambda\beta}{}^\beta\partial_\mu \Lambda^\lambda - C_{\mu\lambda}{}^\beta\partial_\beta \Lambda^\lambda. \tag{A.23}$$

After some calculations, we obtain

$$(C^{\mu\alpha}{}_\alpha)^\$(C_{\mu\beta}{}^\beta)^\$ = C^{\mu\alpha}{}_\alpha C_{\mu\beta}{}^\beta - \Lambda^\lambda \partial_\lambda(C^{\mu\alpha}{}_\alpha C_{\mu\beta}{}^\beta)$$
$$= L_{\phi 2} - \Lambda^\lambda \partial_\lambda L_{\phi 2} = (L_{\phi 2})^\$. \tag{A.24}$$

Thus, we have seen that the Lagrangian L_ϕ in (A.12) transforms like a scalar with zero weight $w = 0$.

Mathematically, the Lie derivative of the action S, i.e., $(\mathcal{L}_\Lambda S)$, is the directional derivative of S along the vector $\Lambda^\mu(x)$ and is defined on any differential manifold because it is coordinate invariant. Since the action functional S satisfies $\mathcal{L}_\Lambda S = 0$, S does not change when the gauge function $\Lambda^\mu(x)$ changes and, hence, the change of the gauge function $\Lambda^\mu(x)$ leaves the action S invariant.

From a physical viewpoint, the vanishing of the action S under the Lie derivative (or the Pauli–Ślebodziński variation) is equivalent to the invariance of quantum Yang–Mills gravity under arbitrary space-time translations and arbitrary coordinate transformations with the same vector gauge function $\Lambda_\mu(x)$. In this sense, the T_4 gauge invariant quantum Yang–Mills gravity provides a physical interpretation of the Lie derivative in the coordinate expression (or the Pauli–Ślebodziński variation).

We have proved in (A.11) that, based on the Lie derivative and Cartan formula, the action S of Yang–Mills gravity is invariant under the space-time T_4 gauge transformation. Thus, the Lie derivative (or the Pauli–Ślebodziński variation) is the mathematical manifestation of the space-time T_4 gauge transformation in quantum Yang–Mills gravity.

The T_4 gauge transformation and the external translation group with generators $p_\mu = i\partial_\mu$ in Yang–Mills gravity are involved in the Lie derivative in the coordinate expressions, as discussed in Section 7.6. But the T_4 gauge covariant derivative $\Delta_\mu = \partial_\mu + g\phi_{\mu\nu}\partial^\nu$ and the T_4 gauge curvature $C_{\mu\nu\alpha}$ are also naturally associated with fiber bundles. These properties seem to suggest that both the Lie derivative and fiber bundles may be considered as the mathematical foundations of quantum Yang–Mills gravity based on flat space-time.

References

1. W. Pauli, *Theory of Relativity*, Trans. by G. Field (Pergamon Press, London, 1958), p. 66. Pauli's book was originally published in 1921.
2. W. Ślebodziński, *Bull. Acad. Roy. Belg.* **17**, 864 (1931).
3. K. Yano, in *The Theory of Lie Derivatives and Its Applications* (North-Holland, 1955), p. 14.
4. Wikipedia, Lie derivative. See also ion.uwinnipeg.ca, Mathematics Stack Exchange (differential geometry, Lie derivative of volume form).

Appendix B

The Fock–Hilbert Approach to Local Symmetry and Conservation Laws in General Frames of Reference

In Chapter 6, we discussed the conserved and symmetric energy–momentum tensor $t_{\mu\nu}$ of a physical system in the context of a flat space-time framework with external and local translational symmetry, involving arbitrary vector functions. This approach for deriving the symmetric energy–momentum tensor based on the variation of metric tensors was first discussed by Hilbert in 1915, before Noether's theorem was published in 1918.[a] Hilbert obtained the symmetric energy–momentum tensor in a paper discussing Einstein's theory of gravity based on a new action (the Hilbert–Einstein action) with general coordinate invariance in curved space-time.

Somewhat later, Fock proved that the covariant continuity equation in curved space-time, which is satisfied by Hilbert's energy–momentum tensor, implies the genuine conservation of energy–momentum if and only if space-time is endowed with a constant curvature (i.e., a maximally symmetric space-time).[b] Since taiji space-time has a vanishing space-time curvature and can accommodate both inertial and non-inertial frames of reference, Fock's proof implies that within the taiji framework, the conservation laws for energy–momentum hold in all physical reference frames. The purpose of Appendix B is to show how the "local" gauge symmetries of internal and external (space-time translation) groups lead to conservation laws based on the Fock–Hilbert approach.

---

[a]See Ref. 1 in Chapter 6.
[b]See Refs. 2–4 in Chapter 6.

B.1.   Conserved Charges in Inertial and Non-Inertial Frames

The usual derivation of the conservation law for electric charge in inertial frames is based on the global $U(1)$ symmetry. However, in order to see the implication of the local symmetry on conservation laws in general frames of reference, let us consider the gauge invariant actions of quantum electrodynamics (QED) and gauge theories.[1] Using arbitrary coordinates with the metric tensor $P_{\mu\nu}$ in the taiji space-time framework, we have

$$S_{\text{QED}} = \int \left[-\frac{1}{4}F_{\mu\nu}F^{\mu\nu} + i\{\overline{\psi}\Gamma^\mu(D_\mu - ieA_\mu)\psi\} - m\overline{\psi}\psi \right]\sqrt{-P}\,d^4x,$$

$$\tag{B.1}$$

$$S_{\text{YMU}} = \int \left[-\frac{1}{4}f_{k\mu\nu}f_k^{\mu\nu} + \overline{\Psi}[i\Gamma^\mu(D_\mu - igA_{k\mu}G_k) - m]\Psi \right]\sqrt{-P}\,d^4x,$$

for the QED action and the Yang–Mills–Utiyama (YMU) action, respectively. The Yang–Mills–Utiyama action is taken to have a Lie group (or an internal gauge group) with N generators G_k ($k = 1, 2, \ldots, N$) that have constant matrix representations in non-inertial frames. We have used the following relations in the action (B.1):

$$F_{\mu\nu} = D_\mu A_\nu - D_\nu A_\mu = \partial_\mu A_\nu - \partial_\nu A_\mu,$$

$$f_{i\mu\nu} = \partial_\mu A_{i\nu} - \partial_\nu A_{i\mu} + gC_{ijk}A_{j\mu}A_{k\nu},$$

$$\tag{B.2}$$

$$\Gamma^\mu = \gamma^a e_a^\mu, \quad P_{\mu\nu}e_a^\mu e_b^\nu = \eta_{ab}, \quad \eta^{ab}e_{\mu a}e_b^\nu = \delta_\mu^\nu,$$

where γ^a are the usual constant Dirac matrices and e_a^μ are tetrads.[1] The symbol D_μ is the covariant derivative associated with the Poincaré metric tensor $P_{\mu\nu}$ for non-inertial frames. The gauge variations of the electromagnetic potential fields A_μ and the gauge fields $A_{i\mu}$ are given by

$$\delta A_\mu = \partial_\mu \Lambda, \quad \delta A_{\mu i} = -(1/g)\partial_\mu\omega_i + C_{ijk}\omega_j A_{\mu k}, \tag{B.3}$$

where Λ and ω_i are arbitrary infinitesimal functions, and C_{ijk} are the completely antisymmetric structure constants of a Lie group.[1] These structure constants, which are associated with the N generators G_k and the Jacobi identity, are necessary for the formulation of the Yang–Mills theory and have been discussed by Utiyama.[2] Since we are interested in the continuity equations involving fermion currents, it suffices to consider the variation of the actions in (B.1) with respect to gauge fields

A_μ and $A_{i\mu}$. We have

$$\delta S_{\mathrm{QED}} = \int [\partial_\mu(\sqrt{-P}F^{\mu\nu}) + j^\nu\sqrt{-P}]\delta A_\nu d^4x, \tag{B.4}$$

$$\delta S_{\mathrm{YMU}} = \int [(\delta_{ik}\partial_\mu + gC_{ijk}A_{j\mu})(\sqrt{-P}f_k^{\mu\nu}) + J_i^\nu\sqrt{-P}]\delta A_{i\nu}d^4x, \tag{B.5}$$

$$j^\nu = e\overline{\psi}\Gamma^\nu\psi, \quad J_i^\nu = g\overline{\Psi}\Gamma^\nu G_i\Psi.$$

From (B.3)–(B.5), the variations of the gauge invariant actions lead to the following covariant relations involving the electric and Utiyama currents, j^ν and J_i^ν, in non-inertial frames:

$$\partial_\nu[\partial_\mu(\sqrt{-P}F^{\mu\nu}) + j^\nu\sqrt{-P}] = 0, \tag{B.6}$$

$$(\delta_{ij}\partial_\nu + gC_{inj}A_{n\nu})[(\delta_{ik}\partial_\mu + gC_{imk}A_{m\mu})(\sqrt{-P}f_k^{\mu\nu}) + J_i^\nu\sqrt{-P}] = 0, \tag{B.7}$$

where Λ and ω_i are arbitrary functions. Note that the variations of the fermion fields $\delta\psi$ and $\delta\overline{\psi}$ do not affect the results (B.6) and (B.7). With the help of $F^{\mu\nu} = -F^{\nu\mu}$ and gauge field equations

$$(\delta_{ik}\partial_\mu + gC_{imk}A_{m\mu})(\sqrt{-P}f_k^{\mu\nu}) + J_i^\nu\sqrt{-P} = 0, \tag{B.8}$$

(B.6) and (B.7) can take the form of covariant continuity equations in non-inertial and inertial frames

$$D_\mu j^\mu = \frac{1}{\sqrt{-P}}\partial_\mu\left(\sqrt{-P}\,j^\mu\right) = 0, \tag{B.9}$$

$$\frac{1}{\sqrt{-P}}\partial_\nu(\sqrt{-P}[J_i^\nu + gC_{ijk}A_{j\mu}f_k^{\mu\nu}]) = 0. \tag{B.10}$$

As shown in Section 6.5, these continuity equations lead to the conservation of charge in general frames of reference in flat space-time,

$$Q_e = \int \sqrt{-P}\,j^\mu dS_\mu, \tag{B.11}$$

$$Q_{\mathrm{U}i} = \int \sqrt{-P}\,[J_i^\nu + gC_{ijk}A_{j\mu}f_k^{\mu\nu}]\,dS_\nu. \tag{B.12}$$

The vector dS_μ is dual to the antisymmetric tensor $dS^{\nu\gamma\sigma}$ (which is an infinitesimal element of a hypersurface) and $dS_0 = dxdydz$ is the projection of the hypersurface on the hyperplane $w = x^0 = $ constant. The electric charge Q_e in (B.11) characterizes the electromagnetic interaction and Utiyama's charges $Q_{\mathrm{U}j}$ in (B.12) characterize the interaction of the gauge fields $A_{j\mu}$.[2] The values of these constant charges are the same in both inertial and non-inertial frames and are truly universal constants. Such results indicate that the coupling constants of gauge fields in quantum chromodynamics and electroweak theory are truly universal.

B.2.　Energy–Momentum Tensor of Fermions

Since there is no GL(4) representation for spinors under the Lorentz group, the Hilbert energy–momentum tensor (6.2) can be defined only for tensor objects under the Lorentz transformations. For fermions, we must introduce tetrads or vierbeins e_μ^a or e_b^ν, i.e., four covariant or contravariant vector functions. By definition, $e_b^\nu \equiv \eta_{ab} P^{\mu\nu} e_\mu^a$, and they satisfy the identities $e_\mu^a e_a^\nu = \delta_\mu^\nu$, etc. Under the T_4 gauge transformations, these quantities transform as 4-vectors in (7.4),

$$\delta e_\mu^a = -\Lambda^\lambda \partial_\lambda e_\mu^a - e_\lambda^a \partial_\mu \Lambda^\lambda, \quad \delta e_\alpha^\mu = -\Lambda^\lambda \partial_\lambda e_a^\mu - e_a^\lambda \partial_\lambda \Lambda^\mu. \quad (B.13)$$

For the energy–momentum tensor of fermions, the variation δe_μ^a (or δe_μ^a) in the invariant action S_ψ involving pure fermions[c] (i.e., the Lagrangian in S_{YMU} with $f_{k\mu\nu} = 0$ in (B.5)) leads to

$$\delta S_\psi = \int d^4x \sqrt{-P} T_\mu^a \delta e_a^\mu = 0,$$

$$\delta S_\psi = \int d^4x \sqrt{-P} T_a^\mu \delta e_\mu^a = 0. \quad (B.14)$$

Because Λ^λ are arbitrary functions, from equations (B.13) and (B.14) we have

$$\frac{1}{\sqrt{-P}} \partial_\sigma (\sqrt{-P} T_\lambda^a e_a^\sigma) + T_\mu^a \partial_\lambda e_a^\mu = 0,$$

$$\frac{1}{\sqrt{-P}} \partial_\sigma (\sqrt{-P} T_a^\sigma e_\lambda^a) - T_a^\nu \partial_\lambda e_\nu^a = 0, \quad (B.15)$$

where we have used integration by parts.

---

[c]For fermion Lagrangians with external space-time gauge groups, see Ref. 3.

We define a new symmetric energy–momentum tensor by the relation in taiji space-time with arbitrary coordinates,

$$T^{\mu}_{.\nu} = \frac{1}{2}(e^{\mu}_a T^a_\nu + e^a_\nu T^{\mu}_a) = T^{.\mu}_{\nu.}, \tag{B.16}$$

which is a mixed tensor.[4] We have used a dot to mark the position of the indices for clarity. For symmetric tensors, the dots will eventually be unnecessary. From (B.16), we have

$$Q_\nu \equiv \frac{1}{\sqrt{-P}}\partial_\mu(\sqrt{-P}T^{\mu}_\nu) = (-e^a_\sigma e^\sigma_b T^b_\mu \partial_\nu e^{\mu}_a + e^b_\sigma e^\sigma_a T^{\mu}_b \partial_\nu e^a_\mu), \tag{B.17}$$

where we have used (B.15) and the identity $e^a_\nu e^\nu_b = \delta^a_b$. Differentiation of this identity gives

$$e^a_\sigma \partial_\nu e^{\mu}_a = -(\partial_\nu e^a_\sigma)e^{\mu}_a. \tag{B.18}$$

We also have the relations

$$\partial_\nu P_{\alpha\mu} = -P_{\alpha\rho}P_{\mu\sigma}\partial_\nu P^{\rho\sigma} = -P_{\alpha\rho}P_{\mu\sigma}(-P^{\gamma\sigma}\Gamma^\rho_{\gamma\nu} - P^{\gamma\rho}\Gamma^\sigma_{\gamma\nu})$$
$$= P_{\alpha\rho}\Gamma^\rho_{\mu\nu} + P_{\mu\sigma}\Gamma^\sigma_{\alpha\nu}, \tag{B.19}$$

$$\Gamma^\lambda_{\mu\nu} = \frac{1}{2}P^{\lambda\alpha}\left(\partial_\nu P_{\mu\alpha} + \partial_\mu P_{\nu\alpha} - \partial_\alpha P_{\mu\nu}\right).$$

It follows from (B.17)–(B.19) that

$$0 = Q_\nu - T^\alpha_\rho e^\rho_a \partial_\nu e^a_\rho = Q_\nu - T^{\alpha\sigma}e_{\sigma a}\partial_\nu e^a_\alpha = Q_\nu - \frac{1}{2}T^\alpha_\rho P^{\rho\mu}\partial_\nu P_{\alpha\mu}$$

$$= \frac{1}{\sqrt{-P}}\partial_\mu\left(\sqrt{-P}T^{\mu}_\nu\right) - T^\alpha_\rho\Gamma^\rho_{\alpha\nu} = D_\mu T^{\mu}_\nu. \tag{B.20}$$

Thus, in non-inertial (and inertial) frames, we have the conservation law of the fermion energy–momentum tensor $D_\mu T^{\mu\nu} = 0$, where $T^{\mu\nu} = T^{\mu}_\sigma P^{\sigma\nu}$. In taiji space-time, equation (B.20) implies the true conservation of the energy–momentum tensor.[d]

---

[d]If a theory is invariant under a local Lorentz transformation, the quantity $e^{\mu}_a T^{a\nu}$ can be shown to be symmetric in μ and ν (see Ref. 4).

References

1. K. Huang, in *Quarks Leptons & Gauge Fields* (World Scientific, 1982), Chapter 4.
2. C. N. Yang and R. L. Mills, *Phys. Rev.* **96**, 191 (1954); R. Utiyama, *Phys. Rev.* **101**, 1597 (1956).
3. J. P. Hsu, *Phys. Rev. Lett.* **42**, 934 (1979); *Phys. Rev. Lett. E* **42**, 1720 (1979); *Phys. Lett. B* **119**, 328 (1982).
4. S. Weinberg, in *Gravitation and Cosmology* (John Wiley & Sons, 1972), pp. 370–371.

Appendix C

Calculations of $H^{\mu\nu}$ in the Gravitational Field Equation

C.1.   Evaluations of the Component H^{00}

To see and appreciate the wonderful physical implications of the new T_4 gauge curvature in (8.14) and the invariant action (8.16), one must explore the gravitational equation (8.26) in detail. The new gravitational field equation (8.26) and the "effective" metric tensors embedded in the eikonal equation (8.36) and the classical Hamilton–Jacobi equation (8.40) are key properties of Yang–Mills gravity in flat space-time.

In this section, we carry out explicitly the detailed calculations for the perihelion shift of Mercury and the deflection of light. For this purpose, it suffices to solve for the static and spherically symmetric solutions for $J_0^0(r)$, $J_1^1(r)$, and $J_2^2(r) = J_3^3(r)$ in the tensor field equation (8.26). It is convenient to use the spherical coordinates $x^\mu = (w, r, \theta, \phi)$ with the metric tensor $P_{\mu\nu} = (1, -1, -r^2, -r^2 \sin^2\theta)$. The tensor $H^{\mu\nu}$ is given in (8.26) and contains six terms:

$$H_1^{\mu\nu} = D_\lambda(J_\rho^\lambda C^{\rho\mu\nu}), \quad H_2^{\mu\nu} = -D_\lambda(J_\alpha^\lambda C^{\alpha\beta}{}_\beta P^{\mu\nu}), \quad H_3^{\mu\nu} = D_\lambda(C^{\mu\beta}{}_\beta J^{\nu\lambda}),$$
$$H_4^{\mu\nu} = -C^{\mu\alpha\beta} D^\nu J_{\alpha\beta}, \quad H_5^{\mu\nu} = C^{\mu\beta}{}_\beta D^\nu J_\alpha^\alpha, \quad H_6^{\mu\nu} = -C^{\lambda\beta}{}_\beta D^\nu J_\lambda^\mu. \tag{C.1}$$

With the help of the following relations of covariant derivatives:

$$D_\lambda C^{\rho\mu\nu} = \partial_\lambda C^{\rho\mu\nu} + \Gamma^\sigma_{\lambda\sigma} C^{\sigma\mu\nu} + \Gamma^\mu_{\lambda\sigma} C^{\rho\sigma\nu} + \Gamma^\nu_{\lambda\sigma} C^{\rho\mu\sigma}, \tag{C.2}$$

$$D_\lambda J_\beta^\alpha = \partial_\lambda J_\beta^\alpha + \Gamma^\alpha_{\lambda\sigma} J_\beta^\sigma - \Gamma^\sigma_{\lambda\beta} J_\sigma^\alpha, \tag{C.3}$$

$$D_\nu J_{\alpha\beta} = \partial_\nu J_{\alpha\beta} - \Gamma^\sigma_{\nu\alpha} J_{\sigma\beta} - \Gamma^\sigma_{\nu\beta} J_{\alpha\sigma}, \tag{C.4}$$

and $D_\lambda P_{\alpha\beta} = 0$, we can express $J^{\mu\nu}$ and $H^{\mu\nu}$ in terms of three functions, S, R and T,

$$J_0^0 = S(r), \quad J_1^1 = R(r), \quad J_2^2 = J_3^3 = T(r),$$

$$J^{00} = J_{00} = S(r), \quad J^{11} = J_{11} = -R(r), \tag{C.5}$$

$$-r^2 J^{22} = \frac{J_{22}}{-r^2} = -r^2 \sin^2 \theta J^{33} = \frac{J_{33}}{-r^2 \sin^2 \theta} = T(r).$$

All other components of $J^{\mu\nu}$ vanish.

Let us calculate the first term H^{00}:

$$H_1^{00} = D_\lambda(J_\rho^\lambda C^{\rho 00}) = (\partial_\lambda J_\rho^\lambda - \Gamma_{\lambda\rho}^\sigma J_\sigma^\lambda + \Gamma_{\sigma\lambda}^\lambda J_\rho^\sigma) C^{\rho 00}$$

$$+ J_\rho^\lambda(\partial_\lambda C^{\rho 00} + \Gamma_{\lambda\sigma}^\rho C^{\sigma 00} + \Gamma_{\lambda\sigma}^0 C^{\rho\sigma 0} + \Gamma_{\lambda\sigma}^0 C^{\rho 0\sigma})$$

$$= (\partial_1 J_1^1) C^{100} - (\Gamma_{21}^2 J_2^2 + \Gamma_{31}^3 J_3^3) C^{100} + (\Gamma_{12}^2 J_1^1 + \Gamma_{13}^3 J_1^1) C^{100}$$

$$+ J_1^1 \partial_1(J^{11} \partial_1 J^{00}) + J_2^2 \Gamma_{21}^2 C^{100} + J_3^3 \Gamma_{31}^3 C^{100}$$

$$= \left[\left(\partial_1 R + \frac{2}{r} R\right) C^{100} - R\partial_1(R\partial_1 S), \tag{C.6}$$

$$\partial_1 R = \frac{dR(r)}{dr}, \text{ etc.}, \tag{C.7}$$

where we have used (C.2)–(C.5) and

$$C^{100} = J^{11} \partial_1 J^{00} = -R\partial_1 S, \quad C^{000} = C^{200} = C^{300} = 0, \tag{C.8}$$

$$C^{\alpha\beta\lambda} = -C^{\beta\alpha\lambda}, \quad P_{\mu\nu} = (1, -1, -r^2, -r^2 \sin^2 \theta), \quad P_{\mu\nu} P^{\nu\alpha} = \delta_\mu^\alpha, \tag{C.9}$$

$$\Gamma_{\alpha\beta}^\mu = \frac{1}{2} P^{\mu\sigma}(\partial_\alpha P_{\sigma\beta} + \partial_\beta P_{\sigma\alpha} - \partial_\sigma P_{\alpha\beta}), \tag{C.10}$$

$$\Gamma_{22}^1 = -r, \quad \Gamma_{33}^1 = -r \sin^2 \theta, \quad \Gamma_{12}^2 = \frac{1}{r}, \quad \Gamma_{33}^2 = -\sin \theta \cos \theta,$$

$$\Gamma_{13}^3 = \frac{1}{r}, \quad \Gamma_{23}^3 = \cot \theta, \tag{C.11}$$

where all other components of the Christoffel symbols $\Gamma_{\alpha\beta}^\mu$ vanish.

For the second term H_2^{00}, we find

$$
\begin{aligned}
H_2^{00} &= -(\partial_\lambda J_\rho^\lambda - \Gamma_{\lambda\rho}^\sigma J_\sigma^\lambda + \Gamma_{\sigma\lambda}^\lambda J_\rho^\sigma) C^{\rho\beta}{}_\beta \\
&\quad - J_\alpha^\lambda (\partial_\lambda C^{\alpha\beta}{}_\beta + \Gamma_{\lambda\sigma}^\alpha C^{\sigma\beta}{}_\beta + \Gamma_{\lambda\sigma}^\beta C^{\alpha\sigma}{}_b - \Gamma_{\lambda\beta}^\sigma C^{\alpha\beta}{}_\sigma) \\
&= -(\partial_1 J_1^1) C^{1\beta}{}_\beta + J_\sigma^\lambda \Gamma_{\lambda\alpha}^\sigma C^{\alpha\beta}{}_\beta - J_\alpha^\sigma \Gamma^l d_{\sigma\lambda} C^{\alpha\beta}{}_\beta - J_1^1 \partial_1 C^{1\beta}{}_\beta \\
&\quad - J_2^2 \Gamma_{2\sigma}^2 C^{\sigma\beta}{}_\beta - J_3^3 \Gamma_{3\sigma}^3 C^{\sigma\beta}{}_\beta \\
&= -\left[\frac{dR}{dr} + \frac{2R}{r} + R\frac{d}{dr}\right] C^{1\beta}{}_\beta,
\end{aligned}
\tag{C.12}
$$

where

$$
C^{1\beta}{}_\beta = -R\left(\frac{dS}{dr} + 2\frac{dT}{dr}\right) - \frac{2T^2}{r} + \frac{2TR}{r},
\tag{C.13}
$$

$$
C^{0\beta}{}_\beta = C^{2\beta}{}_\beta = C^{3\beta}{}_\beta, \quad P^{00} = 1.
\tag{C.14}
$$

For other components of H^{00}, we obtain

$$
H_3^{00} = D_\lambda(C^{0\beta}{}_\beta J^{0\lambda}) = 0, \quad H_4^{00} = -C^{0\alpha\beta} D^0 J_{\alpha\beta} = 0,
\tag{C.15}
$$

$$
H_5^{00} = C^{0\beta}{}_\beta D^0 J_\beta^\beta = 0, \quad H_6^{00} = -C^{\lambda\beta}{}_\beta D^0 J_\lambda^0 = 0,
\tag{C.16}
$$

where we have used $D^\lambda = P^{\lambda\sigma} D_\sigma$. From previous results, we obtain H^{00} as follows:

$$
\begin{aligned}
-H^{00} &= -\left(\frac{dR}{dr} + \frac{2R}{r}\right) C^{100} + R\frac{d}{dr}\left(R\frac{dS}{dr}\right) \\
&\quad + \left[\frac{dR}{dr} + \frac{2R}{r} + R\frac{d}{dr}\right]\left(-R\frac{dS}{dr} - 2R\frac{dT}{dr} - \frac{2T^2}{r} + \frac{2TR}{r}\right) \\
&= \frac{d}{dr}\left(R^2\frac{dS}{dr}\right) + \frac{2R^2}{r}\frac{dS}{dr} + \left(\frac{dR}{dr} + \frac{2R}{r} + R\frac{d}{dr}\right) \\
&\quad \times \left(-R\frac{dS}{dr} - 2R\frac{dT}{dr} - \frac{2T^2}{r} + \frac{2TR}{r}\right),
\end{aligned}
\tag{C.17}
$$

where C^{100} is given by (C.8). This result is consistent with equation (9.4).

C.2.   Evaluations of the Component H^{11}

Next, we calculate H^{11}. Its first term is

$$H_1^{11} = (D_\lambda J_\rho^\lambda) C^{\rho 11} + J_\rho^\lambda (D_\lambda C^{\rho 11} = J_2^2 \Gamma_{22}^1 C^{212} + J_3^3 \Gamma_{33}^1 C^{313}. \qquad (C.18)$$

The T_4 gauge curvature $C^{\mu\nu\sigma}$ in H_1^{11} can be evaluated:

$$C^{212} = J^{22}(\partial_\sigma J^{12} + \Gamma_{2\sigma}^1 J^{\sigma 2} + \Gamma_{2\sigma}^2 J^{1\sigma}) - J^{11}(\partial_1 J^{22} + \Gamma_{1\sigma}^2 J^{\sigma 2} - \Gamma_{2\sigma}^1 J^2 \sigma)$$

$$= -r J^{22} J^{22} + J^{22}\frac{1}{r}J^{11} - J^{11}\partial_1 J^{22} + \frac{2}{r}J^{22}$$

$$= -\frac{T^2}{r^3} + \frac{1}{r^3}RT - R\partial_1\left(\frac{T}{r^2}\right) - \frac{2}{r^3}TR$$

$$= -\left[\frac{R}{r}\frac{d}{dr}\left(\frac{T}{r}\right) + \frac{T^2}{r^3}\right] = -C^{122}, \qquad (C.19)$$

$$C^{313} = \frac{C^{212}}{\sin^2\theta} = -C^{133}. \qquad (C.20)$$

Thus, we have

$$H_1^{11} = 2rTC^{122} = 2rT\left[\frac{R}{r}\frac{d}{dr}\left(\frac{T}{r}\right) + \frac{T^2}{r^3}\right]. \qquad (C.21)$$

Similarly, we obtain

$$H_2^{11} = \left[S' + \frac{2}{r}R + R\frac{d}{dr}\right]C^1{}_\beta{}^\beta, \quad S' = \frac{dS}{dr}, \qquad (C.22)$$

$$C^1{}_\beta{}^\beta = J^{1\sigma}D_\sigma J_\beta^\beta - J^{\beta\sigma}D_\sigma J_\beta^1$$

$$= J^{11}\partial_1(J_\beta^\beta - J_1^1) - J^{\beta\sigma}\Gamma_{\rho\sigma}^1 J_\beta^\rho + J^{\beta\sigma}\Gamma_{\sigma\beta}^\rho J_\rho^1$$

$$= -R(S' + 2T') - \frac{2}{r}T^2 + \frac{2}{r}TR, \quad T' = \frac{dT}{dr}, \qquad (C.23)$$

$$H_3^{11} = \left[\frac{dJ^{11}}{dr} + \frac{2T}{r} - \frac{2R}{dr} + J^{11}\frac{d}{dr}\right]C^1{}_\beta{}^\beta, \qquad (C.24)$$

$$H_4^{11} = -C^{1\alpha\beta}p^{1\sigma}(\partial_\sigma J_{\alpha\beta} - \Gamma^\rho_{\sigma\alpha}J^{\rho\beta} - \Gamma^\rho_{\sigma\beta}J^{\alpha\rho}) = -C^{100}p^{11}\partial_1 J_{00}$$

$$-C^{122}p^{11}\partial_1 J_{22} - C^{133}p^{11}\partial_1 J_{33} + 2[C^{122}p^{11}\Gamma^2_{12} + C^{133}p^{11}\Gamma^3_{13}]J_{\rho 3}$$

$$= -R\left(\frac{dS}{dr}\right)^2 - 2C^{122}(Tr^2)' + 4rTC^{122}, \tag{C.25}$$

$$H_5^{11} = C_\beta^{1\beta}p^{1\sigma}D_\sigma J_\alpha^\alpha$$

$$= -C^1_{\beta\beta}\partial_1 J_\alpha^\alpha - \left[-R(S' + 2T') - \frac{2}{r}T^2 + \frac{2}{r}TR\right](S' + R' + 2T'), \tag{C.26}$$

$$H_6^{11} = -C_\beta^{\lambda\beta}p^{1\sigma}(\partial_\sigma J_\lambda^1 - \Gamma^\rho_{\sigma\alpha\lambda}J_\rho^1 + \Gamma^1_{\rho\sigma}J_\lambda^\rho)$$

$$= \left[-RS' - 2RT' - \frac{2}{r}T^2 + \frac{2}{r}TR\right]R', \quad R' = \frac{dR}{dr}, \text{ etc.} \tag{C.27}$$

The first two terms in (9.5) can be identified with

$$-(H_1^{11} + H_4^{11}) = -2rTC^{122} + R\left(\frac{dS}{dr}\right)^2 + 2C^{122}\frac{d(Tr^2)}{dr} - 4rTC^{122}$$

$$= R\left(\frac{dS}{dr}\right)^2 + 2r^3\frac{d(T/r)}{dr}\left[\frac{R}{r}\frac{d}{dr}\left(\frac{T}{r}\right) + \frac{T^2}{r^3}\right]. \tag{C.28}$$

The third term in (9.5) is

$$-(H_2^{11} + H_3^{11} + H_5^{11} + H_6^{11}) = \left(S' + 2T' - \frac{2T}{r}\right)C^{1\beta}_{\ \beta}, \tag{C.29}$$

where $C^{1\beta}_{\ \beta}$ is given by (C.13).

C.3.   Evaluations of the Component H^{22}

Now let us consider H^{22}. The first term H_1^{22} is

$$H_1^{22} = D_\lambda(J_\rho^\lambda C^{\rho 22}) = (D_\lambda J_1^\lambda)C^{122} + (D_\lambda J_3^\lambda)C^{322}$$

$$+ J_\rho^\lambda(\partial_\lambda C^{\rho 22} + \Gamma^\rho_{\sigma\lambda}C^{\sigma 22} + \Gamma^2_{\sigma\lambda}C^{\sigma 22} + \Gamma^2_{\sigma\lambda}C^{\sigma 22})$$

$$= (\partial_1 J_1^1 - \Gamma^2_{21}J_2^2 - \Gamma^3_{31}J_3^3 + \Gamma^2_{12}J_1^1 + \Gamma^3_{13}J_1^1)C^{122} + J_1^1\partial_1 C^{122}$$

$$+ J_2^2\Gamma^2_{12}C^{122} + J_3^3\Gamma^3_{13}C^{122} + J_1^1\Gamma^2_{21}C^{122} + J_2^2\Gamma^2_{12}C^{212} + J_1^1\Gamma^2_{21}C^{122}$$

$$= \left(\frac{dR}{dr} - \frac{T}{r} + \frac{4RT}{r} + R\frac{d}{dr}\right)C^{122}, \tag{C.30}$$

where

$$C^{122} = \frac{R}{r^2}\frac{dT}{dr} - \frac{RT}{r^3} + \frac{T^2}{r^3}. \tag{C.31}$$

Similarly, we also have

$$H_2^{22} = -P^{22}D_\lambda(J_\alpha^\lambda C^{\alpha\beta}{}_\beta) = \frac{1}{r^2}[(D_\lambda J_0^\lambda)C^{0\beta}{}_\beta + (D_\lambda J_1^\lambda)C^{1\beta}{}_\beta + (D_\lambda J_2^\lambda)C^{2\beta}{}_\beta$$

$$+ (D_\lambda J_3^\lambda)C^{3\beta}{}_\beta + J_0^\lambda D_\lambda C^{0\beta}{}_\beta + J_1^\lambda D_\lambda C^{1\beta}{}_\beta + J_2^\lambda D_\lambda C^{2\beta}{}_\beta + J_3^\lambda D_\lambda C^{3\beta}{}_\beta]$$

$$= \frac{1}{r^2}[(\partial_\lambda J_1^\lambda - \Gamma_{\lambda 1}^\rho J_\rho^\lambda + \Gamma_{\lambda\rho}^\rho J_1^\lambda)C^{1\beta}{}_\beta + J_1^1 \partial_1 C^{1\beta}{}_\beta + J_2^2 \Gamma_{21}^2 C^{1\beta}{}_\beta + J_3^3 \Gamma_{31}^3 C^{1\beta}{}_\beta]$$

$$= \frac{1}{r^2}\left[\frac{dR}{dr} + \frac{2R}{r} + R\frac{d}{dr}\right]\left[-R\frac{dS}{dr} - 2R\frac{dT}{dr} - \frac{2T^2}{r} + \frac{2RT}{r}\right], \tag{C.32}$$

where we have used $C^{0\beta}{}_\beta = C^{2\beta}{}_\beta = C^{3\beta}{}_\beta = 0$.

The rest of terms in H^{22} can also be obtained in the same way:

$$H_3^{22} = (D_\lambda J^{\lambda 2})C^{2\beta}{}_\beta + J^{\lambda 2}D_\lambda C^{2\beta}{}_\beta = J^{\lambda 2}\Gamma_{\lambda\rho}^2 C^{\rho\beta}{}_\beta$$

$$= \frac{1}{r}J^{22}C^{1\beta}{}_\beta = -\frac{T}{r^3}\left[-R\frac{dS}{dr} - 2R\frac{dT}{dr} - \frac{2T^2}{r} + \frac{2TR}{r}\right], \tag{C.33}$$

$$H_4^{22} = C^{2\alpha\beta}P^{2\sigma}D_\sigma J_{\alpha\beta} = C^{2\alpha\beta}P^{22}(\partial_2 J_{\alpha\beta} - \Gamma_{2\alpha}^\rho J_{\rho\beta} - \Gamma_{2\beta}^\rho J_{\alpha\rho}$$

$$= -C^{2\alpha\beta}(-\Gamma_{2\alpha}^1 J_{1\beta} - \Gamma_{2\alpha}^2 J_{2\beta} - \Gamma_{2\alpha}^3 J_{3\beta} - \Gamma_{2\beta}^1 J_{\alpha 1} - \Gamma_{2\beta}^2 J_{\alpha 2} - \Gamma_{2\beta}^3 J_{\alpha 3})$$

$$= \frac{1}{r^3}C^{212}J_{22} - \frac{1}{r}C^{212}J_{11} = -\frac{1}{r^3}C^{122}(J_{22} - r^2 J_{11})$$

$$= \frac{1}{r}(R - T)\left(\frac{R}{r^2}\frac{dT}{dr} - \frac{RT}{r^3} + \frac{T^2}{r^3}\right), \tag{C.34}$$

$$H_5^{22} = C^{2\beta}{}_\beta D^2 J_\alpha^\alpha = 0, \quad C^{2\beta}{}_\beta = 0, \tag{C.35}$$

$$H_6^{22} = -C^{\lambda\beta}{}_\beta P^{2\sigma}D_\sigma J_\lambda^2 = C^{\lambda\beta}{}_\beta P^{22}\Gamma_{2\lambda}^\rho J_\rho^2 + C^{\lambda\beta}{}_\beta P^{22}\Gamma_{2\rho}^2 J_\lambda^\rho$$

$$= \left(-\frac{T}{r^3} + \frac{R}{r^3}\right)\left[-R\frac{dS}{dr} - 2R\frac{dT}{dr} - \frac{2T^2}{r} + \frac{2TR}{r}\right], \tag{C.36}$$

where we have used $C^{233} = -C^{323} = 0$.

Collecting all six terms H_1^{22}–H_6^{22} together, we have

$$
H^{22} = \left(R\frac{d}{dr} + \frac{dR}{dr} + \frac{5R}{r} - \frac{2T}{r} \right) \left(\frac{R}{r^2}\frac{dT}{dr} - \frac{RT}{r^3} + \frac{T^2}{r^3} \right)
$$

$$
+ \left(\frac{R}{r^3} - \frac{2T}{r^3} \right) C^{1\beta}{}_\beta + \frac{1}{r^2} \left(\frac{dR}{dr} + \frac{2R}{r} \right) C^{1\beta}{}_\beta
$$

$$
= \left(R\frac{d}{dr} + \frac{dR}{dr} + \frac{5R}{r} - \frac{2T}{r} \right) \left(\frac{R}{r^2}\frac{dT}{dr} - \frac{RT}{r^3} + \frac{T^2}{r^3} \right)
$$

$$
+ \left(\frac{1}{r^2}\frac{dR}{dr} + \frac{R}{r^2}\frac{d}{dr} - \frac{2T}{r^3} + \frac{3R}{r^3} \right) \left(-R\frac{dS}{dr} - 2R\frac{dT}{dr} - \frac{2T^2}{r} + \frac{2RT}{r} \right).
$$

$$
\tag{C.37}
$$

This result is consistent with equation (9.6), whose first two terms are just H^{22}. Note that there is no minus sign associated with H^{22} in (C.37), in contrast to the previous terms in H^{00} and H^{11} in (C.17), (C.28) and (C.29).

The tensor $A^{\mu\nu}$ in the field equation (8.28) can also be calculated by the same method discussed above.

Appendix D

Tensor Properties of Physical Quantities
in Taiji Space-Time

The mathematical structure of taiji space-time is neither that of the Minkowski space-time of special relativity nor the Riemannian space-time of general relativity, but something in-between the two. Its metric tensors are more general than the Minkowski metric tensors so that it can describe physics and the absolute space-time properties in non-inertial frames, and are simpler than the metric tensors of curved space-time so that all conservation laws (including the conservation of energy–momentum) and quantum gravity and its unifications with all other forces can be formulated. Similar to the concept of the golden mean, taiji space-time appears to be the most suitable framework for describing physics that is supposed to be covered by special relativity and general relativity.

Taiji space-time or the taiji symmetry framework is a mathematical framework with arbitrary coordinates and Poincaré metric tensors $P_{\mu\nu}$ that are consistent with vanishing Riemann–Christoffel curvature tensors. The differential coordinate $dx^\mu = (dw, dx, dy, dz)$ in general frames is, by definition, a contravariant vector. The coordinate x^μ is also a contravariant vector in inertial frames with the Minkowski metric tensor $\eta_{\mu\nu} = (1, -1, -1, -1)$. However, x^μ is not a contravariant vector in, say, non-inertial frames (or in spherical coordinates).

In taiji space-time, the coordinate transformations are obtained on the basis of the principle of limiting continuation of physical laws, which plays the same fundamental role as the principle of relativity in special relativity. To discuss experiments involving non-inertial frames, we often use coordinate transformations between an inertial frame $F_I(x_I^\mu)$ and a non-inertial frame $F(x^\nu)$,

$$x_I^\mu = x_I^\mu(x^\lambda), \quad \text{or} \quad dx_I^\mu = T^\mu{}_\nu dx^\nu, \quad T^\mu{}_\nu = \frac{\partial x_I^\mu}{\partial x^\nu}. \tag{D.1}$$

One can obtain the covariant metric tensor $P_{\mu\nu}$ from (D.1) using

$$ds^2 = \eta_{\mu\nu} dx_I^\mu dx_I^\nu = P_{\mu\nu} dx^\mu dx^\nu. \tag{D.2}$$

Based on the known covariant metric $P_{\mu\nu}$ in (D.2), one can derive the contravariant metric tensor $P^{\mu\nu}$ using the relation $P^{\alpha\lambda} P_{\lambda\mu} = \delta_\mu^\alpha$. The principle of limiting continuation of physical laws requires the accelerated transformations (D.1) and the associated Poincaré metric $P_{\mu\nu}$ to reduce to the Lorentz transformations and the Minkowski metric tensor $\eta_{\mu\nu}$, respectively, in the limit of zero acceleration.

The Poincaré metric tensors $P_{\mu\nu}$ and $P^{\mu\nu}$ can be used to raise or to lower the indices of vectors (e.g., V^μ) and tensors (e.g., $\phi_{\mu\nu}$, $C_{\mu\nu\alpha}$),

$$V_\mu = P_{\mu\nu} V^\nu, \quad V^\nu = P^{\nu\lambda} V_\lambda, \quad \phi_\mu^\lambda = P^{\alpha\lambda} \phi_{\mu\alpha}, \quad \text{etc.,} \tag{D.3}$$

where V_μ may be the covariant coordinate differential dx_μ, the covariant momentum p_μ, the covariant wave vector k_μ, etc. All these covariant vectors have the same transformation properties, which can be derived from the transformations of the corresponding contravariant vector. For example, any contravariant vector V^ν or k^ν transforms like dx^μ in (D.1),

$$V_I^\nu = T^\nu{}_\lambda V^\lambda, \quad k_I^\nu = T^\nu{}_\lambda k^\lambda. \tag{D.4}$$

The transformation of a covariant wave vector k_μ can be obtained by using $V_I^\nu = \eta^{\nu\mu} V_{I\mu}$, (D.3), and (D.4),

$$\eta^{\nu\alpha} k_{I\alpha} = T^\nu{}_\lambda P^{\lambda\mu} k_\mu, \tag{D.5}$$

where (D.4) is the transformation of a contravariant vector and $\eta^{\nu\alpha} k_{I\alpha} = (k_{I0}, -k_{I1}, -k_{I2}, -k_{I3})$. The transformations (D.5) may be useful if the inverse transformations are difficult to obtain.

The components of a contravariant (covariant) vector or tensor are denoted by a superscript (subscript). A tensor such as ϕ_μ^ν is a mixed tensor. This distinction between contravariant, covariant, and mixed tensors is necessary to maintain, especially in discussing experiments involving non-inertial frames, as we shall see below.

The physical momentum of a particle is a covariant vector p_μ because it is originally introduced in, say, the Lagrangian formulation of mechanics through the relation,

$$p_\mu = \frac{\partial S}{\partial x^\mu} \equiv \partial_\mu S, \quad \left[\frac{\partial S}{\partial x_I^\mu} = \frac{\partial S}{\partial x^\nu} \frac{\partial x^\nu}{\partial x_I^\mu} \right], \tag{D.6}$$

which holds in inertial and non-inertial frames because S is a scalar action for the motion of a particle.

The contravariant momentum vector is defined by $p^\mu = P^{\mu\nu} p_\nu$ and, hence, depends on the Poincaré metric tensors. Other physical functions such as the covariant electromagnetic potential $A_\mu(x)$ or the mixed gravitational potential ϕ_μ^ν were originally introduced through the replacement $\partial_\mu \to \partial_\mu + ieA_\mu$ in electrodynamics and $\partial_\mu \to \partial_\mu + g\phi_\mu^\nu \partial_\nu$ in Yang–Mills gravity. These functions A_μ and ϕ_μ^ν are, by definition, independent of the Poincaré metric tensors. Thus, the contravariant electromagnetic potential $A^\mu = P^{\mu\nu} A_\nu$ and the covariant tensor $\phi_{\mu\alpha} = P_{\mu\nu}\phi_\alpha^\nu$ depend on the metric tensors. These properties are useful when one performs variation of a Lagrangian with respect to the metric tensors in the Fock–Hilbert approach to local symmetry and conservation laws. One must know which functions are dependent on the metric tensor.

To illustrate the experimental difference between a covariant and a contravariant tensor, let us consider the rotational transformations of wave vectors between an inertial frame $F_I(x_I^\lambda) \equiv F_I(x_I)$ and a rotating frame $F(x)$, which can be derived from the principle of limiting Lorentz–Poincaré invariance, as discussed in Chapter 5. The frequency shift of electromagnetic waves emitted from an orbiting radiation source has two relations,

$$k_{I0} = \gamma^{-1}[k_0 + \Omega y k_1 - \Omega x k_2], \dots \tag{D.7}$$

for the covariant wave vector $k_\mu = (k_0, k_1, k_2, k_3)$, and

$$k_I^0 = \gamma^{-1}[k^0 + \gamma^2 \Omega^2 wx k^1 + \gamma^2 \Omega^2 wy k^2], \dots \tag{D.8}$$

for the contravariant wave vector $k^\mu = (k^0, k^1, k^2, k^3)$. These two relations predict different frequency shifts, as measured in an inertial frame. Such an ambiguity can be resolved by a proper understanding of the mathematical nature of physical quantities. Since the physical wave vector must have the same transformation properties as the covariant momentum, $p_\mu = \hbar k_\mu$, one should use the relation (D.7) to predict the observable frequency shift of an orbiting radiation source, and not (D.8).

Author Index

313

Subject Index

Printed in the United States
By Bookmasters